#8

Environment of Life

Environment of Life

Second Edition

Kenneth E. Maxwell

California State University, Long Beach

Dickenson Publishing Company, Inc.

Encino, California and Belmont, California

ISBN-0-8221-0167-X

Library of Congress Catalog Card Number: 75-7293

Printed in the United States of America

Printing (last digit): 9 8 7 6 5 4 3 2 1

Interior design by Jim Beggs
Interior artwork by Francie Rozycki, Bob Swingler, and Sam Sanders
Cover photo by Dr. Lee Kronenberg
Cover design by A. Marshall Licht

This book is dedicated to

EDMUND C. JAEGER

teacher, philosopher, and long-time student
of mountain and desert wildlife, who opened
doors to a world of wonders.

CONTENTS

x   CONTENTS

FOREWORD

A generation ago, when I was an undergraduate taking my first biology·
course, it did not occur to me or to my teachers that the external environment served as anything more than a playground for life. We were
aware, of course, that evolution required a rigorous environment to
shape the adaptations that made current life different from the past. But
this generalized environment was remote from our familiar life of smoke,
smog, pollutants, additives, wastes, and synthetics, which we took for
granted as symbols of a powerful industrial nation.

There were occasional "voices in the wilderness" who cautioned
against radiation damage or industrial health hazards. There were also a
few conservationists who saw how rapidly our natural resources were
being depleted. And there was a tiny number of Malthusians who
warned that the world's population was getting out of hand. Somehow
we did not heed these voices. It was easier to enjoy the products of our
industry than to worry about their ultimate cost or the priorities for living. It was fashionable then to reject these warnings as the sour reflec-
tions of gloomy pessimists. Besides, real scientists did not concern
themselves with society or values or the consequences of research. Only
facts, theories, and our perception of reality seemed to matter.

That was a false perception of life. Organisms live in an environment
and cannot be isolated from it. Our knowledge of that environment now
extends to our cells and the molecules that compose these cells. We
should no longer tolerate ignorance of how living things work, behave,
and survive because we are part of that life, and much of what applies to
viruses, bacteria, fungi, fruitflies, and mice applies to us. As scientists
we must consider the consequences of our work; we must speak out
when life is exposed to hazards; we must educate ourselves and our
students so that we can set priorities, argue values, and demand enough
scientific literacy so that our law makers, business executives, and
consumers can shape an environment suitable for life.

Kenneth Maxwell's *Environment of Life* is well suited for this
task. I have admired Maxwell's pioneering efforts to bring things together. He selects freely from the world of inorganic matter and the
biologist's multitude of living beings; he reaches into the industrial,

xi

urban, and agricultural practices of society and shows us what takes place in the living environment. He uses genetics, development, reproduction, and physiology to focus on man. This human concern permeates the book. Maxwell sifts through the layers of gases, liquids, and minerals which constitute our familiar environment and surprises us with the toxic and beneficial effects we derive from them. He satisfies our curiosity about the biological effects of antibiotics, tobacco smoking, drugs, radiation, preservatives, and numerous other products familiar and unfamiliar to our environment. Most of all, without polemics, he evaluates the biological risks and benefits of our natural and synthetic environments.

A major objective of undergraduate education is the stimulation of the students' interest in the world they will try to alter. Maxwell's approach provides a sound knowledge of the components of our environment and how these effect our own lives. His skill in providing the science that is hidden from our view and relating it to our concerns makes *Environment of Life* a valuable asset for the liberal arts education of college students.

Elof Axel Carlson

PREFACE

Concern about man's tinkering with the environment has stimulated
an evolution in the science of ecology. The relationship between living
organisms and their natural environment remains as important as ever,
but more emphasis is now placed on the disrupting effects of man. There
is a growing awareness of the effects of human activity on the environment. We are taking a closer look at the by-products of technology and
their impact on man himself as well as on other forms of life. We are
more concerned about human values. We are attempting to reappraise
past failures, predict the effects of present trends, and take precautionary measures against future mistakes.

The environment is a variegated pattern of biological and physical
components. Nature is not a stack of boxes labeled with the various
sciences, but an integrated whole. I have not hesitated to draw on the
various specialties when necessary to impart an understanding of the
biological importance of a principle, a process, or a product. By bringing
the one-ness of nature into sharper focus, I hope that the reader can gain
a keener understanding of the profoundly enormous biological effects
which are sometimes caused by seemingly small interactions in the environment.

In the chaotic state of the technological society, perspective may well
be the key to survival. For this reason, I have included examples and
historical background of important environmental problems. I have discussed the steps that were necessary to achieve some of the scientific
advances which relate importantly to environmental biology. Readers
who are unfamiliar with science and its methods may have expectations
of "pat" answers to environmental problems. They may be puzzled
when they come face-to-face for the first time with the fact that in biology, as in all of science, there are more questions than there are
answers. It is important that they not misunderstand this, lest
they be misguided to the conclusion that science cannot be helpful in
answering these questions.

Many of the crucial decisions affecting life on earth are being
made by people with a limited knowledge of science, and of environmental biology in particular. A purpose of this book is to help the

xiii

reader to prepare for the gigantic steps that must be taken—and quickly—toward the goal of reasonable but effective solutions to the most threatening and difficult problems ever experienced by man.

Since the first edition of this book, great progress has been made toward solving environmental problems, but the advances seem small in comparison to the magnitude of the job that has to be done. While it is not the purpose of the book to present a running account of current events, this edition updates many of the advances and setbacks experienced in our rapidly changing ability to deal with the environment. I have included more material on the basic principles of ecology and consolidated the material into one chapter near the beginning of the book. I have also more broadly related "environmental biology" to the problems that are enmeshed in the complex web of human activities that impede or enhance environmental improvement. This revision is reflected in an expanded discussion of energy and its consolidation into a new chapter, the addition of a new chapter on conservation of natural resources, especially land and wildlife, and the addition of *problem areas* at the end of most chapters.

The overall purpose and thrust of the book continues to be primarily informative and authoritative to serve as useful background for those who by experience, training, or education may not be as familiar with the environment and its biological impact as they would like to be. Many people feel an urgent need for solutions to environmental problems, and they are desperately looking for ideas to deal with them. Unfortunately, ideas alone fall short, because answers grow and thrive only in the soil of experience, experimentation, and knowledge. If you remove their roots from this support, they become infertile or distorted and wither away. But if ideas and embryonic answers are nourished on scientific principles and basic facts, they may flourish and become fruitful. It is my hope that this book will provide some of that nourishment.

K. E. M.

ACKNOWLEDGMENTS

The ideas and facts in *Environment of Life* are from many
sources. I cannot adequately express my indebtedness to the numerous
research workers and authors whose contributions to environmental
biology enabled me to prepare a distillation of their work and thoughts,
though I alone must bear the responsibility for the selection and interpretation of the concepts.

I am grateful for the help provided by those who reviewed the outline
or the manuscript and who pointed out deficiencies and needed revisions. Among those who made helpful comments and suggestions during
preparation of the first edition were Dr. Travis W. Brasfield, Dr. Gene
Bozniak, Dr. Guy N. Cameron, Dr. Rosalyn Kane, Dr. Donald E.
Landenberger, and Dr. Donald R. Scoby. Dr. Elof Carlson, who critically reviewed both the outline and the manuscript, was especially helpful in preparation of both the first and second editions. I am deeply
grateful for the help of my colleague Dr. Greayer Mansfield-Jones, who
meticulously read the manuscript for factual content and made invaluable suggestions for both editions.

Especially helpful by making innumerable suggestions for revisions in
preparation of the second edition were Dr. LaMont C. Cole, Dr. Theodore G. Daub, Dr. Charles S. Galt, Dr. Raymond E. Hampton, Dr.
Charles A. M. Meszoely, Dr. Carl H. Reidel, Dr. C. Lee Rockett, Dr.
Emory F. Swan, Dr. Thomas B. Widdowson, and Dr. Milton J. Wieder.
Dr. Donald Shipley kindly allowed me to draw on useful information
and references from his material.

Much of the typing of the manuscript for the first edition was capably
done by Bette Alexander. All of the charts, graphs, and artwork for the
first edition were done by Bob Swingler and Sam Sanders. Many fine
illustrations for the second edition were drawn by Francie Rozycki.

This work was made possible by the editorial cooperation of John
Crain during the initial planning, and the stimulating influence of John
Lawrence, who attentively helped in planning and organizing the subject
matter and to whom I am indebted for putting a burr under my saddle
and for appreciating the agony of saddle sores. Michael Snell gave
generously of his editorial time and attention to the organization and

content of the book, and was indispensably helpful in the preparation of the second edition. I am especially indebted to Janet Greenblatt who, with diligence and perseverance, succeeded in mitigating my extravagances, perceiving inconsistencies, and sharpening my presentation.

Finally, to the many sympathetic people who helped in the physical preparation of the manuscript and provided spiritual support, I am deeply grateful.

K. E. M.

Environment of Life

TECHNOLOGICAL MAN: THE SOCIAL DILEMMA

> We travel together, passengers on a little space ship, dependent on vulnerable supplies of air and soil . . . preserved from annihilation only by the care, the work, and I will say the love, we give our fragile craft.
>
> Adlai Stevenson

Society has come to realize slowly that the earth—immense in extent though it be and unimaginably rich in natural resources—is limited in the amount of abuse that it can take without disastrous effects on the biosphere. Perhaps we failed to perceive this earlier because the technological revolution descended upon us so precipitately, and we have been so busy basking luxuriantly in the profusion of its benefits that we have not had time to think, nor have we wanted to think, about the morning after.

Modern thinking about science is one of suspicion and distrust. We have seen that science can be used to make the world a better place to live in. But we question that the benefits are worth the price that we are paying in the various forms of environmental deterioration. Man has acquired the ability to make bigger changes in the environment, and to make them immeasurably faster, than the changes brought about by natural causes. Lakes and rivers that existed yesteryear are no longer there; tidelands, valleys, meadows, and hills disappear; clean air, water, and soil become fouled with the ever-mounting excrement of human pursuits faster than we can evaluate them. The terrifying thought occurs that the momentum of technological innovation may be beyond control. We fear that runaway technology has set in motion a trend toward environmental deterioration that may be irreversible. And we wonder whether we have the capacity to extricate ourselves from the technological web that we have cunningly spun to control the forces of nature, but in which we have become inadvertently ensnared by our own fumbling and bumbling.

Where did we go wrong? Is it the fault of science, which is knowledge, or of technology, which is the way we use knowledge? It may help to consider where science fits into society's scheme of things.

1

Photo courtesy of E. I. du Pont de Nemours & Company.

Chemical plants, complex monuments to today's technology, are vast and complicated networks of pipes, towers, and automatically controlled units.

SCIENCE AND SOCIETY

Someone has said that science is knowledge of the way the world is. This is not a bad definition, but it is incomplete. Another description is that science is a body of knowledge tested against the way nature is. That is better, but still does not go far enough because it does not explain how such knowledge is acquired, which is certainly part of science, nor how we learn to use such knowledge, which is also part of science. And, most important of all, it does not explain how we evaluate the validity of the reputed ''knowledge'' and the way it is used.

There is an insatiable human need for consistency, the lack of ambiguity. This quality of human nature is the driving force in science—the need to know. Paradoxically, it is incompatible with another face of science. Science abhors rigid conclusions. It is intolerant of the positive flat-out answer ''yes or no,'' ''it is thus and so.'' It is ever probing, finding alternatives, seeking inconsistencies, overturning cherished beliefs, and making enemies of those who do not like to change their views—which is nearly everyone. This clash between science and society is resolved not by changing human nature or science but by understanding society's desires and science's methods.

Let us look at an essential natural resource. Everyone would like to have water that is pure and free of microbes, but for what society thinks are practical reasons, it accepts some bacterial pollution. In a number of states including New York, for example, the permissible bacterial contamination of surface waters used for public water supply is an average of 5,000 coliform bacteria per 100 milliliters of water after complete treatment and 50 coliform bacteria per 100 milliliters after disinfection.[1] Clearly, some pollution of drinking water with sewage is tolerable to some people. This means that by society's decree, a little pollution is more tolerable than the sacrifices in energy, time, and money that would be required to make the water 100 percent pure. Society has made a compromise between what it would like to have and what it considers to be infeasible because of the unacceptable cost in time and energy that would be required to achieve it.

The problem, then, is to evaluate the contamination, hazard, or environmental modification and to determine whether the amount that must be tolerated is an acceptable price to pay for the particular human activity in question. To do so requires much more than the application of science and technology. It also involves a socioeconomic evaluation of what has come to be called *benefits versus risks*.

Benefits versus risks

Most of the benefits and some of the major injurious effects of any technological development can be determined quantitatively. The side effects of medicines, the number of fatalities to be expected from the use of automobiles, the causation of cancer by cigarette smoking, the toxic effects of air pollutants on the intensification of emphysema, damage to the nervous system by mercury and lead, the disappearance of species of wildlife—these are things about which scientific data can be collected and, over a period of time, valid conclusions can be made concerning the beneficial and damaging effects. Some of the information will not come soon enough and there will be disagreements among scientists as to details, but theoretically, a consensus will emerge that in the long run should give a reasonably clear picture of the good and bad features of the product or the process. Unfortunately, there is a complication.

There is a saying that everything has its price. But that depends on the personal values of those who are paying. Many people think that the price of some things are too high; others eagerly pay almost any price for the same "benefits," no matter what the coin. No two people see a benefit as having the same value; no two people see a particular risk in the same light. This characteristic of our nature is so inherent that we may have to examine it closely to recognize it. Some people can hardly wait to take a chance on the wheel at Monte Carlo or the slot machines at Las Vegas; others think that the transient fun is not worth the probable cost. Most of us are willing to gamble our lives on the highway or in an airplane. But some of us are less eager and some are not willing to do it at all. For them the benefits do not justify the risks. Anyone can think of innumerable examples. They are such a part of the fabric of our daily life that we are scarcely conscious of applying the process to our decisions concerning all technological hazards.

This leads us to the next problem. Who will answer the questions "What is an acceptable risk?" and "How safe is safe enough?" Here we are dealing with another aspect of the human element. While each of us can give answers readily enough, agreement among groups of people is

---

[1] 100 milliliters equals slightly less than a cupful.

more difficult. But the answers cannot be given by a few people. The questions must be answered by society at large, and to do so, society must have the facts. This usually means some knowledge of technology, almost always a knowledge of biology, and more than a little general science.

Knowledge and perspective

Almost every comprehensive study of environmental quality has contained strong appeals for more scientific information, in depth and on a wide front. A single technological advance takes an average of at least seven to ten years to bring to fruition. In the aggregate, present technology is based on a foundation of many years of accumulated knowledge. Correction of mistakes also takes time and this can be expected to require even more intensified scientific effort than in the past. However, neither science—which is knowledge about nature—nor technology—which is the application of knowledge aimed at fulfilling human aspirations—possesses any moral or ethical qualities of its own. They can be used for either constructive or destructive purposes depending on the competence, or the intent, of the users. That is why science and technology can either create problems or be used to solve them.

Knowledge and how we use it are two separate things. Knowledge does not guarantee wisdom. Therefore, we cannot blame science for its misuse. We can build energy-consuming engines, powerful bulldozers, and polluting factories, but it would be a contortion of reason to blame the machinery for what we do with it. Nobel laureate Peter Medawar put it this way: " . . . in the management of our affairs we have too often been bad workmen, and like all bad workmen we blame our tools."

TECHNOLOGY AND GROWTH

There is nothing inherently virtuous about growth. Animals grow, plants grow, and crops grow. So do bacteria, parasites, cancers, and city dumps. Growth is one aspect of nature, but so is decay. The ideal over a long period of time is the *steady state*—a balance between biological productivity and biological degradation—in which wide fluctuations that might produce catastrophe are avoided. Neither population growth nor economic growth can be evaluated independently. Their beneficial or detrimental effects depend on a combination of circumstances that may change with circumstances and time.

Since the beginning of the century, technological growth has about doubled every twenty years. While it has been presumed that this exponential growth in technology is associated with a parallel growth in socioeconomic benefits—health, education, income, cultural life, and material welfare—the socioeconomic disadvantages are less readily acknowledged—congestion, higher costs of government, transportation, and public services, technological unemployment, reduced physical and mental health, increased crime, and higher taxes. To these must be added the losses in personal convenience, recreational opportunities, open spaces, and an untrammeled environment. A new concept of the value of "growth" is in order.

Population growth

There are strong indications that as the population growth curve bends ever more sharply upward, the people-related problems increase at an even faster rate of acceleration. The race to make resources keep pace

USDA photos.

Left photo: We have the knowledge, from research and experience, of how to manage forests and keep them productive. Felling a ponderosa pine with a power saw in the Lassen National Forest. *Right photo:* Burned forest is followed by soil erosion.

with human needs may, within a few generations, see the capabilities of mind and body and the reserves of the human spirit overwhelmed and trampled in the reproductive stampede. People may not come to realize that the human population is on the verge of outbreeding its world until it is too late to do anything about it.

Malthus had urged sexual restraint and families of fewer children. This idea was revived in modified form nearly two centuries later as zero population growth. But the world was not ready in Malthus's time, nor is the world at large ready now, for such strictures on its reproductive freedom. We have the knowledge to control human reproduction. But we do not have, as yet, either the sense of urgency or a global plan to curb human fertility. Philosophical and religious viewpoints as well as technological and economic considerations are involved. Thus, while the onslaught of human flesh threatens to nullify the best efforts to control and clean up the debris of civilization, the most basic problem of all—population growth—appears to be more in need of socioeconomic solutions than technological answers.

It is clear that the future of mankind may depend on how well those of us who are now living can come to understand the population problem and whether, understanding, we can generate the unity of purpose to find an answer to the seemingly irresistable human reproductive urge.

THE WORLD OF NATURE

The dilemma of present-day society is depicted allegorically in the ancient Sumerian *Story of the Deluge* and again in the Book of Genesis. The earth had become too corrupt for man and beast. So God washed away the

FDA photo.

Hurricane Camille struck the Gulf Coast in August 1969 with winds of 200 miles an hour. This beached shrimp trawler is a dramatic example of the forces of nature beyond man's control.

corruption from the land with a mighty flood and gave life on the planet a fresh start.

Man has become a god in the ancient sense. Our power over nature has become so great that primitive man would have accepted our divine magic without question. But by playing at being deities, perhaps we have come to believe too much in our omnipotence. Devotion to the magic wand of technology has replaced the instinctive reverence for primeval nature and its immutable laws for survival. Our power over other forms of life and our capability to modify the biosphere impart a self image that fits imperfectly with the reality of our position as part of the ecosystem. Historically, man is the measure of all things. We have judged man by his domination over nature; and we have judged the world by the impact that man can make on it.

This view of the natural world as an object for exploitation is not confined to any one group of people, nation, or historical time period. Pliny spoke of the deforestation in Greece and its effect on local climate. The early Chinese used up their trees, and the barren soil aggravated the annual flooding. Roman colonists in the Mediterranean area similarly abused the countryside. They so decimated the animal life to provide carnage for their bloody circuses for the entertainment of the populace that many of the large animals permanently disappeared from the European and North African scene. Callous disregard for the welfare of man's companions on the planet is evident in many places and at many times.

In too many instances we have gone past the point of no return. The impairment of the ecosystem can no longer be looked upon as isolated, temporary disarrangements resulting from occasional overindulgence by exuberant humans. The accelerating pace of technological exploitation of natural resources and the growth of populations to unmanageable proportions leaves us in the rubble of an ecological crisis too vast to be ignored.

For want of other recreational facilities, hundreds of thousands of drivers of motor vehicles have now left the highways and taken their steeds of steel onto the deserts of the Southwest. The soil is misplaced, vegetation is trampled, nesting birds are driven away, eggs are exposed, and food is destroyed.

Drained swamps, bulldozed hillsides, concrete highways, mutilated landscapes, and the ever-widening encroachment of the pavement of urban sprawl require, today, planning for conservation on a dimension not heretofore contemplated.

Wildlife in distress or in danger of extinction can be viewed as warnings of disarrangements in the biosphere upon which man, too, must rely. Coal miners used to carry a caged canary into the mine to warn by its behavior of poisonous coal gas; submariners are said to have taken caged mice with them to warn by their behavior of the depletion of oxygen and dangerously high concentrations of carbon dioxide; today, we observe damage to plants at concentrations of air pollutants below those which have any readily detectable effect on humans. It is becoming increasingly evident that wildlife species of plants and animals can often be monitors of man's environment.

There are satisfactions to be derived from living in harmony with our co-occupants on the planet. The Sequoia and bristle cone pine are national treasures. If lost, they are lost for all time. Many animals that are thought of as nuisances by some people have their rightful place in the preservation of man's welfare. The grizzly, grey wolf, and coyotes; the abalones and sea otters; the osprey, condor, pelican, and peregrine falcon; and even the rare black toad, along with a hundred odd species on the ragged edge of extinction, would become irreparable losses to mankind unless we defend and nurture them.

GOALS AND VALUES

The solution to our environmental problems resides partly in our ability to shift our goals. Until recently technology was concerned with achieving narrowly defined and immediately profitable results at the lowest possible cost. Often, undesirable side effects were evaluated, if at all, after environmental damage became evident. Relatively little consideration was given to pollution, blemishes on the environment, or depletion of natural resources. Even after injuries occurred and could no longer be ignored, the attitude was often one of "progress cannot be impeded." The eternal conflict between those who desire the benefits of a technological innovation and those who object to the side effects has been resolved in the past largely by an evaluation of "economic" benefit. In the future, the preservation of irreplaceable natural endowments must be given first consideration, not last. Goals will have to be shifted to emphasize the *quality* instead of the quantity and the size of things. Precautionary measures will have to be taken before, not after, the potentially injurious effects take place.

Left photo courtesy of New York Convention and Visitors Bureau. Right photo by Bill Beebe.

Left photo: Bronze statue of Prometheus at Rockefeller Center. In Greek mythology, Prometheus stole fire from the gods in heaven and gave it to man. Eventually man would use fire to heat the furnaces of industry and drive the pistons of internal combustion engines, causing profound changes in the environment. *Right photo:* Black-necked stilt.

We tend to appraise all human values in terms of material productivity. But the dollar sign falls short when dealing with things that cannot be purchased with cash or a credit card. How do you put a money value on the worth of an unblemished landscape, the greenery of a hillside, the chromatic splendor of a river valley, the rushing rock-rippled stream, the splash of trout in a sparkling lake, the staccato of quail bursting from cover, the crispness of clean morning air, or the inscrutable depth of a star-stabbed night? These things can be measured only in terms of the satisfaction of living.

In the end, we will have to put a money value even on these more intangible aspects of life, because it is becoming increasingly expensive to protect the environment. The price that we pay—in money, inconvenience, and effort—will be determined by how much society is willing to revise its priorities in the quest for material gain.

Renewal of confidence

Fire was a new technology discovered by *Homo erectus* or his ancestors. He felt its warmth and tasted its gastronomic benefits. Half a million years later modern man used fire to build a technological revolution. But the talisman has been lost. The magic of science has evaporated. The illusion of accomplishment has faded and in its place there is realism—and the awareness of failure. Thus, we are faced with a new danger, a decline in resolve and determination—an affliction that Peter Medawar said might be described by a future historian as "failure of nerve." We can lose perspective in an orgy of self-contempt and despair, diverting our energies to unproductive ends. Or we can cultivate an appreciation of our inherent virtues and in so doing generate the degree of confidence and self respect that we need to cope with the environmental crisis—a crisis that appears to be the most formidable obstacle to survival that our species has encountered in all of evolutionary history.

That is the challenge.

Let others praise ancient times;
I am glad that I was born in these.

Ovid, about 2 B.C.

Yet cease your ire, you angry stars of heaven!
Wind, rain, and thunder, remember earthly man
Is but a substance that must yield to you;
And I, as fits my nature, do obey you.

Shakespeare, *Pericles*

READINGS

ALLEN, D. L., *Population, Resources and the Great Complexity*. PRB Selection No. 29, pp. 1–6. Washington, D.C., Population Reference Bureau, 1969.

BRANSCOMB, L. M., "Taming Technology." *Science* 171:972–977 (1971).

COMMONER, B., *The Closing Circle*. New York, Alfred A. Knopf, 1971.

COMMONER, B., *Science and Survival*. New York, The Viking Press, 1963.

GARDNER, J. W., *The Recovery of Confidence*. New York, Pocket Books, 1970.

GREENBERG, D. S., "British AAS: Counterattack on Gloom about Science and Man." *Science* 165:1239–1240 (1969).

HORSFALL, JAMES G., "A Socio-economic Evaluation," in C. O. Chichester, ed., *Research in Pesticides*. New York, Academic Press, 1965. (Also in Maxwell, K. E., *Chemicals and Life*. Belmont, Calif., Dickenson Publishing Co., 1970, pp. 292–306.)

ODUM, H. T., *Environment, Power, and Society*. New York, John Wiley & Sons, 1971.

STARR, C., "Social Benefit versus Technological Risk." *Science* 165:1232–1238 (1969).

USDA photo.

PART ONE
THE NATURE OF THE ENVIRONMENT

THE BIOSPHERE: AN OVERVIEW

The natural environment of living things is a complex, though fragile, membrane around the earth. If the earth were the size of a grapefruit, sliced in half for a cross-sectional view, the environment of life would be too thin to be seen easily with the naked eye. Yet this epidermis of the earth is abundantly rich in the quantity and variety of animals, plants, fungi, bacteria, and other living things.

All living organisms, together with the physical and chemical components of the total environment, make up the *biosphere,* or what is often called the *ecosphere*. It involves all the interactions between the living and nonliving components that produce, support, and affect life. Thus, the biosphere is more than the abode of iving things—it is a self-sufficient biological system in which the thread of life is sustained in a dynamic state of equilibrium by a flow of energy from the nearest star, the sun. The energy input from the sun is captured, used over and over again by the web of life, and eventually deposited in the earth or emitted back into space.

The biosphere—or ecosphere—is remarkably constant in physical and chemical characteristics. So special are these conditions that perhaps nowhere are they exactly duplicated on any satellite among all the stars in the universe, though there may be many that are similar. The temperatures of bodies in space range over millions of degrees, from indescribable coldness to the inferno of the center of the hottest star.[1] By contrast, the biosphere of the planet earth is an island of equanimity, where the temperature is always within a few degrees of that necessary to maintain water, the giver of life, in its liquid state. Other properties of the biosphere are equally fortuitous for life. The size of the earth and its gravity, rotation, and distance from the sun, as well as the chemical composition of the rocks, the ocean, and the atmosphere, hold within the membrane a delicate balance in which life can flourish. Only slightly beyond these limits, life, as we know it, would perish.

---

[1]The temperature of the sun beneath the surface is estimated to be 30 to 40 million degrees Fahrenheit.

The biosphere is a region of remarkable equanimity compared to the extremes of temperature and other conditions elsewhere in the universe.

Evolution and the variety of life

Some form of life has existed on earth for at least 2 billion, and possibly 3½ billion, years. During that time there have been dynamic upheavals and cataclysmic changes in climate as well as other conditions that affect life. The history of the dynamics of life on earth is permanently written in sedimentary rocks in the form of fossils. The record tells of a parade of living types from early cellular protoplasm to progressively more complicated and highly specialized forms. The record reveals that many of the kinds of animals and plants that flourished in the remote past could not adapt to changes and lost the struggle for survival.

G. G. Simpson, a paleontologist, guessed that no less than 50 million and possibly as many as 5,000 million species of organisms have lived during the entire course of evolution. They evolved, flourished, and perished, except for a few unusually well adapted types and others that evolved relatively recently and still exist. Today, there are probably about 2 million species of animals and plants living. No one knows the exact number. Estimates place the number of known and described species at about 1¾ million, of which there are about 37,000 species of microorganisms—bacteria, yeasts, protozoans, and similar small life; 90,000 species of fungi; 250,000 species of more highly specialized plants—mosses, ferns, palms, pines, and flowering plants of familiar and exotic variety; and 1,360,000 species of animals of 20 or more major types or *phyla* ranging from the primitive hydra, sponges, starfish, insects, and crabs to the vertebrates and other higher animals.[2]

Homo sapiens appeared on the scene only recently. If the existence of life on earth were compressed into a twenty-four hour period, the occu-

---

[2]H. Orians, *The Study of Life* (Boston: Allyn and Bacon, 1969).

The variety of life in the biosphere ranges from microscopic organisms to giants of the plant world. This is a giant sequoia (Big Tree), *Sequoiadendron giganteum*.

pancy of man would be less than a second. During that figurative second, the climate of the earth has been moderately stable; and the environment has been favorable to man and the other forms of life that have been able to adapt to it.

ENERGY OF LIFE

All life is dependent on photosynthesis, a photochemical reaction that takes place in chlorophyll, the green pigment in most plants and a few microorganisms (see Figure 1.1). It brings about the first step in the production of sugars—the most basic of all food substances. All life, therefore, derives from the radiant energy of the sun, which drives the chemical reaction of photosynthesis. The sun is an enormous nuclear reactor utilizing a fusion process that puts out 10 billion pounds of converted energy every second. Yet its mass is so great that it has lost only 0.03 percent of its matter in the 4 or more billion years of its solar existence.

Actually, the sun is a typical star. It is quite average according to all the standard ways of classifying the stars of the universe within the reach

fats and proteins are made by chemical changes
in the sugar and starches in various parts of the plant

blossom uses food in production of seeds

carbon dioxide is absorbed

oxygen is released

plants breathe through leaves

bark or hard case protects plant from bruises and insect
damage

trunk, stem, and tendrils reach up into sunlight and air

chlorophyll, the green substance in the plant, absorbs energy
from light and water, this energy is used to make starch and
sugar

branches and leaves are arranged to expose greatest possible
area to light

foods manufactured in leaves and branches are carried down
stem

roots anchor the plant, absorb minerals and moisture,
and serve as food reservoir

MOISTURE   SOIL   MINERALS

From I. J. Hindawi, National Air Pollution Control Administration.

FIGURE 1.1 Sunlight is the energy source for photosynthesis in the complex system of the living plant.

of human vision—in size, mass, temperature, brightness, and most other measureable qualities. Yet, by earthly standards, the radiation from the sun is fantastic. It is radiated into space in all directions. The earth soaks up only a small fraction—about one part in two billion—equal to 126,000,000,000,000 horsepower. It is enough, for the energy of life and all the energy stored in coal, oil, and other organic matter during the existence of this planet came from the sun's energy.

Even that tiny portion of the sun's radiant energy that strikes the earth is mostly reradiated into space, and only a fraction of that which is absorbed is usable by chlorophyll. The part of the electromagnetic spectrum that activates chlorophyll is in the range of visible light, mainly the blue and red. Green light is incompletely absorbed by chlorophyll and is therefore reflected, which accounts for the green color of succulent foliage. However, some plants and microorganisms contain other pigments that are involved in photosynthetic processes, so that the efficiency of light utilization is variable depending on the species.

The amount of the sun's energy that is effective in sustaining life is surprisingly small. Ordinarily, no more than 5 percent of that which is absorbed can be stored as photosynthetic energy under the most favorable conditions. At least 20 percent of that, and more often about 50 percent, is utilized by the plant itself for its metabolic functions. The net percentage efficiency of photosynthetic food production is estimated to be about 0.5 percent under average favorable conditions, but this can reach a maximum of 4 percent or higher at the peak of the growing season during long summer days. Much higher efficiencies may be achieved under laboratory or other special conditions. Even that is not indicative of the productive efficiency of the biosphere because the surface of the earth is incompletely occupied by chlorophyll-bearing plants. In fact, the portion of land that is usable for high-yielding crops is decreasing rapidly as a result of the encroachment of cities, homes, factories, and highways.

Food chain of life   Organisms that engage in the capture of electromagnetic energy by the process of photosynthesis include chlorophyll-bearing organisms of the seas, which are mostly microscopic algae. As a rule, such photosynthetic plants of the oceans live close to the surface, for sunlight does not penetrate water more than a few feet with the intensity needed to support a high efficiency of photosynthetic activity. Fresh-water algae and shallow, fixed plants are also important in the food chain (see Figure 1.2). But it is on land that photosynthetic plants are overwhelmingly productive. Here the grasses, herbs, shrubs, and trees, which are the food-producing vascular plants, create the sustenance for hundreds of thousands of microorganisms, parasitic plants, and herbivorous animals as well as, ultimately, even the carnivores.

The role of the food chain has greatly increased in importance by the accumulation of waste products and other pollutants in various parts of the environment. Some contaminants, for example mercury, DDT, and other synthetic organic compounds, are absorbed or taken in with contaminated food and passed along to the next higher organism, sometimes becoming highly concentrated in those at the end of the food chain (Figure 1.2). These relationships will be discussed in greater detail later.

THE MACROSPHERE OF MAN

The environment of man is many things, external and internal. Physical, chemical, biological, and cultural conditions, and their ramifications, collectively comprise "the environment." Since the onset of the industrial revolution, there have been subtle but significant changes in the bio-

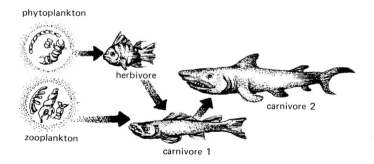

FIGURE 1.2 A toxic substance can be concentrated in a food chain.

sphere. These changes have become accelerated in recent years, so that in the momentary span of less than 100 years many forms of life have become threatened with extinction, and the health and welfare of people themselves are in jeopardy. The United States Fish and Wildlife Service lists more than 100 domestic species and nearly 300 foreign species of endangered wildlife. This can be taken, to some degree, as an indicator of a deteriorating environment that affects man as well.

Some of the most drastic changes in the human environment are taking place as a result of the accelerating pace in the use of energy and other natural resources. The depletion of irreplaceable resources threatens to bring about drastic changes in both the world's economy and the quality of life. Yet this is only one part of the problem. Waste products are produced for which no satisfactory disposal system has been found. The solutions to these and related problems are the most difficult and challenging of human undertakings, for like other organisms, we must live within the ecological limits of the biosphere.

Man in the biosphere Man is a generalized physical animal, remarkably adaptable to changing conditions, and is highly specialized in only one attribute—a well developed central nervous system. The enlargement at the top of the spinal cord, called the brain, has endowed man with a degree of cleverness unapproached by any other living thing. However, the evolutionary selection process by which man acquired cleverness did not, with equal vigor, simultaneously select for wisdom. Also, unfortunately, man was left with many of the primitive attributes that were useful during a more brutal stage in the climb to his precarious perch in the uppermost branches of the evolutionary tree.

Nature seems to have played a grotesque joke on man. Without the wisdom to match his technological brains, he thrashes over and around the face of the earth, wasting, destroying, poisoning, and polluting, seemingly hypnotized by the enormity of his cleverness and obtusely oblivious to the disastrous consequences of his deeds.

READINGS

ADLER, I., *How Life Began*. New York, The New American Library of World Literature, 1957.

"The Biosphere," *Scientific American,* 223, no. 1 (September, 1970).

GROBSTEIN, C., *The Strategy of Life*. San Francisco, W. H. Freeman, 1964.

ECOLOGICAL PRINCIPLES

ORIGIN AND MEANING OF ECOLOGY

Ecology is the study of living organisms in relation to one another and to their environment. Ernst Haeckel, a German biologist who was a champion of Charles Darwin's theory of natural selection, is given credit for first using the word *oecology* in print in 1869. *Ecology* comes from the Greek work *oikos,* meaning "house." It has the same Greek root as the word *economics,* from *oikonomia* or "household manager." Recently, *ecology* has become part of the general vocabulary and carries the broad implication that it encompasses all of nature including man and his activities as they relate to the environment.

Ecology in this broad sense was practiced early in man's history as a way of dealing with the surroundings, especially in relation to health and nutrition, for example, as applied to medicine and agriculture. The Greek scholars Hippocrates, Aristotle, and Theophrastus, and later both Greek and Roman philosophers and writers including Lucretius, Dioscorides, Cato, Varro, and Pliny the Elder, studied and wrote about many things related to ecology. But ecology developed as a science only slowly. Though it began to take shape in its present form during the scientific renaissance of the eighteenth and nineteenth centuries, it did not flourish as a distinct branch of science until about the beginning of the present century. Even then, ecology remained for the most part a branch of observational natural history until adoption of the experimental approach, enabling the new science to provide fresh concepts, especially about animal and plant populations and communities.

For each species within a community,[1] it is theoretically possible to describe its position in relation to the total biotic and abiotic factors to which it is adapted. These include the various aspects of space (loca-

---

[1]A *community,* in an ecological sense, includes all the populations of different species that occupy a given area. The term *population,* from the Latin *populus,* meaning people, originally referred to human populations, but in ecology, the term is used to mean a group of individuals of any one kind of organism.

tion), time, characteristics of the environment, and the functional relationships with other species. The organism's place in the habitat thus identified is called its *ecological niche*. However, the term *niche* is used in a variety of ways, often with limited meaning, referring, for example, more specifically to the function, location, or adaptive strategy of the species.

The ecosystem   The term *ecosystem* was adopted to describe an environmental system consisting of the community of all living organisms in a given area together with the nonliving physical and chemical aspects of the environment. It can easily be seen from the concept of an ecosystem that ecology must involve many other branches of the biological and physical sciences, for they all bear on the environment of life, the biosphere. It can be said, then, that the science of ecology has hardly any limits at all.

The science of ecology has increasingly related to human problems. A new dimension has been introduced by pollution of air, water, and soil; the exploitation of irreplaceable natural resources; destruction of natural habitats; unwise and irresponsible misuse of land and ocean; encroachment on the habitats of forms of wildlife, the protection of which is essential for the welfare of man; and the threat that we are becoming overly successful in that we threaten to overrun the earth with an overwhelming capacity for changing the environment of the planet and its delicate balance. If we accept the idea that ecology can be used to solve problems, it is useful to give serious thought and study to the fundamentals of ecology and related sciences. The enormous complexity of the ecosphere, which makes it possible for living organisms, including man, to survive and flourish, emphasizes the time-tested rule that successful problem-solving starts not with solutions but with close observations of nature and an understanding of natural law.

BIOTIC COMMUNITIES

The biomes   A good perspective of the animal, plant, and microbial life of the biosphere is obtained by relating biotic communities to climatic regions. Anyone who travels cross-country or from lowlands to high elevations cannot fail to be impressed by the transition in plant life that is seen as one moves from a warm area to a cooler one, or vice versa. Similar changes can be seen in animal populations, although they are sometimes more difficult to observe. Ecologists classify the major terrestrial communities into recognizable units, called *biomes*. A biome is a region having a characteristic fauna and flora (animal and plant components). All the communities in a given climatic region are natural parts of the biome (Figure 2.1).

The biome in the northernmost parts of the land masses of America, Europe, and Asia is the *tundra*, a Russian word meaning "north of the timberline." In most of the tundra the subsoil is permanently frozen and is referred to as permafrost. There are many lakes and bogs and few trees, and much of the ground is covered with mosses, lichens, a few grasses, and numerous small perennials. Animal life includes large numbers of birds, many of which nest on the tundra in the summer, reindeer, caribou, polar bear, wolves, foxes, hares, and incredibly numerous insects, nota-

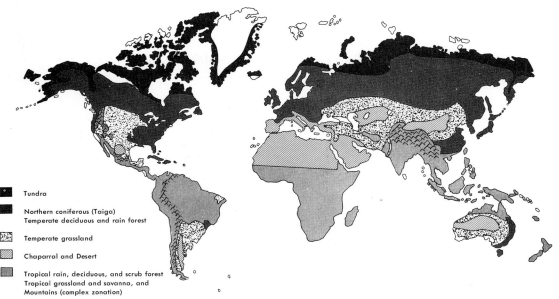

Tundra

Northern coniferous (Taiga)
Temperate deciduous and rain forest

Temperate grassland

Chaparral and Desert

Tropical rain, deciduous, and scrub forest
Tropical grassland and savanna, and
Mountains (complex zonation)

From *Patterns and Perspectives in Environmental Science,* National Science Board, National Science
Foundation, 1972.

FIGURE 2.1 Major terrestrial biomes of the world.

bly mosquitoes. Thus, the tundra is a biome that is teeming with life, though limited in the number of species that it supports.

Moving south from the tundra, one encounters a zone of coniferous forests as the dominant feature of another biome, the *taiga*. The plants and animals associated with the conifers differ in many respects from those in the tundra. There are more species, and those that favor the taiga are more abundant.

Moving still farther south, there is a wide area in the temperate zone where the summers are longer and warmer and there is abundant rainfall. Here, the plant communities are dominated by broadleaved trees, forming the *deciduous forest* biomes,[2] in which are typically found many more species of plants and animals than in the more northerly regions.

Other major biomes of the world include the *tropical rain forests,* which are plant and animal communities of great variety and complexity; the *grasslands,* vast regions of low or uneven rainfall in both temperate and tropical regions; *tropical savannas,* grasslands with scattered trees or clumps of trees in areas of high rainfall but a prolonged dry season; *deserts,* regions of so little rain that not even grasses are abundant; *chaparral,* in temperate regions with high winter rainfall but dry summers; and *scrub forest* biomes, which are intermediate between desert and areas of higher rainfall. Some ecologists make additional classifications and subdivisions.

In general, each biome has its characteristic complement of plant and animal species, with much overlapping, but with population densities de-

---

[2]Deciduous plants are those that drop their leaves seasonally.

A coniferous forest biome, or *taiga*.

termined largely by the climatic effects on the biotic complex. Clearly, the distribution of plants and animals and the abundance of the various species is correlated with the climate of the biomes. The population density of humans has less uniformity and is less predictable.

SUCCESSION

Biotic communities, though they may sometimes appear to be static, are often vibrant with dynamic change. When you observe any natural ecosystem over a period of time, it can be seen that there is an orderly process of community development. Starting with the first organisms to invade an unoccupied habitat, called *pioneer* populations, a series of changes takes place in the kinds of plants and animals that occupy the area. The course of events is influenced by the fact that the community itself causes changes in the physical environment as well as in the biological makeup of the ecosystem. Thus, a sequential replacement of one population or community by another takes place, called *ecological succession*.

One of the most easily seen examples of ecological succession is on abandoned farmland, or even a city lot, cleared of vegetation and left untended. The soil, at first virtually barren, does not remain so for long. Plants and animals appear with amazing rapidity. Seeds are brought in by birds, other animals, and wind. Within a remarkably short time, if moisture is available, the area is populated with many kinds of organisms. The first to appear are usually green plants of various kinds we call weeds. Insects and microorganisms quickly move in on these pioneer invaders, and predators follow, each contributing to the competitive relationships of the biotic community and each bringing about its measure of sequential change. The whole series of transitory communities is called a *sere*, from the Latin word *serer*, meaning to put in a row, and the identifiable tran-

A tropical rain forest.

sient communities of the sere are called *developmental stages,* or *seral stages.* But these should not be thought of as discrete steps. There is gradual evolution toward a comparatively stable community in which the flow of energy maintains the community in a state that changes very slowly thereafter. Such a terminal, nearly stabilized system is called a *climax community.* When the succession of populations leading to such a climax develops on virgin soil, such as newly exposed rock, sand, or eroded land, the result is called *primary succession.* But most successions begin on sites where previous populations have been destroyed or damaged, such

A desert biome.

as clearings, plowed fields, ditches, or abandoned property. These are called *secondary successions*. They usually take place more rapidly than primary successions due to the presence of nutrients and other biological necessities that remain from previous inhabitants.

The process of succession is perhaps best understood if we start with a mature climax such as a forest, in which the populations of plants, animals, and microorganisms may have changed very little for hundreds or even thousands of years (see Figure 2.2). If the trees are cut, the logs hauled away, and the underlying vegetation removed leaving the bare soil, a scenario unfolds that will take many years to run its course. First, the landscape will be changed by erosion and there will probably be deposition of silt in lower areas. Plants from seeds that are left from the forest cover are the first to appear. Other plants spring up from seeds brought in by wind, rodents, and birds, or they may come up from roots left in the ground. The first plants are apt to be grasses and annual broadleaved plants. But as soon as they become established, the environment changes. There is shade for the tender shoots of new plants, succulent foliage for the feeding of insects and fungi, changes in the structure of the soil that improve its capability for absorbing and retaining moisture, the invasion of herbivorous animals and their predators, and the falling of dead vegetation and animal waste that provide nutrients and a congenial habitat for the proliferation of bacteria and the beginnings of a layer of humus. A new microclimate appears because both temperature and humidity are drastically different from those of the bare soil. Eventually woody shrubs appear, and these may force out smaller plants that require more sunshine than they can now get. The shrubs also provide cover for larger animals that may further reduce the more succulent vegetation; and, if the animals include predators, they will change the competitive relationships in the ecosystem that affect both animal and plant populations, including insects and other small organisms. At last, conditions of shade, moisture, and soil nutrients will be created that are favorable for the growth of young trees, but they will almost certainly not be the kind that dominated the original forest. If, for example, the forest was predominantly Douglas fir, the first trees to sprout and grow vigorously may be a species of alder whose vigorous growth is apt to form dense stands of cover that prevent or delay the establishment of fir trees for many years. After 50 to 100 years, more or less, depending on a wide variety of circumstances, the landscape may begin to take on the appearance of the original forest.

Ecological succession, of course, is not confined to areas afflicted with human disruption, nor is it confined to terrestrial ecosystems. Lakes typically have a life history in which succession of plant and animal communities play a prominent role. A newly formed lake will characteristically have clear, rapidly replaced water, and it is apt to be relatively free of vegetation and debris. But as the lake ages, sediment collects first near the inlet and later over the entire bottom. Plant life encroaches around the edges, and as dead organic material accumulates, algae and other microorganisms flourish. These, in turn, provide nourishment for small animals, and an environment is created that is favorable for sustaining populations of fish and other larger organisms. The natural process of enrichment is called *eutrophication,* from the Greek words *eu,* meaning "well,"

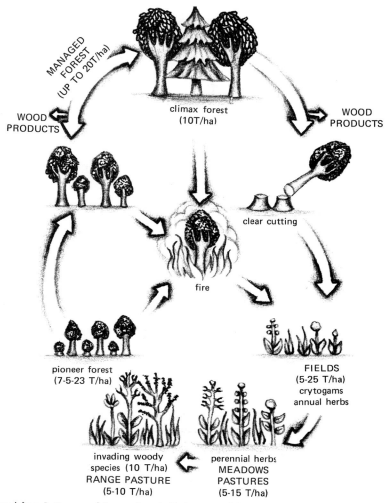

WOOD
PRODUCTS

MANAGED
FOREST
(UP TO 20T/ha)

climax forest
(10T/ha)

WOOD
PRODUCTS

clear cutting

fire

pioneer forest
(7-5-23 T/ha)

FIELDS
(5-25 T/ha)
crytogams
annual herbs

invading woody
species (10 T/ha)
RANGE PASTURE
(5-10 T/ha)

perennial herbs
MEADOWS
PASTURES
(5-15 T/ha)

Adapted from *Patterns and Perspectives in Environmental Science,* National Science Board, National Science Foundation, 1972.

FIGURE 2.2 Seral stages of a deciduous forest, showing the successional sequence. Where man has arrested the successional sequence to create other systems, numbers indicate annual dry-matter production in metric tons per hectare (1 t/ha = 100 g/m²). Cryptogams are plants—ferns, mosses, algae, and the like—which reproduce spores and do not produce flowers or seeds.

and *trophē,* meaning "nourishment." The lake, typically, becomes marshy around the edges, and as it more completely fills with sediment and debris, the entire lake turns into a marsh. Eventually, as terrestrial grasses and moisture-loving plants take over, the marsh turns into a meadow, and in the final act of this scenario the former lake has been so drastically changed that it becomes a congenial environment for woody shrubs and trees. The lake has become a forest! (See Figure 2.3.)

Evolution   We have been looking at changes in biotic communities that take place during relatively short periods of biological time—decades or perhaps

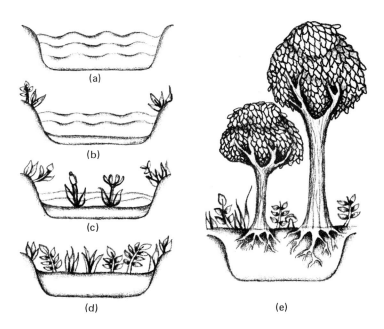

FIGURE 2.3 Ecological succession in a lake. (a) Pioneer stage in newly formed lake: clear water, bare bottom. (b) Sediment and silt, especially near inlet, a few plants around edge, algae and other small aquatic plants and aquatic animals. (c) Marshy at edges, terrestrial plants increase, bottom-rooted plants, larger aquatic animals including fishes. (d) Meadow, with remaining marshy areas. (e) Trees: the lake has become a forest.

hundreds or even thousands of years. But we know that comparable successions have occurred during the long span of the earth's prehistory called *geological time,* extending over billions of years. Many of the most recent successions were similar in their general characteristics to those that we have described, culminating in climax communities such as the deciduous and conifer forests, grasslands, deserts, tropical rainforests, and other types of communities that we see today. But the most dramatic changes in biotic communities during the millions of years of geologic time were those in which entire populations of one or more species became extinct and were replaced by new species of plants and animals. We know from the fossil record of life in sedimentary rocks that there was a continuous succession of populations and communities since life first appeared on the planet probably more than 3.5 billion years ago. Time and again, species were replaced by other organisms that were closely related but more adaptable to the slowly changing environment during geologic time. The emergence of those that were better able to cope with new conditions is one of the key features of the process of *evolution*. It differs, however, in one important respect from ecological succession as we ordinarily think of it. In evolution there is no climax, as indeed there may be no true climax in ecological succession if we extend contemporary time long enough because each evolutionary progression adds to the biological complexity of the ecosystem, and the diversity makes it inevitable that change will continue to take place. How these sequential changes of ecological succession occur over the short term are more easily understood if we consider the nutritional and energy relationships of biotic communities.

BASIC ELEMENTS OF ECOSYSTEMS

Every ecosystem consists of two basic components: the *abiotic,* or nonliving, environment and the *biotic,* or living, part of the system. The living and nonliving components interact in an exchange of nutrients and a dynamic flow of energy that leads to a clearly defined and structured ecosystem.

The abiotic portion of the ecosystem includes all the physical and chemical features of the environment. The word *abiotic* comes from the Greek *a,* meaning "not," and *bios,* meaning "life," therefore "without life." The abiotic factors of the environment include temperature, moisture, light and other forms of radiant energy, altitude (including barometric pressure), latitude (including light incidence, length of day, and seasonal influences), the tides of the oceans and movement in smaller bodies of water, and wind and its effect on evaporation and cooling, as well as more subtle physical influences on living organisms, some of which are discussed later. The abiotic environment also includes the inorganic and organic substances of the air, water, and soil. The most important of these, if in fact any one of them can truthfully be said to be more important than the others, is water itself, for it forms most of living matter. Other chemical substances vital to life include nitrogen, oxygen, and carbon dioxide of the atmosphere, and all the minerals, salts, and chemical elements, especially those that are important in plant and animal nutrition. The latter include nitrates, phosphates, and potassium as well as other nutrient elements that are needed in smaller amounts. These all differ in concentration from place to place and therefore in their availability to living organisms. The presence of acids and bases also directly affects the availability of mineral nutrients and is sometimes critical for the so-called micronutrients, mineral elements that are needed in only very small amounts. All these are part of the abiotic portion of the ecosystem, and together they greatly influence the populations of animals, plants, and microorganisms that occupy a given area or habitat.

The biotic, or living, portion of the ecosystem consists of three broad categories of organisms: *producer* organisms, *consumer* organisms, and *decomposer* organisms. These form the communities of living animals, plants, and microbes of the ecosystem. The total amount of living organic material in the ecosystem is called the *biomass.*

Producers   In nearly all natural ecosystems, the producer organisms are mainly green plants, but a few of the photosynthetic organisms are bacteria. They depend on sunlight for their initial source of energy, which they use by the process of *photosynthesis* (described in more detail in Chapter 3). Other exceptions include types of bacteria that are able to use chemical energy to synthesize organic compounds from inorganic material, by so-called chemosynthesis. These are relatively small in

Predators, called secondary consumers, are an important part of the food web.

number and play a minor role in the economy of nature.[3] The overwhelming majority of producer organisms are those that contain the green pigment chlorophyll, which is able to capture solar radiation and convert it directly to chemical energy in the form of organic compounds. All of the organisms that synthesize their own organic compounds by taking energy from the environment, whether by photosynthetic or chemosynthetic processes, are called *autotrophs* from the Greek words *autos*, meaning "self," and *trophē,* meaning "food" or "feeding." The predominant form of producer organisms on land are the flowering plants, which include many of the familiar trees, shrubs, and grasses. Other familiar terrestrial producers are the somewhat more primitive forms of plant life including ferns and the conifers such as pines and firs.

In bodies of water, algae and other photosynthetic organisms form the producer populations. Collectively, these microscopic free-floating producer organisms are called *phytoplankton*. In the oceans, the most important groups of phytoplankton are single-cell organisms called *diatoms,* which form shells of silica that display fascinating forms, shapes, and colors under the microscope, and the free-swimming organisms *dinoflagellates* and *microflagellates*. Diatoms predominate in northern oceans while the flagellates are found in greater abundance in subtropical and tropical waters. Together, the diatoms and flagellates are the basic source of food that supports the animal life of the sea, though near shore large multicellular algae, commonly called seaweeds, play a producer role in the economy of marine life.

Consumers   Consumer organisms are, generally speaking, animals. All animals obtain their energy from organic materials synthesized by other organisms, either producers or prey. Consequently, consumer organisms are called *heterotrophs,* from the Greek words *heteros,* meaning "other," and *trophē,* meaning "nourishment"—hence, "depending on others for nourishment." Heterotrophic organisms that feed directly on plants are known as *primary consumers*, or *herbivores,* while those that prey on the

---

[3]A group of chemosynthetic bacteria called hydrogen bacteria have been studied for use in life-support systems in spacecraft because of their high efficiency in removing carbon dioxide from the atmosphere.

herbivores are called *secondary consumers,* or *carnivores.* Here we have the fundamental features of a *food chain,* in which the energy stored up by a producer, for example a species of grass, is passed along in chainlike fashion to a herbivore such as a deer and then to a carnivore such as a mountain lion. Thus there is some justification for the aphorism, "all flesh is grass." Many carnivores feed on other carnivores, so these are often divided into first-level, second-level, and third-level carnivores, and so on. The final consumer in the food chain is sometimes called a *top carnivore.* Organisms that are *parasitic* in animals are highly specialized carnivores.

Food chains are sometimes classified as two basic types: the *grazing food chain,* in which the food energy starts with a green plant and goes through the grazing herbivore to the meat-eating carnivore, and the *detritus*[4] *food chain,* which starts with dead organic material and goes through microorganisms, detritus-feeding organisms, and predators. Actually, food chains do not often exist as isolated lines of predator-prey transfer of food energy. There are almost always cross-connecting links between food chains, often of such complexity that they form what is best described as a *food web.*

Decomposers   Decomposer organisms are heterotrophs that perform an essential role in the food web by recycling organic detritus in the ecosystem. The largest ecosystem, the biosphere itself, would collapse of its own weight in the absence of decomposers. Within them, the droppings and carcasses of the carnivores and unconsumed herbivores would accumulate, along with the debris of dead producers, in a gigantic mass of organic tissue and waste products that would eventually choke out most, if not all, forms of producer organisms. Fortunately for life on earth, when an animal collapses or expires in the sea, or when a piece of dead tissue such as a leaf, branch, or tree falls to the ground, it is quickly invaded by decomposing bacteria and fungi. These convert the organic material of the producers and the consumers to inorganic nutrients that can be used by the producers for synthesizing organic compounds. Larger detritus-feeding organisms, especially insects, also enter the detritus food chain.

Trophic level   The relative position occupied by an organism in a food chain is referred to as the *trophic level.* In a typical food chain, as we have seen, the first trophic level is occupied by green plants, the second trophic level by plant-eaters, the third trophic level by the carnivores that eat the herbivores, and the fourth trophic level by the carnivores that prey on the first-level carnivores. In some cases, a particular species can occupy more than one trophic level because of broad tastes and food requirements that form multiple interconnecting links in the food web.

As a practical matter in the economy of nature, the number of trophic levels in a food chain are limited. The food chain can be thought of as a flow of energy through transfer of food energy from one organism to another. At each step in the transfer, as much as 80 to 90 percent of the potential energy of the victim is unavailable to the beneficiary at the next higher trophic level. Most of the energy is dissipated as heat generated by

---

[4]The word *detritus,* borrowed from geology, means "a product of disintegration," from the Latin *deterere,* "to wear away."

the metabolic processes of the organism during the normal activity of feeding, surviving, and reproducing. Thus, the number of steps in a food chain is usually limited to four or five. The shorter the food chain, the greater the amount of available energy at the top of the chain, and the greater the biomass that can be supported at the top trophic level. This explains in part why the oceans in the cold regions, which have short, simple food chains, are so highly productive. For example, baleen whales, which feed on krill and other small organisms, are the end of a short food chain. However, many of the species of ocean fish are at, or near, the top of a relatively long food chain, involving a progressive loss of available energy. Thus, there is a sharp limit on the amount of high trophic level food available from the sea.

The loss of available energy at each transfer is of great practical importance in human nutrition. For example, let us take a relatively short food chain—from corn to cows. Only a small part of the earth's share of solar energy is captured by the plant in the first place, and only part of that is converted into food energy because the conversion itself uses up most of the energy and most of it is lost to the environment as heat. From the producer onward, each step in the food chain entails about a ten-fold loss of available energy. (See Figure 2.4.) That is why it costs more for cutlets than corn, and for bacon than beans, of the same caloric content. As the human population soars upward, meat and other animal products will become increasingly scarce and costly. People in the more affluent parts of the world may have to drastically change their diets unless world population growth is blunted. These and related problems are discussed further in Chapter 16.

Ecologic pyramids

The energy relationships between trophic levels are sometimes shown graphically, and because there is a loss at each transfer from one level to the next, the graph takes the shape of a pyramid (see Figure 2.5). An *ecological pyramid* can show any one of several quantitative features. For example, the numbers of individual organisms at each trophic level can be shown as a pyramid of numbers, or the energy flow at each level can be similarly diagrammed. A pyramid of biomass is another way to show one of the more fundamental relationships of an ecological community. Total productivity may be greater than the biomass because biomass is only the amount of living material present at any particular time. If the producer organism such as an alga is reproducing rapidly with a high energy flow, it might be capable of supporting a much larger biomass of herbivore than the biomass of the alga would indicate. The total assimilation of food energy is called the *energy flow* through that level. (See Figure 2.4.) The energy flow equals the production of biomass plus the loss of energy to the environment through metabolism, technically called respiration. The energy inputs and outputs in an ecosystem can be calculated as a mathematical model, but this could become highly complex (see Figure 2.6).

LIMITING FACTORS

One of the most visible facts of nature is that a particular plant or animal can be seen in abundance in one place but can be found only with difficulty or not at all in other places. There are obvious explanations for the

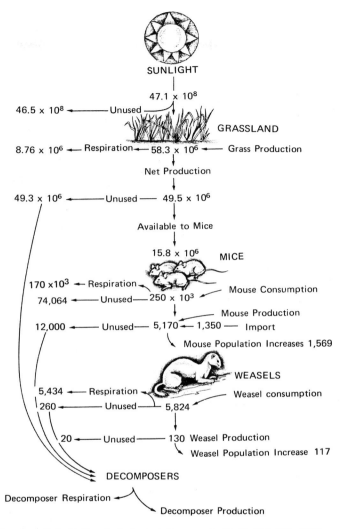

SUNLIGHT

47.1×10^8

46.5×10^8 ◄——— Unused

GRASSLAND

8.76×10^6 ◄— Respiration ◄— 58.3×10^6 ◄——— Grass Production

Net Production

49.3×10^6 ◄——— Unused ——— 49.5×10^6

Available to Mice

15.8×10^6 MICE

170×10^3 ◄— Respiration

Mouse Consumption

$74,064$ ◄——— Unused ——— 250×10^3

Mouse Production

$12,000$ ◄——— Unused ——— $5,170$ ◄— $1,350$ ——— Import

Mouse Population Increases 1,569

WEASELS

$5,434$ ◄——— Respiration

Weasel consumption

260 ◄——— Unused ——— $5,824$

20 ◄——— Unused ——— 130 Weasel Production

Weasel Population Increase 117

DECOMPOSERS

Decomposer Respiration ◄

Decomposer Production

Based on data from F. B. Golley, in *Ecological Monographs* 30 (2): 187–206, 1960.

FIGURE 2.4 An energy-flow budget for a plant–meadow mouse–weasel food chain (energy values are in kilocalories per hectare). The plant converts only about 1 percent of the incoming solar energy to plant tissue, and most of that is lost to respiration (metabolism) and decomposition. Of the available energy that remains, meadow mice consume only about 2 percent. But the weasels, next in turn, use up 30 percent of the available energy in the mouse biomass. Of the total energy used in each step of the food chain, the amounts consumed in respiration are 15 percent by plants, 68 percent by mice, and 93 percent by weasels. This supports the general principle that each successive stage in a food chain shows an increase in utilization of the energy taken up. Because the available energy tapers off, food chains rarely have more than four or five steps, and they usually have fewer. In this particular food chain, so little of the energy that entered the system was converted to weasel biomass that there would be insufficient flesh to support a secondary carnivore preying on the weasels.

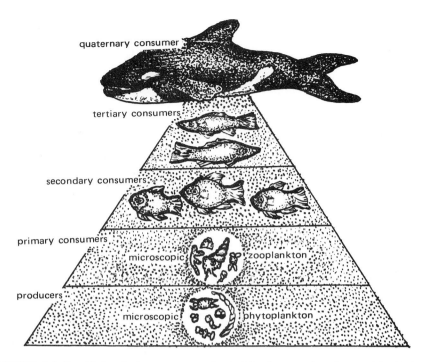

quaternary consumer

tertiary consumers

secondary consumers

primary consumers

microscopic   zooplankton

producers

microscopic   phytoplankton

FIGURE 2.5 Trophic levels shown as an ecological food pyramid in the sea. The higher the step in the pyramid, the fewer the number of individuals and the larger their size. In many ecosystems, the larger animals bypass some of the lower trophic levels. For example, man gets food from all levels below himself, including that of the producers.

variations in abundance and distribution of many familiar plants and animals. You would hardly expect to see palm trees in Alaska or pine trees along the Amazon. Nor are you apt to find wild elephants in North Dakota or polar bears in Yucatan. Sometimes climatic effects, very often temperature or moisture, limit the distribution of an animal or plant to circumscribed areas. But there are many more situations in which the reasons for the distribution and abundance of an animal or plant are subtle, complex, and often frustratingly difficult to identify. To do so is not only of scientific interest but of great practical importance as well, especially with respect to public health, agriculture, and conservation of natural resources. Thus, some of the most challenging problems in ecology today are those that deal with the relationships between environmental conditions and the abundance of plants and animals. Ecological conditions that restrict or control the distribution or abundance of plants or animals are called *limiting factors*. (See Figure 2.7.)

The concept of limiting factors originated in 1840 in relation to fertilization of agricultural crops. Justus Liebig, a renowned German biochemist with wide-ranging interests, perceived that the main reason for loss in fertility of agricultural soils was the withdrawal of nutrients by the plants. He was the first to experiment with the addition of mineral fertilizers in place of manure. He noted in his experiments that the size of the crop depended on, and was limited by, the nutrient that was present in smallest amount. This principle came to be known as *Liebig's law of the minimum*. The idea was elaborated by others to include additional abiotic factors, but it is now recognized that the relationship between limiting factors and the productivity or abundance of plants and animals is not a simple one. For example, high concentrations of one substance can sometimes partly compensate for the limiting effect of one in low concentration. Some of the most sophisticated techniques to deal statistically with such interactions were developed originally in experimental work on soil fertility.

Limits of tolerance

The law of the minimum was extended by an American ecologist, Victor Shelford, who pointed out in 1913 that populations can be limited by too much of an environmental factor as well as by too little. This can apply to any one of several factors when they approach either the upper or the lower limit of tolerance for the organism. For example, such factors as heat, light, water, and salts, which may be required by an organism in certain amounts, can become inhibitory or even lethal in amounts that are too high or too low. Thus, each organism has an ecological maximum as well as minimum, with a range in between called the *limits of tolerance*. This idea, known as *Shelford's law of tolerance,* applies to each factor of the abiotic environment. It should be understood, however, that this is

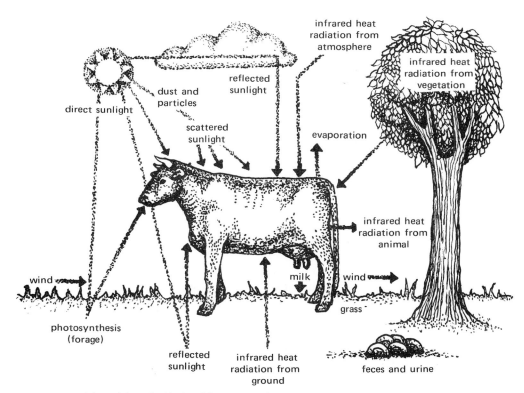

Adapted from the National Science Board, 1972.

FIGURE 2.6 Energy budget of a cow, in which the multiple energy inputs are shown qualitatively. It is possible to determine each input quantitatively, and to express it mathematically as part of a larger model describing in greater detail the energy balance in a field or pasture.

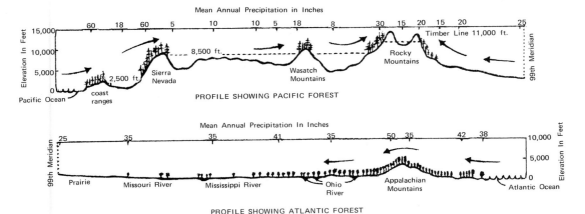

PROFILE SHOWING PACIFIC FOREST

PROFILE SHOWING ATLANTIC FOREST

Redrawn from Raphael Zon, "Climate and the Nation's Forests," *Yearbook of Agriculture,* 1941.

FIGURE 2.7 Moist air currents can be an important limiting factor. These profiles show the influence of moisture on the distribution of forests in North America along the thirty-ninth parallel.

not the whole story. Whether an organism thrives in a particular habitat depends not only on the abiotic factors but also on the biological interrelationships such as competition, available food, and predators.

A great deal of useful information about plants and animals has been obtained from tolerance studies, particularly those which Eugene Odum calls "stress tests," in which a range of conditions is imposed on the organisms. It is known from such experiments, supplemented by field observations, that an organism can have narrow limits of tolerance for one factor and wide limits of tolerance for other factors. The species most apt to be widely distributed are those with a wide range of tolerance for all factors. A set of terms has come into use in ecology to describe the degree of tolerance for a limiting factor. The prefixes *steno-,* from the Greek *stenos,* meaning "narrow," and *eury-,* from the Greek *eurys,* meaning "broad," are used in such terms as stenohydric and euryhydric, referring to a narrow or broad range of tolerance to water; and stenothermal and eurythermal, referring to a narrow or broad range of tolerance to temperature.

If a factor limiting primary producers is increased, the ecosystem may be drastically changed. An example is the stimulation of excessive growth of algae in a lake by the addition of large quantities of phosphates from sewage or agricultural drainage. The condition, called *eutrophication,* often leads to disruption of the ecosystem resulting in serious consequences to recreation and fisheries. (Eutrophication is discussed in detail in Chapter 7.) Often, increasing a limiting factor is beneficial as, for example, the addition of water to desert areas for irrigation of crops. But even in this case, problems involving limiting factors may result from the drainage of irrigation water into ponds or lakes. The Salton Sea, in the desert of California, has become increasingly salty over the years to the point where some of the valuable forms of aquatic life, including sport fish, are in danger. Studies are underway to take corrective measures for this complex problem.

The principles of limiting factors and tolerance levels have a wide variety of practical applications. One objective in ecological management is to improve the productivity of a plant or animal or to increase its geographical range by transporting or "introducing" individuals to a new area that may be deemed favorable for their growth and reproduction. This has frequently been done with desirable plants and game animals. On the other hand, the opposite approach is needed when it becomes necessary to suppress a pest or prevent its spread. Most of the serious insect and weed pests of the United States were accidental importations. Under new ecological conditions, they performed better than expected and escaped from domestication to become pests. The management of pests requires a thorough understanding of their ecological limits of tolerance if efficient methods of control are to be developed without the excessive use of chemicals.

Some of the most troublesome and costly weeds were deliberately introduced under the mistaken notion that they would become valuable farm crops. Johnson grass, one of the sorghums,[5] was introduced to South Carolina in 1830. It had limited usefulness as a forage grass but took to its new environment with vigor and became a widely distributed weed pest. It also contains a cyanide-forming constituent that has been linked to livestock deaths. Another immigrant that many people think was too successful is hemp,[6] better known as marijuana, a native of Asia. The plant was introduced into the United States because of its economic importance for producing fiber (hemp) and seeds, which are used in birdseed mixtures and for oil. Cultivation became unprofitable for these purposes and was discontinued around 1955, but the plant had already escaped beyond the confines of cultivation and is now commonly found growing wild, especially in the alluvial bottom lands of the Mississippi and Missouri Valleys. The federal government regards marijuana plants as weeds and prohibits, except by license, the possession of living or dried *Cannabis,* or any parts of the plant. The raging controversy over the social consequences of the weed go beyond the purpose and scope of discussions of its ecology.

Indicator organisms   Another application of tolerance levels is in the use of *indicator organisms*, that is, those animals, plants, and microorganisms that indicate ecological conditions. If we can determine what kind of organisms are present, we can tell a great deal about the physical environment. *Steno* species usually make better indicators than *eury* species because they are less tolerant of extremes. Indicator organisms are especially useful in determining pollution. For example, the presence of the common colon bacillus, *Escherichia coli*, is the standard for determining whether water is polluted with sewage. One of the more ingenious uses of indicator organisms is based on the discovery that when certain trees such as junipers and pines grow in soil containing uranium, the element is taken up through the roots and deposited in the above-ground plant parts. By analyzing the plants, they can be used as indicators of uranium deposits. Selenium is another poisonous element that is taken up by plants. Some

---

[5]*Sorghum halepense* (L).

[6]*Cannabis sativa* (L).

plants require selenium and will be found only where there is selenium in the soil. Consequently, these indicator plants are useful in mapping geologic formations and soils containing dangerously high amounts of selenium which can be taken up by forage and grain crops in concentrations poisonous to livestock and humans. An interesting sidelight is that selenium is often associated with uranium, making it possible to use plants that require selenium in uranium prospecting.[7]

NUTRIENT CYCLING

We have seen that the energy flow through the community levels of an ecosystem involves both production and decomposition, and that the cycling of nutrients within the food web contributes to maintaining the ecosystem. But important elements that make up the organic matter of living material remain in the biomass only momentarily. They are lost to the environment through oxidation, respiration, and other metabolic routes and are regained in different form at whatever trophic level in the food web furnishes the nutrient supply for any particular organism. This cycling of nutrient materials through the different trophic levels of the biotic community, to and from the abiotic portion of the environment, is known as *biogeochemical cycling,* a term that refers to the biological involvement of "earth" chemicals.

The elements in highest demand among a large number of essential nutrients are nitrogen, carbon, oxygen, hydrogen, phosphorus, and potassium. Nitrogen, phosphorus, and potassium are among the *macronutrients,*[8] so-called because they are important limiting factors in the growth of green plants, and are often added to soils as fertilizers in large amounts. Carbon, hydrogen, and oxygen, although needed in large amounts, are readily obtained by photosynthetic plants from water and atmospheric carbon dioxide.

Since nitrogen is a key constituent in the amino acid building blocks of all proteins, nitrogen follows a crucial pathway through the food web (Figure 2.8). Compounds of the element are used in large quantities as a fertilizer, and because of its importance as a major limiting factor and its usefulness in growing food crops, the nitrogen cycle has been studied extensively. Nitrogen is abundant in the air, which is almost four-fifths nitrogen gas; thus the atmosphere plays a key role in furnishing nitrogen salts for plant nutrition. The routes that this essential nutrient takes are discussed further in Chapter 3.

The nutrient cycles for other macronutrients—phosphorus, potassium, calcium, sulfur, magnesium—are in general simpler than the nitrogen cycle, and they differ in some of their essential details. Phosphorus is apt to be the most important limiting factor among the macronutrients because it tends to be required by organisms in greater ratio to the other elements than it occurs in the environment. This is of great ecological

---

[7]Such plants are a species of poisonvetch, or locoweed, of the genus *Astragalus* (Leguminosae).

[8]From the Greek word *makros,* meaning "large."

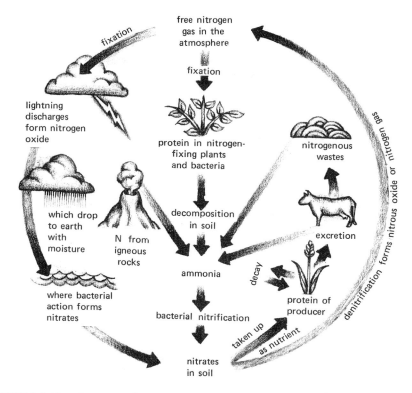

FIGURE 2.8 The nitrogen cycle.

importance. If the soil of some part of the earth's surface is deficient in phosphorus, that region is apt to be severly limited in its productivity. In such areas where food crops are grown, large amounts of phosphates are often required to balance out the nutrient supply.

The carbon cycle is a good example of the dependence of living organisms on inorganic substances in the environment, for all life is dependent on the ability of photosynthetic plants to take carbon dioxide from the air and synthesize carbohydrates (Figure 2.9). Photosynthetic organisms produce about 100 billion tons (10^{17} grams) of organic matter every year and approximately its equivalent is oxidized back to carbon dioxide and water by metabolic activity of living organisms.[9] But it doesn't balance out exactly. Beginning about 600 million years ago, during the period of geological time called the Cambrian, a small part of the organic matter became buried under sediments without being completely decomposed or oxidized. The fossil remains, which were laid down in especially large deposits about 300 million years ago, formed the fossil fuels—coal, petroleum, and natural gas. This locking up of the unoxidized carbon-containing organic matter also resulted in an imbalance in the removal of carbon dioxide and release of oxygen. Thus, over a long period of prehistory, there was a gradual buildup of oxygen in the atmo-

[9]J. R. Vallentyne, "Solubility and the Decomposition of Organic Matter in Nature," *Arch. Hydrobiol*, 58:423–434 (1962). (Also in Odum, E. P., *Fundamentals of Ecology*.)

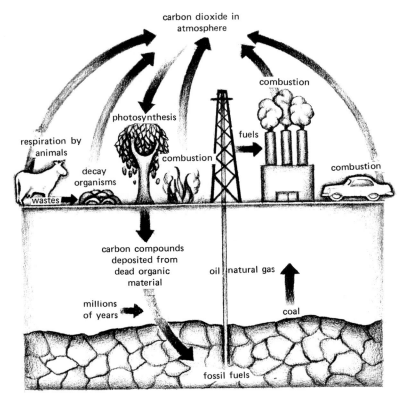

FIGURE 2.9 The carbon cycle in an ecosystem.

sphere that set the stage for the evolution and survival of new forms of life, especially the higher forms of animals. There were times when the balance shifted the other way, probably caused by changes in biological activity associated with solar radiation, volcanic activity, weathering of rocks, and sedimentation. In recent years there has been a sharp increase in the carbon dioxide/oxygen ratio due to the large amounts of carbon dioxide released to the atmosphere from the combustion of fossil fuels. Some of the probable effects are discussed in detail in Chapter 6.

DEFINITIONS

BIOME: A community unit characteristic of a regional climate. The biome is the largest land community unit that is easily recognizable. All the communities in a given climatic region are natural parts of the biome.

COMMUNITY: All the populations of different species that occupy a given area. The community is the living part of the ecosystem.

ECOLOGY: Science of the reciprocal relationship of organisms to their environment.

ECOSYSTEM: The total living and nonliving components of a particular environment as an interacting system.

ENVIRONMENT: All the external conditions, both biotic (living) and abiotic (nonliving), that affect an organism or group of organisms of a given habitat. The term is sometimes used to describe only the external physical and chemical conditions that affect living organisms.

HABITAT: The place where an organism lives.

NICHE: A species' place in the community in terms of space, time, and its functional relation to other species.

POPULATION: A group of individuals of one species or organism.

SUCCESSION: The replacement of one population or community by another.

READINGS

BILLINGS, W. D., *Plants, Man, and the Ecosystem,* Second Edition, Belmont, Calif., Wadsworth Publishing Company, 1970.

BOUGHEY, A. S., *Ecology of Populations.* New York, The Macmillan Co., 1971.

BOUGHEY, A. S., *Fundamental Ecology.* San Francisco, Intext Educational Publishers, 1971.

EMMEL, T. C., *An Introduction to Ecology & Population Biology.* New York, W. W. Norton & Co., 1973.

KORMONDY, E. J., *Concepts of Ecology.* Englewood Cliffs, N. J., Prentice-Hall, 1969.

ODUM, E. P., *Fundamentals of Ecology,* Third Edition. Philadephia, W. B. Saunders Company, 1971.

RODGERS, C. L. and R. E. KERSTETTER, *The Ecosphere. Organisms, Habitats and Disturbances.* New York, Harper & Row Publishers, 1974.

WHITTAKER, R. H., *Communities and Ecosystems.* New York, The Macmillan Co., 1970.

USDA photo.

PART TWO

MAJOR COMPONENTS OF THE ENVIRONMENT

THE ATMOSPHERE

Air does many things to make life possible on earth. It is a reservoir of oxygen needed by animals and of carbon dioxide essential for plants. It serves as a distribution system for moisture. The atmosphere is an insulating blanket around the earth. Without it the temperature at the equator would rise to 180° during the day and drop as low as −220°F at night. It burns up meteors that would bombard the surface of the earth from space. Without the atmosphere there would be no sound and no flight. There would be no conventional long-distance radio communication, for this is dependent on the electrons in the upper atmosphere. Without air there would be no lightning, no clouds, no wind, no rain, no snow, and no fire. The surface of the earth would be as bleak and sterile as the moon. (Figure 3.1.)

The atmosphere shields the earth from lethal concentrations of ultraviolet radiation. It selectively filters the sun's rays so that only two small segments of the electromagnetic spectrum penetrate to the earth's surface in appreciable amounts.[1] One segment is the *optical window,* consisting essentially of the visible spectrum of light, from near-ultraviolet to near-infrared. The other is the *radio window,* consisting of radio waves from about 1 centimeter to 40 meters in length.

It is impossible to define the limits of the atmosphere because the atmosphere becomes progressively tenuous with increasing distance from the earth. There is no boundary between the atmosphere and the void of outer space. However, 75 percent of the earth's atmosphere lies within 10 miles of the surface, and 99 percent of the atmosphere lies below an altitude of 30 kilometers (19 miles). The total mass is estimated at 5,500 trillion tons.

Clean, dry air at sea level contains, by volume, 78 percent nitrogen gas, 21 percent oxygen, 1 percent argon, and 0.03 percent carbon dioxide, plus minute amounts of other gases. The latter, in approximate order of decreasing concentration, include neon, helium, methane, kryp-

---

[1]The electromagnetic spectrum includes the full range of radiation. Radio waves are on the long wave-length end; x-rays and gamma rays are on the short wave-length end. The visible spectrum is in between.

43

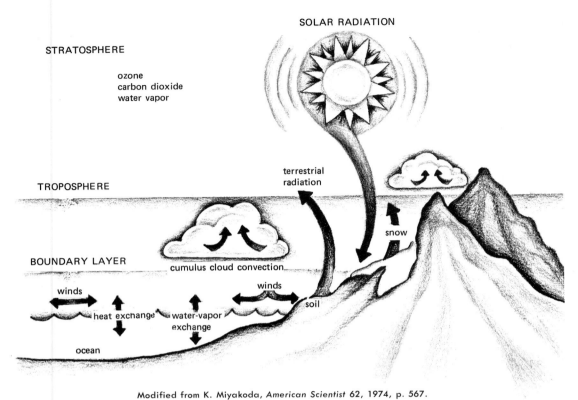

Modified from K. Miyakoda, *American Scientist* 62, 1974, p. 567.

FIGURE 3.1 An atmospheric model.

ton, sulfur dioxide, hydrogen, nitrous oxide, xenon, ozone, nitrogen
dioxide, iodine, ammonia, and carbon monoxide (see Table 3.1).

The nitrogen gas is biologically inert for most organisms, except
under special conditions to be discussed later. It normally takes no part
in the role of animal life except as a diluent for the oxygen and carbon
dioxide, but it is utilized by some bacteria and some plants. The oxygen is
essential to the production of biological energy in the burning of cellular
fuel. Carbon dioxide, despite its low concentration (less than 1/600 that of
oxygen), plays a vital role in the cellular processes of animals and plants.
Most important, it is needed by plants in photosynthesis, by which the
green pigment chlorophyll uses carbon dioxide, water, and sunlight to
synthesize sugars and starches. In animals, carbon dioxide is a waste
product, but it also regulates the rate of breathing. Carbon dioxide also
performs an important function in controlling the acidity-alkalinity balance of the blood by its presence in the form of carbonic acid, identical to
the product in carbonated water.

However, all of the major, and some of the minor, atmospheric components are biologically important, and the participation of each in the
living processes is in some cases critically sensitive to slight changes in
concentration.

OXYGEN

The ancestors of all air-breathing animals had their origins, paradoxically, in the sea. Even in man and other higher animals dedicated to a terrestrial life, evidence of their aquatic ancestry can be seen in the gill crevices in the neck of the fetus.[2] The early embryos of a fish, a bird, a pig, and a man look so much alike during the gill-crevice stage of development that it is difficult for anyone but an expert to tell them apart under a microscope (see Figure 3.2). As in other land-based animals, as well as in those land animals that have returned to the sea, the gill crevices do not develop into gills, which are the respiratory organs of many aquatic animals, but disappear early in embryonic development by transforming into other structures.

The migration from water to air was of tremendous advantage to organisms dependent upon oxygen. Whereas the oxygen content of well-aerated water is no more than 1 percent, the air at sea level contains more than 20 times as much oxygen. Moreover, the diffusion rate of oxygen in air is 45,000 times that of oxygen in water. Air-breathing animals took a giant step, figuratively and literally, when they first crawled out of the slime to set foot on land.

Breath of life   Oxygen is absorbed from the lungs into the bloodstream of man by diffusion through the delicate lining of microscopic sacs called *alveoli* (the singular is *alveolus*). There are about 750 million alveoli in a set of human

Table 3.1   Composition of Clean, Dry Air at or Near Sea Level

Component	Percent by volume
Nitrogen	78.084
Oxygen	20.9476
Argon	0.934
Carbon dioxide	0.0314
Neon	0.001818
Helium	0.000524
Methane	0.0002
Krypton	0.000114
Sulfur dioxide	0 to 0.0001
Hydrogen	0.00005
Nitrous oxide	0.00005
Xenon	0.0000087
Ozone	Summer 0 to 0.000007
	Winter 0 to 0.000002
Nitrogen dioxide	0.000002
Iodine	0 to 0.000001
Ammonia	0 to trace
Carbon monoxide	0.1 ppm—0.2 ppm

From H. J. Sanders, ed., "Chemistry and the Atmosphere," *Chemical and Engineering News Special Report.* Washington, D .C., The American Chemical Society.

---

[2]Gill crevices are vestiges of pharyngeal gill slits present in primitive vertebrates presumed to be similar to the ancestors of higher forms.

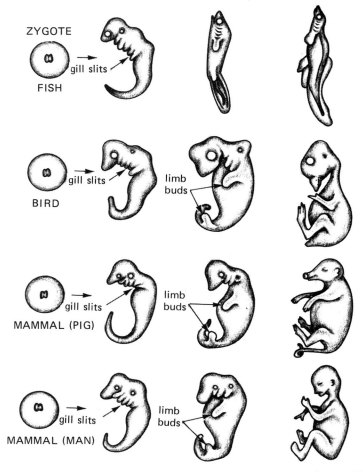

After Haeckel, from C. P. Hickman, *Integrated Principles of Zoology*, Fourth Edition, 1970, by permission of the C. V. Mosby Company, St. Louis.

FIGURE 3.2 Embryos of four vertebrate species in comparable stages of development. Note that the early stages of fish, bird, pig, and man look very much alike. Distinctive characteristics show up in later stages.

lungs and they provide a surface exposed to the air of more than 1,000 square feet. Thus the lungs have 50 times as much surface exposed to the air as the surface of the skin. These tiny bulbous alveolar organs of the lungs are richly lined with blood capillaries so small in diameter that they can barely accommodate a blood cell. Gas exchange takes place in both directions across the alveolar membrane by inward diffusion of oxygen and outward diffusion of carbon dioxide. Other gases and substances that may be in the air naturally or as contaminants also readily pass through the membranes of the alveoli. We will consider examples of these later.

The alveoli are easily and irreparably damaged by dust, smoke, and chemical pollutants. *Emphysema* is a condition caused by a deterioration in the membranous walls of the alveoli. The damage reduces the surface area that is normally available for gas exchange. The result is an impairment of the inward diffusion of oxygen through the alveolar membrane into the blood stream and the outward diffusion of carbon dioxide into the air. Emphysema is a form of suffocation. Smoke, dust, tobacco tars, and smog are contributing causes.

Blood: the oxygen carrier

A major function of the blood is to carry oxygen from the point of entry at the alveolar membranes to the tissues of the body. The blood plasma contains 2 to 3 percent dissolved oxygen. But its capacity is limited, and animals that have to depend on the plasma for oxygen transport must be small or have other mechanisms for transferring oxygen and carbon dioxide within their bodies. In higher animals most of the oxygen is carried in the red blood cells, which are essentially bags of hemoglobin, the oxygen-combining molecule.

The red pigment hemoglobin is a complex protein made up of about 95 percent globin, which is a colorless protein material, and 5 percent hematin, which contains iron and gives the red blood cells their color. Each red blood cell contains about 280 million molecules of hemoglobin (Hb) and every molecule of hemoglobin contains 4 atoms of iron (Fe), each sequestered by 4 nitrogen (N) atoms in a complex structure called a *heme* group (see Figure 3.3). Each heme group can combine with a molecule of oxygen to form *oxyhemoglobin*:

$$Hb + 4O_2 \rightarrow Hb.\ 4O_2$$

A molecule of hemoglobin contains about 10,000 atoms and has a molecular weight of 67,000. Combination with oxygen changes the shape of the hemoglobin molecule in a way that gives it a brighter color, so that arterial blood (from the lungs) is more crimson than veinous blood (returning to the lungs). There are normally about 5 million red blood cells in a cubic millimeter of human blood, equal to about 3 trillion in a cubic inch. In men, 100 milliliters of blood (slightly more than ¼ pint) contain about 16 grams hemoglobin. In women it averages about 14 grams (about ½ ounce) of hemoglobin.

Each red blood cell emerges from the capillaries of alveoli with more than a billion molecules of oxygen. Thus hemoglobin increases the oxygen-carrying capacity of the blood 70 times that of water. Instead of a

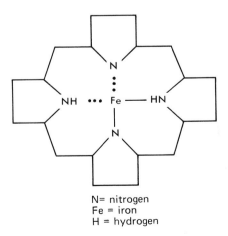

N= nitrogen
Fe = iron
H = hydrogen

FIGURE 3.3 The core of a *heme* group in the hemoglobin molecule. There are four heme groups in the molecule; each heme can carry one molecule of oxygen.

4½-second reserve of oxygen, hemoglobin gives the blood a 5-minute reserve. The bond between hemoglobin and oxygen in oxyhemoglobin is essentially a weak one, and it readily releases the oxygen to the tissues. Whether the hemoglobin takes up or lets go of the oxygen depends on the conditions at the site of action.

Oxygen deficiency

An adult at rest and breathing quietly ordinarily takes into the lungs about ½ liter of air with each breath. At a regular respiration rate of 15 or 16 times a minute, a man (it is higher for women) will inhale and expire 7½ to 8 liters of air a minute, or 475 liters an hour. This is equal to about 125 gallons of air an hour, or 3,000 gallons during a 24-hour day. During strenuous exercise, the higher oxygen demand causes an increase in both the rate and depth of respiration, so that 10 to 20 times as much air will be used in a given period.

The air, like all substances, has weight and is prevented from flying off into space by the pull of the earth's gravity. It is evident that at high altitudes the lungs have to "work" harder because the concentration of oxygen is lower and there is less diffusion pressure to force the molecules through the membrane. Aircraft pilots in nonpressurized cabins at high altitudes must use oxygen masks. At altitudes above 38,000 feet a pressurized cabin is desirable.

The density of the air, that is, the number of molecules in a given volume, at an altitude of 8,800 feet is less than three-fourths the density at sea level. One of the dangers of high-altitude flight is that in spite of the greater demand for air, there is no increase in the carbon dioxide content of the blood, which is the chemical stimulant responsible for inducing the medulla oblongata of the brain to send out the signal for the body to increase the rate of respiration. In fact, the effect may be the reverse of that desperately needed to replenish the dwindling oxygen supply of the tissues, particularly the brain. The reverse effect on the medulla oblongata merely enhances *hypoxia* (deficiency of oxygen), which produces giddiness and eventually irrational behavior and blackout.

A partial safety factor at high altitudes is that the respiratory center in the medulla oblongata also responds to nerve signals from the carotid artery. An increase in the lactic acid content of the blood, such as that resulting from muscular activity or deficiency of oxygen, stimulates the carotid artery whose nerves then send a message to the brain to increase respiration. This, however, does not long overcome the dangerous effect of oxygen-thin ar.

Oxygen toxicity

Too much oxygen, called *hyperoxia,* is as dangerous as oxygen deficiency. Pure oxygen at atmospheric pressure can be tolerated without harm for about 24 hours by the average person. After that time, symptoms of irritation appear followed by fluid in the lungs and other discomforting symptoms. Hospitals often administer oxygen continuously for days or weeks, for which purpose 60 percent oxygen is considered to be the maximum safe concentration, and even lower concentrations are sometimes needed to avoid danger.

Scuba and deep-sea divers would have a similar problem if they were to breathe pure oxygen, because the pressure at a depth of 33 feet is twice that at the surface and increases an additional "atmosphere" each 33 feet. At a depth of 100 feet the pressure is four times that at the surface. A

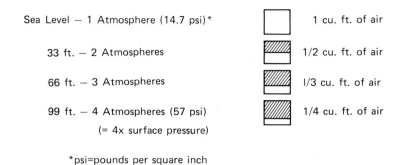

Sea Level — 1 Atmosphere (14.7 psi)*   1 cu. ft. of air

33 ft. — 2 Atmospheres   1/2 cu. ft. of air

66 ft. — 3 Atmospheres   1/3 cu. ft. of air

99 ft. — 4 Atmospheres (57 psi)   1/4 cu. ft. of air
(= 4x surface pressure)

*psi=pounds per square inch

FIGURE 3.4 Effect of compression.

diver breathing pure oxygen will eventually suffer oxygen toxicity. There-fore, compressed air is generally used, although mixtures of oxygen and gases other than nitrogen, such as helium, have been used for deep diving.

According to Boyle's law of gases, the volume occupied by a gas is inversely proportional to the pressure. If the pressure is quadrupled by descending 100 feet beneath the surface of the ocean, the same amount of oxygen now occupies only one-fourth as much space. The concentration of the oxygen molecules is four times what it was at the surface and there is four times as much diffusion pressure against the alveolar membranes (see Figure 3.4). The same is true of the other gases in the air.

The effect of too much oxygen resembles the toxic effect of the power-ful poison strychnine. There is a profound effect on the muscular system and the entire central nervous system. The victim is liable to have headache and nausea recurring perioically, muscular twitching, dizziness, impaired vision, irritability, and possibly convulsions that resemble epileptic seizures. Headache, drowsiness, and dizziness may persist for several hours afterward. The effects appear sooner the deeper the diver goes, although there is individual variation in susceptibility.

World supply of atmospheric oxygen

No change in oxygen content of the air can be detected in records since 1910. According to calculations by Broecker,[3] each square meter (10.76 square feet) of the earth's surface is covered by 60,000 moles (equal to more than a ton) of oxygen gas. Terrestrial, aquatic, and marine plants produce about 8 moles of oxygen gas annually for each square meter of the earth's surface. However, nearly all of this product of photosynthesis is utilized by animals and bacteria with the result that the amount that they consume is practically equal to the oxygen released into the atmo-sphere by plants.

There is a small net addition of oxygen to the atmosphere equal to about 1 part in 15 million parts of the oxygen present. It is probable that much of the excess is removed by oxidation of carbon, iron, sulfur, and other minerals during the normal process of weathering. Unless cata-clysmic changes occur, the oxygen content of the atmosphere is not sus-ceptible to significant change over a geologically short-range time scale of 100 to 1,000 years.

[3]W. S. Broecker, "Man's Oxygen Reserves," *Science,* 168:1537–8 (1970).

NITROGEN

Nitrogen gas is the most abundant component of the atmosphere, comprising 78 percent by volume. In some bacteria and blue-green algae nitrogen fixation is an important means of augmenting the nutrient supply, especially for higher plants when they grow in association with the nitrogen-fixing microorganisms.

Caisson's
disease

Although nitrogen comprises the major bulk of the air, it is almost inert insofar as animals are concerned—almost, but not quite, for it appears to be innocuous until one breathes air under increased pressure, such as when scuba or deep-sea diving, at which time the effects of this treacherous gas can be catastrophic. Nitrogen is a deadly threat to divers, who may fall victim to absorption of the gas into the blood stream. This affliction is known to divers as the "bends," or Caisson's disease.

The high partial pressure of nitrogen during dives forces the gas readily through the alveolar membranes. Under a pressure sometimes several times normal, depending on depth, nitrogen diffuses rapidly into the blood stream. But the tissues have no mechanism for chemically absorbing it as with oxygen and carbon dioxide. Nitrogen can be eliminated only through the lungs. Under high pressure it remains in solution in the plasma and during deep dives the nitrogen in the blood is thus greatly increased. If ascent is too rapid, diffusion through the alveolar membrane does not take place quickly enough to allow the nitrogen to leave the blood and return to the air in the lungs. Instead, the nitrogen may bubble out in the blood vessels like gas from a bottle of soda water when the pressure is suddenly released by removing the cap. The "bends" is a serious and sometimes fatal condition, prevented by rigidly observing the rules for slow return to the surface, for which instruction and practice are required.

Rapture of
the deep

Breathing nitrogen under pressure has an intoxicating effect. At depths of about 100 feet and below, sufficient to increase the pressure to 4 atmospheres or more, the amount of nitrogen forced into the blood stream through the alveolar membrane reaches a concentration at which the nitrogen has a narcotic or anaesthetic action. It is a delightful but dangerous sensation. The condition is called *nitrogen narcosis* or "rapture of the deep."

Nitrogen narcosis is similar in effect to alcohol intoxication. The diver becomes uncoordinated and lacks judgment. The most dangerous part is that under the influence of nitrogen narcosis, the diver may want to go deeper, which could be fatal. Nitrogen narcosis becomes so severe at a depth of 300 feet that a man may be unable to perform even simple tasks. But the affected diver recovers quickly upon return to the surface.

Nitrogen
fixation

All living organisms need nitrogen in large amounts. Nitrogen is a key element in the amino acids as well as in many other dynamic molecules of the protoplasm. The use of nitrogen compounds in the metabolic manufacture of nutrients within the tissues and the fate of nitrogen products during and after decomposition are part of a complicated series of processes referred to as the *nitrogen cycle* (see Figure 2.8). Much of the nitrogen in animals, plants, and their waste products is returned to the soil and reused by plants as part of the cycle. However, microorganisms called *denitrifying bacteria* act on soil nitrates and nitrites and in so doing produce molecular nitrogen (nitrogen gas), which returns to the atmosphere.

Divers are apt to be affected by nitrogen intoxication—"rapture of the deep"—at depths beyond 100 feet.

Probably all of the nitrogen available for food in the soil would be lost if it were not for another part of the nitrogen cycle in which atmospheric nitrogen is eventually returned to the soil.

Nitrogen gas is a chemical substance of moderate chemical stability and does not readily enter into chemical reactions. But its abundance in the atmosphere provides an almost unlimited source of nitrogen. In nature, there are primarily two ways in which the nitrogen is returned to the soil. One is by the energy of lightning which forms small amounts of nitrogen oxides from molecular nitrogen. The nitrogen oxides drop to the earth in moisture and become available for further modification by microorganisms and as plant nutrient. A more important way in which nitrogen becomes "fixed" is by the action of nitrogen-fixing organisms among the algae and bacteria. It is one of the most crucial links in the nitrogen cycle.

Two groups of bacteria are largely responsible for nitrogen fixation: those which live in the soil independently of higher plants, and those which grow in association with higher plants and have a mutual relationship with them. The relationship of two species of organisms living together in an intimate association is called *symbiosis*. The particular kind of symbiosis in which different kinds of organisms living together are of benefit to each other is called *mutualism*.

There are two types of nonsymbiotic nitrogen-fixing bacteria, those which require oxygen and those which do not. *Aerobic* (oxygen-requiring) bacteria called *Azotobacter* do the work in alkaline soil, while *anaerobic* bacteria with the formidable name *Clostridium pasteuranium* work in acid soil. (*Anaerobic* means that it does not depend on oxygen for survival.)

The symbiotic bacteria are the most important nitrogen-fixing organisms. Several kinds of bacteria called *Rhizobium* have a cooperative arrangement with a group of plants called legumes, such as peas, beans, alfalfa, and vetch, whereby the two dissimilar kinds of organisms jointly manufacture nitrate from nitrogen gas. To accomplish this, *Rhizobium* invades the roots of the pea plant or other legume, where it causes innumerable swellings or nodules packed with bacteria. Without *Rhizobium,* the root tissue of the pea plant could not produce nitrate.

Man has devised a way for nitrogen fixation that is probably as important, at least in agriculture, as the sum total of electrical discharges in the atmosphere and biological activity. The most widely used industrial processes for nitrogen fixation use only nitrogen and hydrogen as raw materials. Large amounts of energy are needed to compress the gas mixture to 5,000 pounds per square inch. The crucial step produces ammonia:[4]

$$N_2 \quad + \quad 3H_2 \quad \rightarrow \quad 2NH_3$$

$$\text{nitrogen} \qquad \text{hydrogen} \qquad \text{ammonia}$$

The ammonia is used in anhydrous form as a commercial fertilizer and as a starting point for the manufacture of other nitrogen-containing products for industry and agriculture: aqua ammonia, ammonium salts, urea, and nitrates. Urea is used as a fertilizer, a cattle feed additive, and an industrial chemical.

CARBON DIOXIDE

Carbon dioxide gas is a minor constituent of the atmosphere, comprising about 0.03 percent by volume of the air. By comparison, the air contains 667 times as much oxygen and 2,487 times as much nitrogen. Yet, without carbon dioxide no life could exist, for it is vital to the production of sugars and starches in plants by the process called *photosynthesis* and is a source of carbon, the basic building block for organic compounds in metabolic synthesis. Without photosynthesis there would be no plants, and without plants there would be no animals. Even carnivorous (flesh-eating) animals and parasites must rely for food either on herbivorous animals, which feed on green plants, or on other organisms which derive their carbohydrates (sugars and starches) from photosynthesis.

Light of life: photosynthesis

The sugars and their more complex relatives, the starches, collectively called *carbohydrates,* are produced in green plants from carbon dioxide and water. The reaction is driven by atomic energy—from the fusion fireball, the sun, 93 million miles away, which pumps energy into space in the form of electromagnetic radiation. Packets of light energy called *photons,* traveling at a speed of 186,000 miles a second on their pathway of escape from the sun, penetrate to the earth through the "optical window" of the atmosphere. Energy equal to 100 million Hiroshima bombs reaches the earth every day. Part of the radiation is absorbed by green plants, which use it as the energy for the reaction of photosynthesis.

Photosynthesis takes place in specialized pigments of the plant. The green pigments, in which the reaction mainly takes place, are collectively

---

[4]The first step in the manufacture of ammonia from natural gas is to treat it with steam to convert the natural gas to carbon oxides and hydrogen.

called *chlorophyll,* usually packaged in granules called *chloroplasts,* which can be seen easily under the microscope and are visible in the green coloration of plants. It is a curious fact that the chlorophylls are a type of structure called *porphyrins,* related in structure to the porphyrin in hemoglobin. Instead of iron as in hemoglobin, the porphyrin structure of the chlorophylls contains magnesium.

Light in the green part of the spectrum can be used only slightly by chlorophyll, for most of it is reflected and comprises a large part of the "color" of foliage. The light rays that energize chlorophyll most effectively are in the blue, orange, and red parts of the visible spectrum. Figure 14.1 shows the visible light in relation to the full spectrum of electromagnetic radiation.

The capture of solar radiation by plants and its utilization to energize the synthesis of sugar from carbon dioxide and water by photosynthesis is the most important biological exploit of nature. The biological mechanism of the process has fascinated scientists since the beginnings of plant studies. Yet the process remained one of the most puzzling and stubbornly resistant problems in biological science until a series of brilliant laboratory experiments, made possible by the new techniques of radioactive tracers, removed much of the cloak of mystery from the biochemistry of photosynthesis.

In 1949 Melvin Calvin, a biochemist at the University of California at Berkeley, set out to untangle the confusing knot of knowledge, or the lack of it, concerning the processes by which the green plant manufactures carbohydrates. Calvin and his co-workers used the radioactive isotope of carbon, carbon-14 (usually written ^{14}C), which they incorporated in molecules of carbon dioxide. Their experimental plants were placed in an atmosphere in which part of the carbon dioxide contained carbon-14. Calvin found that the reactions of photosynthesis occurred in split seconds. By stopping the process after exposure of the plants to the radioactive environment for seconds at a time, Calvin was able to follow the very first steps of photosynthesis. For this work Calvin received the 1961 Nobel Prize in chemistry. Subsequent work with both radioactive carbon and a heavy isotope of oxygen, oxygen-18 (usually written ^{18}O), has further elucidated the complicated steps in photosynthesis. For details of the process, the reader is advised to refer to one of the numerous books on biology, many of which contain excellent descriptions of photosynthesis. The following overall equation summarizes the basic process:

$$CO_2 \ + \ H_2O \ \xrightarrow{\ + \text{ Light}\ } \ (CH_2O)_x \ + \ O_2$$

carbon dioxide   water   carbohydrate   oxygen

The oxygen that goes into the makeup of the carbohydrate comes from the carbon dioxide, while the oxygen that is liberated as free gas comes from the "splitting" of water. A more complete overall equation is frequently given, although even this is a gross simplification of an intricate biochemical mechanism:

$$6CO_2 \ + \ 12H_2O \ \xrightarrow{\ + \text{ Light}\ } C_6H_{12}O_6 \ + \ 6O_2 \ + \ 6H_2O$$

carbon dioxide   water   glucose   oxygen   water

The equation represents the conversion of electromagnetic energy (in the form of light rays) to chemical energy (stored in the form of a sugar). Glucose (dextrose), the simplest sugar, contains 670 calories of available energy per 180 grams (¼ pound).

The process of photosynthesis cannot be duplicated artificially. However, changes in the carbon dioxide content of the air can profoundly modify the natural process. Experiments by James Riley and Carl Hodges at the University of Arizona to determine the response of plants to enrichment of the air with carbon dioxide showed that plant growth increased with increasing carbon dioxide concentration. Plant yields were greatly increased at 2,500 ppm (parts per million). The optimum concentration for acceleration of photosynthesis with the species of plants used in the study was around 5 times the normal concentration of carbon dioxide in air. But above an optimum concentration, further increases caused a reduction in growth. The effectiveness of higher carbon dioxide concentrations in bringing about an increase in growth was largely dependent on the degree of turbulence at the surface of the leaves. It is the increase in ventilation resulting from turbulence, rather than greater exposure to light, that causes the "border effect," an increase in growth often observed in plants growing along the border of a field.

Effect on humans   Carbon dioxide is ordinarily not critical insofar as its concentration in man's environment is concerned. The carbon dioxide content of the air is about 320 ppm. A person can tolerate nearly 5,000 ppm, or more than 15 times the normal concentration, without experiencing any seriously adverse effect on respiration, although prolonged exposure to such a high concentration might prove to be deleterious.

During quiet breathing, an adult person puts out 250 to 300 milliliters (about 1 pint) of carbon dioxide a minute. Exercise increases the output. We do not have to consciously and deliberately breathe harder to get rid of it, because carbon dioxide controls the breathing rate. This is brought about by a buildup in the concentration of carbon dioxide in the blood as a result of the concentration gradient across the alveolar membrane and its effect on carbon dioxide diffusion. The respiratory center in the medulla oblongata is extremely sensitive to carbon dioxide in the blood and responds quickly to slight changes in concentration. This can be demonstrated by voluntary rapid breathing to ventilate the lungs, during which the alveolar carbon dioxide is reduced. There will be no impulse to breathe until the carbon dioxide again builds up. The action of carbon dioxide as a respiratory mediator is reinforced by chemoreceptor nerve endings in the carotid artery which are sensitive to oxygen and carbon dioxide in the blood.

The carbon cycle   Carbon dioxide and water are the ultimate waste products from the oxidation of cellular fuel in living organisms. In animals, the elimination of waste carbon dioxide is one of the chief functions of the alveolar membranes through which the gas passes from solution in the blood to the air spaces of the lung sacs and is then expired to the outside. Thus, the molecules of carbon that are taken up by plants in the form of carbon dioxide from the air, and which become the carbon-backbones of organic compounds in living tissue, again return to the atmosphere when the organisms respire or decay. This is part of the *carbon cycle* (Figure 2.9).

Not all the organic material is oxidized and returned to the atmosphere immediately. Much of it becomes imbedded under layers of sediment where it may form coal, petroleum, or natural gas. The carbon in these vast underground graves of long-dead plants or animals may remain entrapped there for hundreds of millions of years before it again returns to the atmosphere in the form of carbon dioxide through the intervention of man when he burns the fossil remains for fuel.

The atmosphere is the repository for vast quantities of carbon dioxide from the burning of fuels. Plants may not be able to withdraw it fast enough to cope with the carbon dioxide expelled into the atmosphere from industrial waste. The dangers of excess carbon dioxide in the atmosphere are discussed in Chapter 6.

READINGS

BROECKER, W. S., "Man's Oxygen Reserves," *Science* 168:1537–8 (1970).

Chemistry and the Atmosphere, C&EM Special Report. Sanders, H. J., Assoc. Ed. *Chem. & Eng. News,* Washington, D.C., American Chemical Society.

GAMOW, G., *A Planet Called Earth.* New York, Bantam Books, 1970.

MACHTA, L. and E. HUGHES, "Atmospheric Oxygen in 1967 to 1970," *Science* 168:1582–4 (1970).

THE HYDROSPHERE

Water is the most useful natural resource on earth, economically, culturally, and biologically. We drink it, eliminate with it, bathe in it, relax in it, fish in it, cook in it, cool with it, irrigate plants with it, and use it for energy, power, transportation, and recreation. Though water is seemingly abundant, the uneven distribution of usable water creates a serious conservation problem in many places where it is vitally needed. In such areas the purity of the water becomes critical. Rivers and lakes in highly industrialized locations may carry an intolerable burden of chemical and human waste products to the point that aquatic life in its natural habitat is obliterated and human health is threatened. Even waters used for irrigation of crops may have excessively high concentrations of salts as the result of leaching or, along coastal areas, from underground seawater intrusion. Wells in such areas become nearly useless as sources of water for agricultural or domestic purposes.

CIVILIZATION AND WATER

About 10,000 years ago or earlier, somewhere in one of the great river valleys of the Nile, the Indus, or more probably along the Tigris and Euphrates Rivers of Mesopotamia, someone hit on an idea that was destined to change the future of mankind. He dug a furrow in the silt and a trickle of water brought the moist spark of life to the parched desert. The man with the wooden hoe did more than make it possible for the seeds of grain to germinate and start the seasonal cycle of crops to lay up a supply of food. He had watered the seed of civilization.

Great irrigation systems eventually spread throughout the reaches of the ancient Mesopotamian plain. They required a highly developed organization for the engineering of a complicated system of river diversions, canals, sluice gates, ditches, and levees and for the systematic dredging of silted-up canals. A central administration for irrigation was needed with the authority of law for the allocation of water and the collection of taxes for maintenance of ditches, canals, and levees as well as for the other functions of the state. Supportive industries became established. A stable food supply was the key to the future. Villages became permanent and some of them grew into towns and cities. The first city-states came into being and kingdoms thrived.

About 300 B.C., Alexander the Great and his army of tough adventurers thrust deeply into Persia and the unknown land beyond. They came to a mighty river with more than twice the flow of the Nile, and on its banks was an ancient civilization. A thousand years before Alexander, the Aryans had invaded the continent and settled on the river. They did not give it a particular name but called it the Indus, the Aryan word for "river." The conquered land became India, the "land of the river." The Indus was destined to irrigate 23 million acres of land, the largest irrigated agricultural region on earth. Civilizations sprang up along the other waterways. In Egypt, the Middle East, India, and China, as well as in South and Central America, ancient civilizations had their origins in the development of the water supply for irrigation and the production of a dependable food supply.

Modern civilization is dependent on water for irrigation, industry, domestic needs, shipping, and, of increasing importance, for sanitation and waste disposal. Most of the areas of the world that are without developed water remain in the hunting and gathering or grazing stage. Civilization's further advance, and possibly the survival of our cultures, will require intensive study of the development of water supplies and careful attention to the protection of water quantity and quality on a worldwide scale.

THE WET PLANET

A man in space can look down on the earth and see that the surface of the planet is mainly an ocean. It is continuous except for interruptions by numerous islands from pinpoint atolls to those the size of continents. If

Right photo USDA photo.

Left photo: Water is an important limiting factor in the growth of plants. Here irrigation makes it possible to grow crops in a desert area. *Right photo:* Water impounded by Coolidge Dam on the Gila River in Arizona supports agriculture of the Salt River Valley.

Photo courtesy of National Aeronautics and Space Administration.

The cloud-whirled blue planet Earth. A traveler arriving from space would observe that its surface is predominantly water, as seen in this view of the Pacific Ocean from Apollo 11. It is late afternoon in western North America (upper right). At the top, the north polar cap gleams white.

Mount Everest, standing at 29,028 feet, were put into the deepest spot in the ocean, its peak would be more than a mile beneath the surface. The ocean occupies 70 percent of the surface and contains 97 percent of all the water on earth.[1] Much of the remainder is frozen in the icecaps and glaciers. By comparison, the water in rivers and lakes is small. Less than 1 percent is in the form of ice-free fresh water in rivers, lakes, and aquifers. Yet this relatively negligible portion of the planet's water is crucially important to all forms of terrestrial and aquatic life. There is also a large underground supply of water. Much of it remains locked deep underground for long periods of time. But the soils near the surface also serve as reservoirs for enormous quantities of water, eventually lost through evaporation or by seepage into underground storage.

The vapor state   Some of the earth's water is in the atmosphere. The warmth of the sun and the air currents evaporate about .033 percent of the water on the surface each year. The vapor condenses and returns as rain and snow and eventually finds its way to the bodies of water to complete the cycle. (See Figures 4.1 and 4.2.) The world annual rainfall is about 480,000,000,000,000 tons of water (see Figure 4.3).

The solid state   Much of the earth's water is in cold storage. Glaciers and the icecaps cover 11 percent of the world's land area; icebergs and pack ice occupy 25

---

[1]The total water on earth is estimated at 326,000,000 cubic miles, equal to 1,440,000,000,000,000,000 tons.

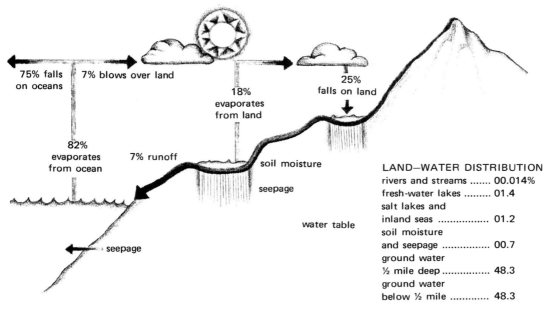

75% falls on oceans

7% blows over land

18% evaporates from land

25% falls on land

82% evaporates from ocean

7% runoff

soil moisture

seepage

water table

seepage

LAND–WATER DISTRIBUTION
rivers and streams 00.014%
fresh-water lakes 01.4
salt lakes and
inland seas 01.2
soil moisture
and seepage 00.7
ground water
½ mile deep 48.3
ground water
below ½ mile 48.3

From the *Conservation Yearbook*, 1967, U.S. Dept. Interior.

FIGURE 4.1 The water cycle.

percent of the ocean area. Permafrost—permanently frozen ground—holds another 10 percent of the land area in its grip, while 30 to 50 percent of the land is covered with snow at any given time. Three-fourths of all fresh water is locked up as ice, mostly in Antarctica and Greenland. Because the cold regions of the earth contain vast resources in minerals, petroleum, timber, and water, the rapidly increasing demand will surely bring more intensive development and industrial activity in those areas. However, environmental problems of the cold regions differ greatly from those in the temperate and tropical parts of the world. Disturbances of the environment that may be mildly disruptive in tropical and temperate areas could be excessively damaging in cold regions. The Soviets are planning a bold and imaginative project using nuclear explosives that would blast out a canal more than 70 miles long across northern Russia. The canal would divert part of the Pechora River, which flows north to the Arctic Ocean, into a tributary of the Volga River and, thence, into the Caspian Sea in order to stabilize the level of that inland sea. But reducing the flow of the Pechora's relatively warm water into the Barents Sea could cause a major climatological backlash, if several theories of American scientists are correct (see Figure 4.4).

Small changes in world climate drastically affect the earth's distribution of water supplies. Glaciers are highly sensitive to climatic trends. Observations indicate that many alpine and valley glaciers have been shrinking during the past 100 years, although some have been increasing. As yet, there is not enough data to prove whether the major ice sheets are shrinking, growing, or in a state of equilibrium. It is important to know. Useful knowledge would be derived from studies of past fluctuations inasmuch as the information might provide a means for predicting future

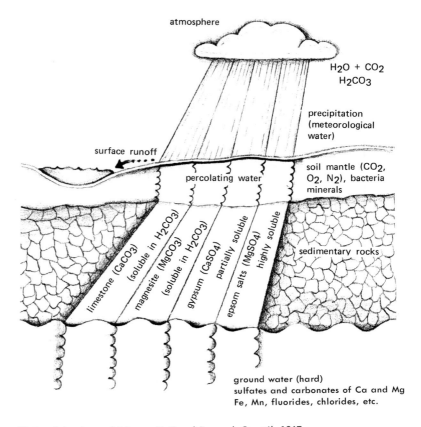

From National Academy of Sciences-National Research Council, 1967.

FIGURE 4.2 Quality relationships between surface waters and ground waters.

trends. Drilling ice cores in glaciers to compare the water, dust, and isotope content of different layers makes it possible to determine precipitation during past centuries. For further discussion of climate and glaciers see Chapter 6.

LIQUID OF LIFE

Water is the essence of the living process. It is the most abundant and the most versatile of the chemicals of life. It is not surprising that in its physical and chemical properties alone, water is the most uniquely remarkable substance known, natural or synthetic. Its importance to life demands an understanding of its role in the process of living organisms and in human life.

Water forms most of our flesh and blood. Protoplasm is mostly water.[2] The content varies in different tissues of the organism and in different plants and animals. In jellyfish, the protoplasm contains 95 percent water. Human blood is 90 percent water. Muscles and nerves contain 80 percent

---

[2]*Protoplasm* is defined as "living substance," usually referring to colloidal material within cells.

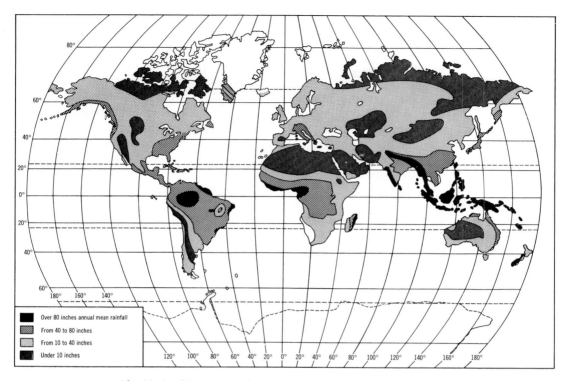

After National Science Board, National Science Foundation, 1972.

FIGURE 4.3 Worldwide annual precipitation.

From Richard Marra, *Sea Secrets*, 1970; courtesy International Oceanographic Foundation.

FIGURE 4.4 Three Russian rivers supply the Arctic Ocean with about half its fresh water. Russian plans to divert that water to the arid Caspian-Aral region might cause enormous changes in the climate of Europe.

or more water. The bones contain smaller amounts of water, and even the enamel of the teeth contains 2 percent water. The total human body consists of 60 percent to 90 percent water, part of the variability being due to the fact that the tissues tend to become dehydrated with age.

There is a similarity between seawater and the dissolved salts in blood (see Figure 4.5). This is not surprising, for many people believe that life was born in the sea. Landlubbers such as man and other terrestrial vertebrates reveal their aquatic ancestry by the presence of gill slits during early fetal development as well as by other embryonic similarities to primitive aquatic organisms more or less representative of the ancestral type. (See Figure 3.2.)

We begin life in a cradle of fluid. Shortly after the fertilized human ovum becomes implanted in the wall of the uterus, membranes appear that are not part of the embryo but form a sort of aquarium for the developing fetus so that it can grow as its fishlike aquatic ancestors did before they abandoned a life in water for the adventure on land. Even the embryo itself is mostly fluid, for water forms most of our body throughout life.

With rare exceptions, animals and plants must have an abundant supply of moisture. Animals and plants that are fugitives from the sea and have managed to escape from the limitations of an aqueous existence cannot survive for long without access to moisture either in the air, the soil, or in food. Some organisms die within seconds if deprived of water. Others must have it within their systems but can survive for years in a dessicated environment. Some animals, for example certain rodents, can manufacture their own water. Humans can withstand a loss of no more than 20 percent of the water in the body without suffering an agonizing death. A healthy person can normally withstand no more than 7 to 10 days without water, but survival time depends greatly on the amount of exertion and loss of water from heat and exposure.

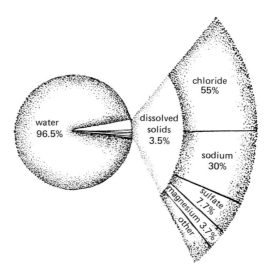

FIGURE 4.5 Composition of sea water.

It is common knowledge that water easily dissolves sugar and salt as well as many other substances. Lakes, streams, rivers, springs, and seas are not pure water but contain a great deal of dissolved substances in solution. The ocean contains an average of about 3.5 percent dissolved solids derived from the land through billions of years of erosion and run-off. The material in solution is mostly common salt (sodium chloride), but there is also an abundance of other mineral constituents, such as calcium, magnesium, sulfates, carbonates, and many metallic elements. Their concentrations are crucially important to the forms of plant and animal life that either select the water as their habitat or merely consume it for survival. The ability to take up and carry materials that are normally solid substances is perhaps the most important property of water in its role as the "solvent of life." Water is the best solvent known. The solvent power of water is due to the ability of its molecules to squeeze into the spaces between the molecules of the solid material and hold them apart (see Figure 4.6). This characteristic enables water to carry the vital chemicals of life—minerals, salts, amino acids, and other organic substances—to the cells of the body. It enables the molecules of water to squeeze their cargo through membranes and deliver them to the sites of action. The return trip may be made with a load of waste products, and the water may be eliminated along with the refuse.

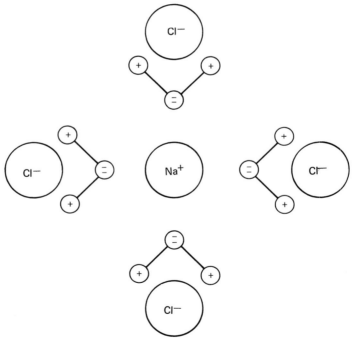

From V. J. Gabianelli, adapted by permission from *Sea Frontiers*, © 1970 by the International Oceanographic Foundation, 10 Rickenbacker Causeway, Virginia Key, Miami, Florida 33149.

FIGURE 4.6 Water molecules are able to "elbow" their way between other molecules, solubilizing them. When less stable compounds are in solution, the water molecules work their way between the atoms and cause them to become ionized. Ionization of common salt is an example.

Terrestrial plants cannot take up mineral nutrients from the soil in the absence of water. Phytoplankton cannot absorb their mineral requirements unless the minerals are dissolved in the water of their environs. Food must be dissolved before it can enter the blood stream of animals. Even the oxygen and carbon dioxide needed by aquatic and marine organisms must be made available to most of them in solution. The products of metabolism are transported within plants in the aqueous solution called sap and within animals they are dissolved in the water of plasma. Many of the waste products of metabolism are carried away dissolved in water.

Transparency of water

In the oceans, as on land, photosynthesis is the first stage in the nourishment of the food chain. Phytoplankton,[3] like land vegetation, thrive on sunlight, carbon dioxide, and mineral nutrients. Fortunately, water is transparent to light. Yet photosynthesis by the producers in the ocean is limited by the relatively poor penetration of light through water compared to air. Some light penetrates clean ocean water to a depth of about 100 fathoms (600 feet). However, most of the photosynthesis takes place in the upper few feet. In clear ocean water, the blue and green portions of sunlight penetrate the best, but these are the wavelengths of visible light that are the least efficient in photosynthesis. Red light is selectively absorbed and therefore is effective only in a narrow zone very near the surface.

Clear water, such as that found in much of the open seas and in many tropical waters, is relatively free of plankton, but there is the compensating factor that light penetration is also greater, resulting in photosynthesis taking place at greater depths. Tropical waters have a high rate of biological turnover because of the higher temperatures. But production of plankton is generally greater in cool waters than in tropical waters. A positive factor is the relatively high solubility of carbon dioxide and oxygen in water of low temperatures.

Lighter than liquid

It is well known that ice floats on water. This is strange behavior for a chemical substance but one that is of great biological importance. Most materials that can be frozen and melted are heavier in their cold or solid state than in their warmer, liquid state. For example, a chunk of iron will sink in a pot of the molten metal.

The ability of ice to float is explained by the molecular behavior of water. The molecules of water are constantly vibrating. When the water is cooled, the vibrations gradually slow down. As cooling proceeds, the molecules crowd together and form an increasingly dense pack. When a temperature of 4°C (39°F) is reached, water is in its most dense condition. Below this point a sudden change takes place. The depressed molecular vibration combined with the strengthened attraction of the *hydrogen bond* causes the molecules to shift their positions to a geometric formation that forms a light, expansive latticework structure. As the water cools further, it continues to expand. When it freezes into ice, it is less dense than water in its liquid form and readily floats instead of sinking to the bottom (see Figure 4.7).

The fact that frozen water floats instead of sinking to the bottom is a

---

[3]*Phyton* (Gr) = "plant"; *plangktos* (Gr) = "wandering." Hence, phytoplankton are marine or fresh-water plants with weak locomotary power that drift with the surrounding water.

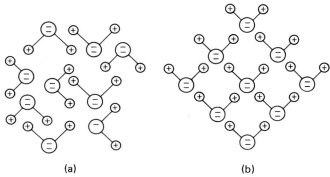

(a)   (b)

From V. J. Gabianelli, adapted by permission from *Sea Frontiers*, © 1970 by the International Oceanographic Foundation, 10 Rickenbacker Causeway, Virginia Key, Miami, Florida 33149.

FIGURE 4.7 As water is cooled, it becomes denser because the motions of the water molecules slow down as the temperature drops. Water reaches its maximum density at 4°C as in (a). But as the temperature drops further from 4° to 0°C, the attraction between the electropositive hydrogen atoms and the electronegative oxygen atoms—called hydrogen bonding—increases, and the water molecules become bonded together in a crystalline lattice, forming ice as in (b). Because of the greater space between the molecules, ice is less dense than water and occupies about 9 percent more volume.

profoundly favorable circumstance for life on the earth. If ice were heavier than liquid water, it would sink to the bottom of rivers, lakes, and frozen seas where it would not receive enough heat from the sun to melt it. Much of the water on earth would be solid ice. Great quantities of the earth's water would be entrapped in an unusable form. Evaporation and precipitation would be greatly reduced. Without moisture in the air, there would be little moderating effect on the sun's radiation and there would be extreme fluctuations in temperature. The world's climate would be drastically altered. Life would be difficult if not impossible. The biosphere as we know it would not exist.

HEATING AND COOLING

When you heat water, it increases in temperature until it boils. Continued heating causes no further increase in temperature. The average velocity of the molecules remains the same no matter how hot the pot. All additional heat is absorbed, and the energy is put to work in breaking up the hydrogen bonds between the molecules, which must take place before the molecules can evaporate.

When the vapor returns to the liquid state it must give up its heat of evaporation. If 1 gram of steam condenses at 100°C, it gives off 540 calories of heat, exactly what it absorbed when it evaporated. This principle accounts for the warmth imparted to objects when vapor condenses on them, and the cooling effect (withdrawal of heat) when moisture evaporates from a surface. This has an important moderating effect in biological systems such as the functioning of the sweat glands; when the temperature rises, the cooling effect of evaporation prevents an excessive increase in body temperature. Evaporation is also important for plants, and accounts in part for the cooling effect in areas of vegetation. It has a cooling effect on moist soil and prevents the surface from becoming as hot as it otherwise would from the direct rays of the sun. It has a similar homeostatic influence on the temperature of bodies of water and thus has an important moderating effect on land temperatures and world climate.

Heat storage   The ability of water to store heat is a characteristic that accounts for much of its biological importance. Water can absorb great amounts of heat while increasing in temperature very little. If you walk in your bare feet over sand, rock, or pavement on a hot day, the heat may soon become unbearable and you will be relieved to step into a pool of water. Although the water has received the same amount of the sun's radiation, it remains refreshingly cool. On the other hand, in the evening the sand and pavement lose their heat but the water stays about the same temperature that it was during the heat of the day. Upon cooling, the temperature of sand drops five times faster than that of water. This great capacity of water to absorb heat, the slowness of water to warm up and cool off, and its ability to give up great quantities of stored heat is summed up in a property called its *specific heat*. The specific heat of a substance is the number of calories required to raise the temperature of 1 gram (about 1/28 of an ounce) of the substance 1°C.

The water in a lake or in the sea gives up 5 times as much heat as the same amount of soil, sand, or rock. This is why a large body of water has a moderating effect on the temperature of the surrounding area. During hot weather when the water receives large amounts of the sun's radiation, the water absorbs great quantities of heat. Equally large quantities of heat are given back to the air during cold weather. Much of the life on earth is dependent on the moderating effect produced by the three-fourths of the globe that is covered by water.

READINGS

GABIANELLI, V. J., "Water—The Fluid of Life," *Sea Frontiers* 16(5): 258–270 (Sept.–Oct., 1970).

NACE, R. L., "Water Resources: A Global Problem With Local Roots," *Env. Sci. & Tech.* 1(7): 550–560 (1967).

REVELLE, ROGER, "Water," *Scientific American* 209(3): 93–108 (1963).

THE BIOGEOSPHERE

THE BIOGEOPHYSICAL ENVIRONMENT

The nature of the biosphere and its relationship to living organisms are determined in large measure by the physical characteristics of the earth. The earth is the fifth largest planet in the solar system, having a diameter of slightly less than 8,000 miles. Its mass (comparable to weight) would be 5.52 times as great as an equal volume of water. This is called its *density*. The surface of the earth is slightly irregular, although this would be hard to discern by an observer in space. For if the earth were reduced to the size of a basketball, the mountains and valleys could hardly be seen with the naked eye and would look almost smooth. The oceans would appear as thin films, barely wetting the surface. It is only when one views these irregularities from the vantage of a comparatively microscopic speck somewhere on the 197 million square miles of the earth's surface that they appear as lofty mountains, deep chasms, and churning oceans.

The land part of the earth is called the *lithosphere,* from the Greek word *lithos,* meaning "stone." The solid part, literally "stonesphere," extends 20 or 30 miles beneath the surface. Beneath that, a zone called the *mantle* consists of heavy rock-like material that flows like sticky gum when under pressure. The deeper the rock, the hotter it is. The core at the center is either molten or, as believed by some geophysicists, so highly compressed that it consists of solid metal, mainly iron and nickel.

The outer portion of the earth is sometimes called the *crust*. It consists of two layers, the *subcontinental* or inner layer and the *continental* or outer layer. The subcontinental layer lies beneath the continents and makes up the ocean floor. It consists mainly of a type of rock called basalt. The continental layer forms all the continents. It is mostly a type of rock called granite.

About 70 percent of the earth's surface is water and about 30 percent is land. Slightly less than 10 percent of the land area is in Antarctica, too cold to be populated by man. The soils of the crust were made from erosion of the rocks by the action of water, ice, and wind. Glaciers tear away pieces of rock, winds literally sand-blast the surfaces, and rivers wear down rock and soil to deposit millions of tons of rocks, mud, sand, and silt in the oceans and other bodies of water every day.

The earth is constantly changing, not only by the erosion of mountains but also by internal distortions that lift parts of the crust to higher levels. Thus, while the material of the mountains is slowly ground away and deposited in the oceans, some of the same material again becomes part of the dry land of mountains and plains.

The history of the earth is written in its crust. Twisting, tilting, upheaval, and distortion clearly show the release of tensions in the earth that changed the surface drastically over millions of years. Volcanic eruptions repeatedly laid new rock over the old. The rocks show the record of glaciers that advanced and receded several times and covered large parts of North America and Europe. There is strong evidence that entire continents, once part of the same land mass, drifted apart over a long period of time. This is called *continental drift*. The history of life on earth from the emergence of primitive plants and animals to the appearance of advanced forms contemporary with man is imbedded in the rocks that were formed by sediments containing the remains of living forms.

GRAVITATION

Gravitation is such an all-pervasive component of the environment that its influence on life is almost taken for granted. Yet its effects are subtle, profound, and far-reaching. Indeed, the existence of life depends on gravitation.

One of the reasons why mice and men have different shapes and structures is the need to overcome the effect of gravity. An upright stance and

Distortion of the earth's crust over millions of years changed the surface drastically.

Volcanic eruptions lay new rock over the old, wiping out plant and animal life, eventually to be replenished, often after erosion and further distortions of the landscape. This is the Kilauea Caldera on the island of Hawaii.

heavy bulk requires a different arrangement of organs, bones, and muscles. The semicircular canals together with their related structures in the inner ears of humans are sensitive to gravitation. They signal their orientation to the brain, enabling the individual to maintain balance. Many lower as well as higher animals have organs of equilibrium of various designs.

Gravitation is the force that causes particles of matter to be attracted to each other. When we speak of *gravity* we refer to the earth's attraction for objects at or near its surface. This is one aspect of gravitational force. But gravitation operates throughout the universe. It is said that an apple falling from a tree aroused Isaac Newton's interest in explaining the attraction of heavenly bodies for each other. Newton, an English scientist and mathematician born in 1642, gave us his law of gravitation, which explains the motion of bodies in the entire physical universe insofar as we know it. Gravitation determines the physical behavior of the stars, it keeps the planets in their solar orbits, and holds the particles of gases together in the sun. Without it, objects would drift around in space in a chaotic state with no direction and no particular destination.[1]

Gravitation and the existence of life

If the earth were slightly less massive than it is, most of the gases of the atmosphere and most of the water would have long ago disappeared in space. The forms of life that we know would probably be much less abundant, perhaps even nonexistent. On the other hand, if the earth were slightly more massive than it is, the increased gravitational force would

---

[1]Newton said that the attraction increases in proportion to the mass and decreases in proportion to the square of the distance. This is called the *inverse square law*. Albert Einstein believed that gravitation does not pull objects toward each other but that, instead, the force is caused by a curvature of what he called *space time*, greatest around large bodies, that pushes other objects toward them.

cause larger quantities of gases to be retained in the atmosphere. More importantly, the delicate balance between the gases of the atmosphere would be upset. Nitrogen, oxygen, carbon dioxide, and other gases would be present in ratios that would not be conducive to life in its present form.

Direct biological effects of gravity

Organisms respond in different ways to the force of gravity. The way in which they react is often a matter of survival. When a seed falls on the ground or in a crevice, its orientation is random. But when the seed sprouts, the structures must grow in predetermined directions. No matter what the original orientation, the root grows downward and the shoot grows upward. This is called *geotropism*. The root is *positively geotropic;* the shoot is *negatively geotropic.*

Many organisms respond directly to gravity. The single-cell protozoan animal *Paramecium caudatum* thrives in fresh water that contains a great deal of decaying organic matter. Despite its unicellular structure, both its internal mechanism and its behavior are amazingly complex. It will head upstream if the current is not too swift and will avoid either total darkness or light that is too bright. It will also respond to chemical, thermal, and electrical stimuli. Its response to gravity is mostly negative; it tends to gather with others of its kind at the surface with its anterior (front) end uppermost. This type of response to the stimulus of gravity among animals is called *geotaxis.*

The tides

Strong gravitational forces distort the earth in its daily rotation on its axis. The moon tends to pull the earth out of shape so that bulges on both sides sweep around the earth, resulting in a twice-daily cycle. The effect is noticeable to us in the ocean tides, which may reach heights of more than 50 feet in some inlets. The sun's gravitational effect is slightly less than half that of the moon, so the tides vary depending on the relationship between the sun and moon. But because the greater force is exerted by

The tides bring a rhythmic surge of nutrients, oxygen, and carbon dioxide to animal and plant life.

Photo by Bill Beebe

Grunion, a small, silvery fish that lives along the coast of Southern California and Baja California, lays its eggs on sandy beaches only on nights of the highest tide. Two weeks later, the young fish burst out of the eggs as the waves wash them from the sand.

the moon, the effects are often referred to as the lunar tides. The succession of biological events in much of marine life is geared to the tides. They bring a rhythmic surge of nutrients, oxygen, and carbon dioxide and they flush the bays and inlets of decaying vegetation and waste materials. Even the sexual life of some marine organisms is geared to the tides.

Weightlessness   The biological effects of weightlessness on astronauts are of major concern in space travel. Changes in blood and bone tissues and in the calcium-phosphorus metabolism of the body give an ominous hint of some of the problems associated with prolonged space flights. The weightlessness experienced in a space capsule causes an abnormally large amount of blood to collect in the thorax (chest area); the muscles tend to lose their tone and become flabby; subtle changes, probably hormone-controlled, take place in the cell membranes that alter their permeability for substances moving inward and outward; there is a drop in the number of red corpuscles; and the bones lose some of their calcium, causing them to become porous and therefore brittle. Astronauts David Scott and James Irwin had periods of irregular heart beat during their twelve-day moon walk in July of 1971. Irwin was afflicted with mild dizziness both during the mission and afterward, but not while on the moon. Both men lost about 15 percent of their total body potassium, an element that is important in maintaining the body's chemical balance. It took Scott thirteen days and Irwin nine days to return to normal after their return to earth. Clearly, gravitation is more important to the welfare of the human body than life on the earth's surface would indicate.

The earth's magnetic field   The earth acts as though it were a large magnet. The north and south magnetic poles are about 1,000 and 1,600 miles, respectively, from the north and south geographic poles. The space in which the magnetic force operates is called the *magnetic field*. The directions along which the magnetic force acts, called *lines of force,* make it possible to use the magnetic compass, an instrument known in Europe and the Orient almost 1,000 years ago. Christopher Columbus carried with him a mariner's compass on his famous voyage in 1492.

A great mystery that has been only partly solved is how migratory birds are able to fly over thousands of miles of land and water and still find their way to exactly the same spot where they spent the previous summer and winter. What is it that gives them a "sense of direction"? There is evidence that some migratory birds are able to "navigate" by using the earth's magnetic lines of force.

The idea that birds have a magnetic sense was considered as early as 1855 when experiments were conducted in Russia and later in France. The results of subsequent studies also suggested that some birds have an organ that is sensitive to the magnetic field. For example, Henry Yeagley attached magnets to the wings of homing pigeons and released them about 65 miles away from their loft. Only 2 out of 20 "magnetized" birds found their way home, whereas all of those carrying unmagnetized rods of the same weight returned. Experiments by other investigators gave contradictory results, showing that when the birds had to navigate across the gradient of the earth's magnetic field, the attached magnets made no difference. However, studies on European robins and ring-billed gulls showed that they used magnetic cues. And Martin Lindauer recently showed that honeybees also respond to the earth's magnetic field.

The most recent evidence is that birds have several ways of finding direction. Experiments show that some birds are able to use the sun and stars for celestial navigation. In some cases navigation appears to be dependent on an internal clock functioning in conjunction with exterior physical forces. We will have to wait for more complete knowledge on the subject before we can give the full answer to this fascinating biological mystery.

Another biological puzzle that, according to one theory, may be related to the earth's magnetic field is the frequency of catastrophic extinctions of animals during the past 500 million years. The evidence is in the fossil record. Eight species of microscopic marine animals called Radiolaria became extinct during the last 2½ million years, and six of the extinctions occurred following reversals in the north and south poles of the earth's magnetic field. Laboratory experiments in which bacteria were kept in a lowered magnetic field for 72 hours showed a reduced reproductive rate. Mice in similar experiments showed changes in enzyme activity. Prolonged exposure to a reduced magnetic field shortened the life span and caused changes in tissues and fertility. Thus it appears that low magnetic fields can be potentially destructive to populations of organisms, possibly by acting directly on biological molecules or ions.[2] Whether this action

---

[2] A theory for the action of a reduction or reversal in the magnetic field is that there is a realignment of biological molecules which are paramagnetic (partially susceptible to magnetism) or diamagnetic (magnetically repellent) or that there is an interference with charged ions in cell membranes. It is also postulated that magnetic reversals cause catastrophic climatic changes, with periods of high magnetic intensity bringing cold climatic conditions by shielding out some of the solar radiation. Still another theory is that changes in the magnetic field permit increases in the penetration of cosmic rays to the earth's surface.

accounts for some of the catastrophic extinctions that are evident in the fossil record is a nagging question that requires further investigation.

RHYTHMS

Rhythms pervade the world of nature. Night and day, light and dark, the phases of the moon, the ebb and flow of the tides, the wet and dry periods, and the seasons of the year are some of the cyclical features of the environment. Other, less manifest, rhythmic components of the physical environment are the pulsations in solar and cosmic radiation, gravity, the earth's electromagnetic field, and barometric pressure. The quality and quantity of plant and animal life are intimately related to the rhythms of the environment, some with obvious cause and effect, others having obscure and mysterious relationships.

Rhythmic oscillations in the functions of animals and plants are so widespread that it is almost a law of nature. The seasonal cycles caused by the inclination of the earth's axis in relation to its revolution around the sun profoundly changes the environment on all parts of the globe. This angle between the earth's axis and its orbit, about 23½ degrees, has an overwhelming influence on most forms of life. Among many small organisms and some large ones reproduction occurs once a year coinciding with the annual cycle. In nearly all mammals the estrous (sexual) cycle follows a rhythmic pattern that is characteristic of the species. In human females, the menstrual cycle coincides closely with the period of the moon's revolution around the earth. Night and day caused by the earth's rotation is related to behavioral or physiological cycles in all organisms that are responsive to light and dark.

The exact cause of the rhythms in biological systems is obscure in the case of many organisms. Some biologists believe that many of the daily rhythms are controlled by built-in biological mechanisms. This view is supported by the fact that the biological cycles of many organisms continue on approximately the same time schedule when they are placed in either continuous light or total darkness in the laboratory and are otherwise protected from the observable fluctuations of night and day. Daily cycles occurring on approximately 24-hour schedules are called *circadian rhythms,* and the term *biological clock* was coined to describe the metabolic mechanisms that supposedly control them.

The existence of a rhythmic mechanism within living organisms was recognized as early as 1902 when Wilhelm Pfeffer, a scientist at the University of Leipzig in Germany, conducted some revealing experiments on bean plants. The leaves of bean seedlings normally stand somewhat upright during the day and droop at night. But when Pfeffer kept his bean seedlings in constant darkness, they continued to elevate their leaves in the daytime and to drop them at night. Since Pfeffer's observations, numerous organisms have been found to maintain their normal cyclic activity despite their removal from the conditions that we ordinarily identify as day and night.

Of course, not all organisms continue their normal cycles when put under artificial conditions. Many of them respond to the new conditions in ways that can be predicted from their behavior in their natural environments. Warming and cooling, freezing and thawing, light and dark can be arranged in ways that completely alter the timing of their cycles in nature. However, there are enough examples of organisms which do not so change, but continue on their apparently predetermined course, to suggest that the living process is under the guidance of something that is not yet fully understood. The trigger of the rhythmic mechanism may be the action of external forces not yet discovered, or it may be some means within the physiology of the organism for continuing the responses that have proved for generations to be beneficial to the species.

There is evidence that in some organisms the force that drives the "clock" is an external geophysical entity. For example, oysters are known to open and close their shells in harmony with the lunar tides. Frank A. Brown of Northwestern University set out to determine whether the cyclic behavior of oysters was actually caused by the tides or whether a biological clock or some other mechanism was involved. He collected oysters on the seacoast of Connecticut and removed them to a laboratory in Illinois, 700 miles inland. For about two weeks, the oysters continued to open and close on schedule, according to the tides of the Atlantic seacoast. But gradually they changed and became synchronized with the time when the tides would occur 700 miles inland if that were the seacoast. Although the biological clock of the oyster could function on a built-in rhythmic schedule for a limited period, it appeared that its timing was set by external forces.

In humans, the performance of sleeping, eating, excretion, physical exercise, work, social activity, and mental states and mood are traceable to rhythmic cycles. But cyclic characteristics of human functions extend also to the physiological and cellular levels, although usually we can neither see nor feel them taking place. Body temperature rises and falls each day by about 1½ to 2 degrees. Hormones of the adrenal cortex decline at night and reach a peak at about the time we rise in the morning. These substances, classified as corticosteroids, moderate the nervous system as one of their functions. The male hormone testosterone is at its daily high in the morning. Protein and amino acid levels in the blood are higher in the morning than at night. There is a daily rhythm in the rate of cell division, in the chemical neurotransmitters of the brain and spinal cord, and in the activity of the liver enzymes, which are necessary for biochemical conversions and the destruction of toxins and other substances no longer needed.

"Jet fatigue" is an experience common to many travelers whose destination is in another time zone. It probably accounts in the large part for the exhaustion often experienced by travelers. Since it would be dangerous to allow the pilot's alertness and efficiency to be adversely affected by jet fatigue syndrome, airline pilots are generally restricted in the number of long-distance flights they can take within a short period of time.

Certain diseases show rhythmic patterns, as does immunity to infection. The incidence of human deaths is greatest during late night and early morning hours. Weekly reversal of the light and dark cycles for laboratory rodents caused a significant reduction in the average life span. Experiments on mice showed that they were more susceptible to pneumonia infection at the end of their period of activity during darkness. The results suggest that the effectiveness of vaccinations and other medications might be increased by proper timing. Better knowledge of circadian rhythms in metabolism might make it possible to better appraise an individual's response to stress and to predetermine periods of maximum strength and endurance. It might help to minimize nervous strain and susceptibility to disease and enhance such desirable responses as patience, keener perception, and mental and physical vigor.

The response of many organisms to toxins is related to circadian rhythms. In experiments with insects, boll weevils were least susceptible to the insecticide methyl parathion at dawn. For some unexplained reason, the period of high resistance recurred at six-hour intervals throughout the 24-hour cycle. When the weevils were subjected to a photoperiod of ten hours of light per 24-hour cycle, the insecticide killed only 10 percent of the weevils at dawn but nearly 90 percent of the weevils treated only three hours later. The fact that insecticides are often applied at dawn because the most favorable atmospheric conditions usually prevail at that time of day needs reevaluating in view of these findings. The influence of external forces on the effects of toxins in the ecosystem is largely unknown.

The time of day when drugs are administered can mean the difference between life and death. In laboratory animals, the effects show pronounced rhythms. When drugs are administered at intervals around the clock, the mortality difference is as much as 60 to 70 percent. Amphetamine (Dexadrine) can kill 78 percent of laboratory test animals at one hour but only 8 percent at another time. Thus, much of the data on the efficacy as well as the injurious effects of drugs and of environmental contaminants and toxicants is meaningless unless the biological time is indicated.

With mice that were subjected to alternating light and dark periods, the response to several kinds of drugs varied depending on the time during a 24-hour period. For example, the stimulating effects of the drug lidocaine hydrochloride were greatest during the activity phase of the animal's cycle. Cyclic variations in the effects of other drugs have been detected.

In a diurnal animal such as man,[3] the time of day should be considered in relation to the individual dose of a medicine that is administered. This important aspect of medicine needs further investigation.

It seems probable that living things are continuously sensitive to forces in the geophysical environment even though we may not be aware of their existence. There is also the possibility that the genes carry signals for the rhythmic functioning of the organism and that the interaction of the biochemical processes with geophysical and other natural forces produce the puzzling phenomena described as biological clocks. The concept of biological clocks is widely accepted among biologists. Whether the clocks are driven from within, by external forces, or both is in many cases an unanswered question.

Photoperiodism

Light has predictable effects on many plants and animals. Knowledge of this makes it possible to manipulate the environment with respect to the light period in ways that are useful for growing some plant and animal

---

[3]Diurnal means active during the daytime.

Left photo: Daffodils are day-neutral plants, more responsive to temperature than to photoperiod. *Right photo:* Marijuana is a short-day plant.

cultures, particularly in commercial horticulture. For example, the physiology of some flowering plants is greatly influenced by night and day. The onset of flowering is related to the timing and duration of light and dark periods. The response is called *photoperiodism*.

Flowering plants can be classified according to their response to the photoperiod. Plants that generally flower in either early spring or late fall can produce flowers only when the day length is shorter than a critical period that varies with the species. For example, cockleburs flower only when the day length is 16 hours or less. Such plants are called *short-day* plants. Among other short-day plants are asters, chrysanthemums, poinsettias, dahlias, goldenrod, ragweed, and violets. Another group of plants, which usually flower later in the spring or earlier in the fall, can produce flowers only when the day length is longer than a period that is critical for the species or variety. For example, spinach flowers only when the day length is 10 hours or more. Plants in this group are called *long-day* plants. Examples of long-day plants are beet, clover, gladiolus, and larkspur. In still another group of plants, flowering is not determined by day length. These are called *indeterminate* or *day-neutral* plants. They include such plants as sunflower, tomato, dandelion, and carnation. Both short-day and long-day plants occur naturally in the temperate zones, but long-day plants predominate in the higher latitudes north and south of about 60°. Predictably, day-neutral plants are widely distributed.

Short-day plants, which would not ordinarily flower during the long days of summer, can be made to do so by shading them during part of the day. Long-day plants, on the other hand, can be made to flower during the short days of winter by the use of artificial light. There is evidence of a relationship between light and the production or release of a hormone within the plant tissues which, in turn, directly controls flowering in some plants. The hormone has been called *florigen*.

Growers of cut flowers make practical use of photoperiodism to induce flowering at times of the year when demand is high. Chrysanthemums can be grown under nearly constant light to force vegetative growth. Then, at the desired time, the light period can be shortened to induce the development of flower buds. People are sometimes amazed at the profusion of

USDA photo.

Effect of day length on the flowering of petunias. Plant scientist Henry M. Cathey demonstrates the effect of chemicals and day length on petunias. The petunia at left was given short days to make it late flowering and to produce a short plant. The petunia at center was given long days to make it early flowering and tall. The petunia at right was sprayed with growth retardant to make it short, and given long days to make it bloom early.

poinsettias on the market at Christmas time. This is another short-day plant that is brought to flower at exactly the right time by controlling the photoperiod.

Some species of animals are sensitive to the light environment. In fish hatcheries, the breeding season of brook trout can be controlled by artificial manipulation of the photoperiod. Trout normally breed in the autumn. But when the day length is artifically increased in the spring and then decreased in the summer to simulate the early arrival of fall, the trout will spawn in the summer, as much as four months early.

THE BIOGEOCHEMICAL ENVIRONMENT

The earth's crust contains all of the 88 natural elements in various combinations. The most abundant elements are oxygen and silicon. They comprise 46 percent and 28 percent, respectively, of the total composition of the crust. The other elements are present in smaller amounts (see Figure 5.1). Analysis of the average granite rock shows that it consists of 67 percent silicon dioxide (SiO_2), 15 percent aluminum oxide, and about 14 percent oxides of iron, calcium, sodium, and potassium. The remaining elements make up less than 4 percent of the rock.

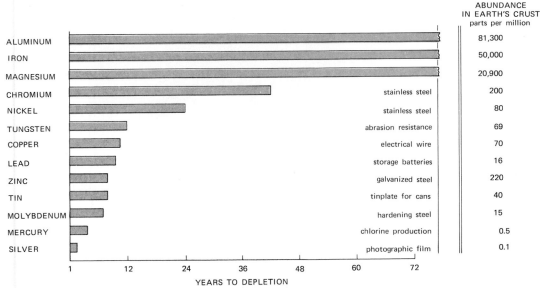

From W. C. Gough, *Why Fusion?*, U.S. Atomic Energy Commission, 1970.

FIGURE 5.1 Depletion of world reserves of commercial grade ores if world population had U.S. living standards.

The soils of the earth are created by the decomposition of rocks. They furnish much of the nourishment for all life on land. Most of the green plants are dependent on the soil for the primary nutrients: nitrogen, phosphorus, and potassium. The abundance of these minerals in any particular place determines in large measure the productivity of the soil. A critical factor in the growth of plants is the availability of the so-called micronutrients, such as zinc, iron, and manganese. When even one of these essential elements is present in insufficient amounts or in a form that is not readily available, plants do not thrive. When deficiencies exist, all the living components of the ecosystem are affected.

Organisms are not only dependent on the geochemical environment; they modify it in many ways. For example, plants growing in a square-foot plot of soil build it into a microcosm entirely different from its original state. Small plants grow in the shelter of the larger ones; fungi and insects proliferate in the discarded plant parts; worms find a congenial home in the organically impregnated soil; and bacteria and other microorganisms multiply. The soil changes in structure and in nutrient content. The decomposed rock becomes a biological metropolis teeming with visible and invisible life. Similar changes take place in forests and valleys but on a grander scale. Conversely, the removal of plant growth by whatever means produces drastic changes. The soil becomes compacted and hard. Its capacity to absorb moisture is reduced, increasing run-off and erosion. Nutrients are lost, some of them irretrievably. Plants can be reestablished only with great difficulty. Having no plants, the land is shunned by animals and microorganisms that depend on plant life for sustenance.

The effects of soil nutrients, water, and climate on soil fertility are discussed further in Chapter 16.

TRACE ELEMENTS

Many of the elements are present in rocks in small quantities and are, therefore, usually components of soils and water in only trace amounts. For this reason, they are often called *trace elements,* even though they may occasionally be found concentrated in large deposits (see Table 8.4). A number of the trace elements are essential to the growth, development, and health of plants. Some of them are needed in such minute amounts that they are called micronutrients, discussed in Chapter 16.

Several of the trace elements are also essential for the growth and development of animals. Iron, manganese, and iodine are needed in small amounts. Other are needed in even smaller quantities. These include copper, zinc, molybdenum, and cobalt. Elements such as vanadium, barium, strontium, silicon, chromium and nickel are required by some species of animals. Still other elements, although present in plant and animal tissues, do not have any known biological function. Such elements are lead, mercury, cadmium, aluminum, and tin.

Some of the elements that are essential for the growth and development of plants and animals can be highly injurious when they are present in the environment at excessively high concentrations, and there is often a narrow margin between the beneficial and injurious amounts. Important examples are discussed in Chapter 8.

READINGS

Biological Rhythms in Psychiatry and Medicine. Public Health Service Publ. No. 2088, 1970. National Institute of Mental Health, Chevy Chase, Md.

BROWN, F. A., JR., *Biological Clocks.* Boston, Mass., D. C. Heath and Company, 1962.

GAMOW, G., *A Planet Called Earth.* New York, Bantam Books, 1970.

PICKERING, J. S., *Captives of the Sun.* New York, Dodd, Mead, 1961.

SNIDER, A. J., "The Rhythm of Life," *Science Yearbook. The World Book Science Annual,* 1968, pp. 112–125. Field Enterprises Educational Corporation.

Photo by John R. Shrader, courtesy Environmental Protection Agency.

PART THREE — ENVIRONMENTAL QUALITY

AIR POLLUTION

The air is one of the most used and abused parts of the biosphere. Industrial, automotive, and domestic activities have resulted in increasingly outrageous insults to the atmosphere. Yet the air is finite. It cannot be manufactured and replenished as the need for it increases. It is at the same time our most precious and most fragile resource. Even animals and plants that live in the sea depend upon dissolved gases from the atmosphere. Our enormously accelerated abuse of the atmosphere has become a health hazard and a threat to life, damaging both plants and animals in areas polluted with poisonous fumes, dust, and smoke.

An ancient problem Early man no doubt suffered from smoke irritation as soon as fire was discovered. The stoic endurance of the American Indian in the smoky atmosphere of his tepee is legendary. Soot stains and the remnants of charred wood in caves are circumstantial but convincing evidence that the prehistoric residents suffered smarting eyes and the sting of wood smoke to the respiratory tract.

Air pollution sometimes has natural causes, as from forest, brush, and grass fires caused by lightning. Volcanoes are a source of air pollution with ash, dust, sulfur compounds, and other gases that can be annoying or even fatal. Tacitus told how Pliny the Elder lost his life near Pompeii during the eruption of Mt. Vesuvius in 79 A.D. while, as commander of the fleet, he was attempting to rescue victims of the disaster. The ash from smoking volcanoes often becomes distributed over wide areas and causes distress due to irritation of the respiratory tract and possibly aggravation of respiratory inflammations. The eruption in 1883 of the volcano on Krakatoa, an island between Sumatra and Java, blew most of the island away and started a tidal wave that killed 36,000 people. Volcanic dust from the explosion spread around the earth and was visible for many months. Albert Schweitzer commented on the beautiful sunsets made by the dust in the air two years after the eruption.

The danger of spoiling the air was recognized in ancient civilized times. The patricians of Rome complained about the smoke that smudged their togas. Pliny, in the first century A.D., grumbled:

Even the air itself wherein and whereby all things should live, we corrupt to their mischief and destruction.

Left photo courtesy of Environmental Protection Agency. Right photo courtesy of U.S. Geological Survey.

Left photo: Mid-Manhattan, New York, during the thick smog of October 27, 1963. Right photo: Volcanic eruptions pollute the air with particulates and gases.

More than 100 organic compounds have been isolated from the combustion of wood; but the wood smoke of earlier times seems a minor nuisance compared to the combustion products of "the black rock that burns" described by Marco Polo upon his return from Asia in 1295. The new source of energy was destined to revolutionize Western industry and culture. Though slow to take hold, coal was abundant in Europe, and its use, though disagreeable, was hastened by population growth and rapid depletion of the forests.

The industrial revolution   Coal smoke was recognized as a hazard in London as early as 1306. In that year a man was executed for burning "sea coal" instead of oak. The residents and rulers of the manufacturing areas in Europe, especially Germany and England, became increasingly alarmed about the injurious and unhealthful effects of the sulfurous smoke from burning coal. King Edward I (1272–1307) issued a proclamation in 1306 prohibiting the use of coal in London. During the reign of Edward II (1307–1327) a man was put to torture for filling the air with coal smoke. But industrialists persisted in the more convenient use of coal and eventually first offenders were punished merely with "great fines" and ransoms. The second offense could result in confiscation and destruction of the coal-burning furnaces.

In 1661, John Evelyn, a leading scientist and founder of the Royal Society, published a small book entitled *Fumifugium; or the Inconvenience of the Aer and Smoke of London Dissipated; together with Some Remedies Humbly Proposed*. Other studies by individuals and commissions followed, and engines were designed to combust their own smoke; but nothing very effectual was accomplished due to the enormous increase in the use of coal.

The New World was not immune to air pollution. This was evident in California before it was occupied by white men. Spanish explorers called the Los Angeles basin the "Valley of Smokes" because the inversion layer that is characteristic of the area during much of the year created a lingering pocket of smoke and haze. An inversion layer is a layer of cool air with warmer air above; the denser cool air tends to remain near the ground. (See Figure 6.1.) A smoky atmosphere is evident today in many areas of the American tropics where "slash and burn" agriculture involves burning large quantities of wood and brush.

Today, the cost of air pollution in the United States falls within a range of $6.1 to $18.5 billion, including estimates of the economic value of air pollution damage to human health, man-made materials, and vegetation. Nationwide emissions of the five principal air pollutants total an estimated 267 million tons per year (see Table 6.1).

AIR CONTAMINANTS

CARBON DIOXIDE

All organic substances used for fuel—petroleum, coal, natural gas, wood—had a living origin. They are the products, directly or indirectly, of photosynthesis. Petroleum, natural gas, and coal are the fossil remains of ancient living organisms, modified by the conditions to which they were subjected under the pressure of sediment and rock deposited by prehistoric seas. Since carbon forms the structural skeleton of the molecules of life, every gallon of gasoline, every chunk of coal, and every stick of wood contains an abundance of carbon that was once part of the atmosphere or the sea. In the form of carbon dioxide, it was taken up initially by plants in the process of photosynthesis. Since carbon dioxide

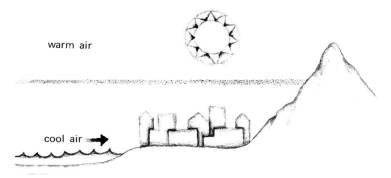

FIGURE 6.1 An inversion layer is formed when cool air moves in under a layer of warm air, or when the air near the ground cools more than the air above. The cool air is denser; therefore it does not rise, and remains stagnant, accumulating atmospheric toxins.

TABLE 6.1   Nationwide Emissions of Major Air
Pollutants, 1970

	Million tons per year
Carbon monoxide	149
Hydrocarbons	35
Nitrogen oxides	23
Sulfur oxides	34
Particulates	26
Total	267

From *Compilation of Air Pollution Emission Factors, Second
Edition.* U. S. EPA Publication AP-42. 1973.

and water are the principal products of the complete combustion of fuel,
the organic carbon is returned to the atmosphere in its ancient form (see
carbon cycle, Figure 2.9).

Carbon dioxide comprises a minute fraction of the gases in the
atmosphere—about 3 parts in 10,000. Yet it has a profound effect on the
temperature of the atmosphere, and hence, on climate. This is discussed
further under the topic Greenhouse Effect. The gas is substantially soluble in water, so that the ocean contains about 1 part in 10,000, a fortunate
circumstance for the photosynthetic activity of marine and aquatic
phytoplankton. Because of the small amount of carbon dioxide in the
biosphere, and its relatively low toxicity, it would not ordinarily be
thought of as a pollutant. However, carbon dioxide does not enter into the
photochemical reactions in the atmosphere and its removal is dependent
on biological and chemical deposition. The equilibrium brought about by
the normal uptake and output from natural sources during the present and
recent past may be upset drastically by the enormous increase in the
combustion of fuels during this century.

*Origin and fate
of carbon
dioxide*

In the geologic past, there were large fluctuations in atmospheric carbon dioxide. Volcanic eruptions injected enormous quantities of carbon
dioxide into the atmosphere. The total amount during several billion years
is estimated to have been at least 40,000 times the amount now in the air.
Most of the carbon dioxide was precipitated from the ocean and other
bodies of water as salts of calcium or magnesium in the form of limestone
or dolomite (magnesium-calcium carbonate). About one-fourth of the
total carbon dioxide was taken up by plants and was buried in the sediments. A small fraction of this was concentrated in deposits of coal, oil
shales, sands, and gas pockets that became the fossil fuels of our industrial age.

As long as man continued to use mainly wood for fuel, the contribution
of carbon dioxide to the atmosphere was negligible and burning affected
the normal carbon dioxide cycle only slightly. Now, however, man is
burning fossil fuels that were deposited by living organisms during the
past several hundred million years. These geologic reserves of locked up
carbon fuel are being oxidized for industrial purposes at a rate that is
causing measurable and significant changes in the carbon dioxide content
of the atmosphere.

*Amount and
concentration of
carbon dioxide*

About 10 times as much carbon dioxide is injected into the atmosphere
from coal fires and furnaces as from the process of breathing. These
sources alone would double the carbon dioxide within 500 years if there

were no natural process of removal. If the recoverable reserves of fossil fuels—coal, lignite, petroleum, natural gas, tar sands, and oil shales—are 2,971 billion metric tons, then the carbon dioxide equivalent of these reserves is more than 300 percent of all the carbon dioxide in the atmosphere present in 1950.

The production of carbon dioxide since 1950 has averaged 9 billion tons per year, resulting in an increase in concentration of about 1.6 ppm per year. At this rate, it is estimated that by the year 2000, there will be about 25 percent more carbon dioxide in the atmosphere than at present. But the use of fossil fuels is increasing. Estimates of the rate of increase in combustion vary between 3.2 percent and 5 percent per year. If the higher rate continues without further acceleration, the quantity of carbon dioxide spewed into the atmosphere during the next 30 years will be an astounding 60 percent of all the carbon dioxide that was in the atmosphere in 1950. And if the current rate of combustion were continued, about 13,000 billion metric tons of fossil fuel would be used up during the next 1,000 years, and would add 17 times as much carbon dioxide as the present amount.

Indirect effects on animals and plants

Although carbon dioxide is essential to the physiology of animals, and more especially to that of plants, the concentration is normally not critical. For example, uncontaminated air has a carbon dioxide content of about 300 ppm (parts per million), yet man can tolerate nearly 5,000 ppm without drastically adverse effects on respiration. Moderately higher concentrations than normal bring about increases in the rate of photosynthesis in plants, but an increase of several fold is required before deleterious effects are noted. However, carbon dioxide has a physical role in the stability of the biosphere that may be as important, indirectly, as the direct effect on animal and plant physiology. Carbon dioxide forms a blanket of insulation around the earth. Changes in temperature caused by an increase in the carbon dioxide content of the atmosphere could alter world climate in ways that would drastically affect all living organisms. It is theorized that these changes can be caused by a greenhouse-like effect.

The greenhouse effect on climate

Much of the sun's energy is emitted as heat rays, consisting of radiation in the infrared portion of the electromagnetic spectrum. The temperature at the surface of the earth is determined by the energy balance between the sun's rays that strike the planet and the heat that is radiated back into space. The near-infrared rays of sunlight penetrate the earth's atmosphere relatively unimpeded, and some of the heat is absorbed and retained by the earth or objects on the surface. The heated earth then reradiates this absorbed energy as radiations of longer wavelength, mainly in the middle-infrared portion of the spectrum. Much of this does not pass through the air envelope to outer space but is absorbed by the carbon dioxide and water vapor in the atmosphere and adds to the heat that is already present. Though carbon dioxide is almost completely transparent to visible light, and partially to near-infrared, it strongly absorbs and reradiates the rays of longer wavelength, especially heat rays having wavelengths of 12 to 18 microns. Thus carbon dioxide acts like the glass of a greenhouse, and on a global scale, tends to warm the air in the lower levels of the atmosphere. This is called the *greenhouse effect* (see Figure 6.2).

Water vapor and ozone are also absorbers of infrared radiation and

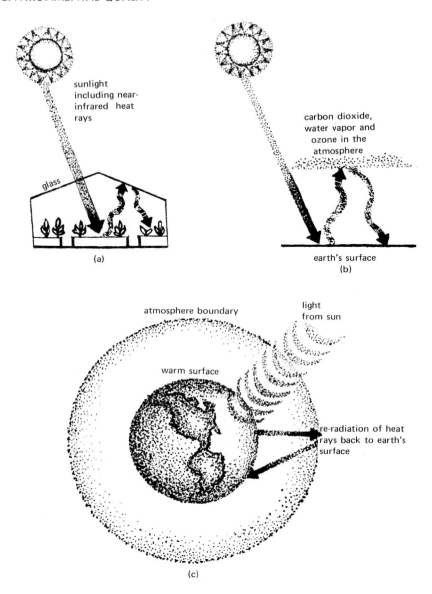

FIGURE 6.2 Greenhouse effect. (a) Heat from the sun readily penetrates the glass roof of a greenhouse in the form of near-infrared rays. These are absorbed by soil, plants, and other objects and reradiated as longer wavelength heat rays. However, the glass is not as transparent to these radiations of longer wavelength; thus much of the heat is held inside the greenhouse. (b) In a similar way, the atmospheric carbon dioxide, water, and ozone are transparent to the near-infrared rays of the sun but absorb the reflected longer wavelength heat rays, thus (c) holding the heat within the boundaries of the earth's atmosphere and having a warming effect.

help to keep the earth warm. Water vapor is an efficient absorber of infrared of around 15 microns and of another band of wavelengths around 6.3 microns. Ozone, an absorber of infrared of around 9.6 microns, is relatively minor in its effect.

It has not actually been proved that carbon dioxide will increase the global temperature to any significant extent. A mathematical analysis of the earth's heat balance shows that either a decrease or an infinite increase in atmospheric temperature can be derived from a variety of assumptions concerning the water content of the atmosphere and its dependence on temperature.[1]

Some students of world climate believe that major fluctuations have been caused by dust in the atmosphere from volcanic eruptions. Volcanic activity during the historical period is known to have thrown out dust and ash which have spread throughout the entire atmosphere and lasted for years. The eruption in 1883 of Krakatoa, an island in the Sunda Strait between Sumatra and Java, is an example of such an occurrence within historic times. Dust in the sky was evident for at least two years afterwards. Some scholars believe that the prehistoric eruption of Thira in the Mediterranean destroyed the Minoan culture of Crete and possibly Troy in about 1400 B.C. If the meteorological effects of the ancient Mediterranean catastrophe were known, the knowledge would be useful in determining the causes of climatic changes and possibly future trends.

There is evidence that the temperature of the entire earth has risen slightly during recent decades. Glaciers in both hemispheres are receding. The occurrence of climatic cycles is well recognized and the causes should be critically analyzed. Between 1885 and 1940, atmospheric warming was worldwide. There was an increase in mean annual air temperature of about 0.5°C (0.9°F) and the mean winter temperature rose even more—by 0.9°C (1.6°F). The greatest increase in annual mean temperature was in the Northern Hemisphere between 40°N and 70°N latitudes, where the average rise was 0.9°C (1.6°F) and where there was an astonishing increase in the average winter temperature of 1.6°C (2.8°F).

After 1940, global warming subsided. Warming of Northern Europe and North America continued between 1940 and 1960, but the mean annual air temperature worldwide as well as for the Northern Hemisphere decreased slightly (0.1°C or 0.2°F). More than 40 percent of the total increase in carbon dioxide content of the atmosphere from combustion occurred during that period. It is evident that for the period that we have studied, other causes of climatic change are at least as important as the release of carbon dioxide by combustion of fossil fuels and may nullify or mask any effect on temperature from carbon dioxide content of the atmosphere, but this is not at all certain.[2]

Antarctic ice cap   Continental ice caps have formed and melted several times during geological history. There is evidence that there have been at least four major glaciations at intervals of about 250 million years, plus several minor glaciations. Each of the frigid spells was short in geologic time, consisting of brief disruptive episodes in the climatic geniality that characterized the normal condition. Probably less than 1 percent of geologic time experienced glacial climates.

---

[1] F. Moller, "On the Influence of Changes in the CO_2 Concentration in Air on the Radiation, Balance of the Earth's Surface and on the Climate," *Journal of Geophysical Research,* 68 : 3877–86 (1963).

[2] A related climatic problem stems from the use of large quantities of gases called flourohydrocarbons, used as refrigerants and propellents in aerosol sprays (see p. 104).

The most recent Ice Age occurred during what is known as the Quaternary, a period spanning the last 1 or 2 million years. The Quaternary includes two epochs: the Pleistocene, during which there was extensive glaciation and cold climate, and the Recent epoch, during which we are now living. Glacial sheets have advanced and receded four times during the Quaternary. We are now in the fourth recession, characterized by a sharp climatic shift to a somewhat milder condition that reached its midpoint of change about 11,000 years ago. It is debatable whether we are now experiencing an interglacial recession or the end of the whole Ice Age. Some of the interglacial stages of the Quaternary were milder than the present.

At present, the Antarctic ice cap extends over 14 million square kilometers (5.4 million square miles) and is about 3 kilometers (1.9 miles) thick. However, a 2-percent change in the global radiation balance could occur by the year 2000 as a result of the estimated increase in carbon dioxide content of the atmosphere. If half of this caloric energy were concentrated in Antarctica, the ice would melt in 400 years. If the ice caps were to melt over a period of 1,000 years, the level of the sea would rise 4 feet every 10 years. The melting of the Antarctic and Greenland ice caps would raise sea level by possibly 400 feet. The inundation would wipe out nearly all major coastal metropolitan centers.

There is a crucial relationship between the presence of the polar ice cap and global temperatures. The ice drastically reduces ocean temperatures, especially at great depths. The polar ice lowers the temperatures in Canada and the United States by several degrees. According to calculations of C. E. P. Brooks, a 5°F drop in polar temperature below the freezing point causes a drop of 50°F in polar winter temperatures. The first drop causes the formation of the ice, whereas the cooling effect of the ice thus formed causes the remainder of the temperature to fall. Conversely, a rise of 2° in the temperature of the earth would at the present time be enough to melt all the polar ice. It is clear that the world is in delicate balance between a normally temperate earth and a glacial climate.

The entire history of modern man and his immediate evolutionary predecessors is confined to the exceptionally harsh and changeable climate of the Ice Age. Man has experienced only the violent period of geologic time when the delicate climatic balance alternated between advance and retreat of the frozen barrier. While the climatic upheavals of the great advances and retreats were but momentary episodes in geologic time, they are extremely slow with reference to the span of human experience. A man-made acceleration in the giant swing of the climatic pendulum conjures possibilities of climatic catastrophes that must be evaluated before irreversible changes take place.

Other biological effects   The effect of carbon dioxide on global temperatures can indirectly affect the fishing industries because the fisheries are sensitive to changes in temperature of the ocean currents. The temperature of the ocean in the polar regions tends to be stabilized by freezing and melting. However, if the global air temperature rises, the temperatures of the surface waters in temperate and tropical parts of the world would rise accordingly. From 1880 to 1940 an increase of 1° to 2°C (about 2°F) occurred in North Atlantic waters, causing a shift in the cod fishery toward Greenland and other northern waters. The warming of the ocean also caused a retreat of the sea ice at the edge of the Arctic Ocean.

A rise in temperature associated with carbon dioxide concentration would affect plant productivity favorably in some areas and unfavorably in others. Plant growth and the physiology of cold-blooded animals are, in general, speeded up by increases in temperature. The activity of warm-blooded animals would be affected in ways that are difficult to predict.

The expected carbon dioxide in the atmosphere by the year 2000 would increase the photosynthetic rate of plants. But limiting factors in plant growth, such as water temperature and nutrient supply, may largely mask the effect of an increase in the concentration of carbon dioxide. Therefore, the effect on a global scale, such as in forests, deserts, and oceans, may be minor. But in those agricultural areas where the factors affecting plant growth can be favorably controlled, the expected increase in carbon dioxide may result in a measurable improvement in productivity.

Another effect of carbon dioxide is to increase the acidity of water. An atmospheric increase in carbon dioxide of 25 percent would cause an increase in the acidity of fresh water by about 0.1 pH. This amount of change would probably not be enough to modify the biota significantly in most bodies of water.

Physical effects of carbon dioxide

Building stones, especially limestone materials such as marble, are susceptible to acid-forming substances. The calcium carbonate of limestone is converted by carbon dioxide and water to bicarbonate, which is water soluble. "Stone cancer" is a term that has been used to describe the erosion of buildings and art objects in the polluted atmosphere of Venice, Italy. Much of the loss there, however, may also be the result of sulfur and sulfide combustion products in the industrial effluent.

CARBON MONOXIDE

The silent killer

Carbon monoxide is the most important product of combustion. The gas is colorless, odorless, tasteless, and nonirritating. It is highly toxic. Prolonged breathing of low, commonly experienced concentrations is dangerous. But carbon monoxide is not, in itself, directly toxic. Poisoning is due solely to its interference with the normal functioning of hemoglobin as a transporting agent for oxygen in the blood. The oxygen-carrying capacity of the blood is due to an affinity of the hemoglobin in the red blood cells for oxygen. The hemoglobin and oxygen combine, forming what is called *oxyhemoglobin*:

$$O_2 \quad + \quad Hb \quad \rightleftharpoons \quad HbO_2$$
$$\text{oxygen} \qquad \text{hemoglobin} \qquad \text{oxyhemoglobin}$$

However, hemoglobin has 204 times the affinity for carbon monoxide as for oxygen. The product formed is *carboxyhemoglobin*, a conversion product that does not have the ability to combine with and carry oxygen:

$$CO \quad + \quad Hb \quad \rightarrow \quad COHb$$
$$\text{carbon monoxide} \qquad \text{hemoglobin} \qquad \text{carboxyhemoglobin}$$

Because the reaction and the asphyxiating effect take place quickly, carbon monoxide is a dangerous and deadly poison. The reaction is reversible, so if the victim survives, the hemoglobin is eventually made available again for carrying oxygen, but the recovery of the hemoglobin is slow. Deficiency of oxygen (hypoxia) due to high altitude adds to the effect of carboxyhemoglobin. An altitude of 335 feet is equivalent to the effect of 1 percent carboxyhemoglobin.

Any confined space in which carbon monoxide is apt to accumulate presents dangers. The toxicant gas is a potential hazard in space flights and undersea operations. If the nuclear submarine *Nautilis* were submerged for 5 days during which smoking and other forms of combustion were prohibited, the CO produced by its crew would give a concentration of over 25 ppm in the atmosphere of the vessel. Nuclear submarines are equipped with catalytic burners to convert carbon monoxide to carbon dioxide, which is easily scrubbed from the system.

Chronic poisoning

The most serious aspect of carbon monoxide for most people is chronic poisoning from prolonged exposure to low concentrations. Experiments on physiological response to low levels of carbon monoxide show that there is an effect on pulse rate, respiration, blood pressure, and neurological reflexes. There is disagreement on the nature of chronic carbon monoxide poisoning. Some investigators contend that the toxin is not cumulative and is released from the system in 3 to 4 hours. American investigators, in particular, have claimed that low concentrations are not injurious. European toxicologists, on the other hand, take a gloomier view of extended exposure to low concentrations and say that it is deleterious. The weight of evidence favors the latter position, and suggests that chronic carbon monoxide poisoning may be one of the most important modern-day problems in public and personal health. No threshold limit for toxicity of carbon monoxide has been demonstrated and it must be assumed that even very small sublethal concentrations are injurious if exposure is constant or prolonged.

Cigarettes and carbon monoxide

The concentration of carbon monoxide in the expired air of smokers is higher than with nonsmokers. Regular cigarette smokers have blood concentrations of carboxyhemoglobin in the range of 5 percent to 10 percent. A blood level of 5 percent carboxyhemoglobin and possibly 2 percent carboxyhemoglobin affects performance. (The background, or normal, level in nonsmokers who have not been exposed to carbon monoxide is about 0.4 percent.)

Tests on mental performance show that there is impairment in arithmetic capability at about 5 percent carboxyhemoglobin. There were differences in perception of time intervals at a concentration of carboxyhemoglobin 2 percent above the background level. It is apparent that the effect of carbon monoxide intoxication (as measured by carboxyhemoglobin level) is dependent to a great extent on the susceptibility of the individual to cigarette smoking and other sources of carbon monoxide exposure.

The birth weights of babies whose mothers smoke cigarettes are lower than those of nonsmoking mothers. Also, the birth weights of babies born at high altitudes are less than the weights of babies born at low altitudes. These findings, taken together, suggest that birth weight is related to the availability of oxygen.

It has long been known that coal miners exposed to carbon monoxide have an increased hematocrit (percent volume of red blood cells). Cigarette smokers were also found to have an increase in hematocrit within minutes of smoking. There is also a long-term increase in hematocrit which is only partially reversible. Heavy smokers can reduce the effect when they stop smoking, but there is theoretically an increased possibility of clot growth with increased hematocrit, which may explain the high cardiovascular mortality of cigarette smokers.

Cigarette smoking involves hazards apart from that of carbon monoxide, for example lung cancer and the possibility that nitrous acid, a known mutagen, may be an injurious component of cigarette smoke (Chapter 13).

Carbon monoxide from combustion

Virtually all combustion sources give off carbon monoxide: stoves, furnaces, open fires, forest and brush fires, burning structures, waste disposal, burning coal mines, factories, power plants, and internal combustion engines (see Table 6.2).

Carbon monoxide is the major pollutant from automobiles, comprising 80 percent of all automobile emissions (see Figure 6.3). In Los Angeles County, the carbon monoxide emission from automobiles in 1971 was estimated at 8,960 tons per day, which equaled 98 percent of the carbon monoxide from all sources. The total from all sources was 9,105 tons per day and comprised 70 percent of all major atmospheric pollutants.

The average driver in the United States uses 2 gallons of gasoline per day, and the combustion of each gallon gives off about 3 pounds of carbon monoxide. The consumption of gasoline is increasing at an alarming rate. There were 40 billion gallons used in 1950, and the consumption by passenger cars alone had risen to 73 billion gallons by 1972. This amount is equivalent to more than 100 million tons of carbon monoxide per year.

The mean annual concentration of carbon monoxide in cities in 1966 ranges from a low of 3 ppm measured in Washington to a high of 100 ppm or more in Chicago, Detroit, and London.

There is evidence that the concentrations of carbon monoxide in the atmosphere in and around cities affect the accident rate. Studies of automobile accidents in Paris showed that the people who were involved had higher levels of carboxyhemoglobin in the blood than that found in policemen and others who were also regularly exposed. One study showed that the air in 10 percent of the automobiles in heavy traffic contained more than 30 ppm carbon monoxide, a concentration that approaches the threshold for acute poisoning. Underground garages have

TABLE 6.2 Emissions of Carbon Monoxide in the United States, Estimated for 1970.

	Carbon monoxide million tons per year
Motor Vehicles	111.0
Stationary Combustion	0.8
Solid Waste burning	7.2
Industrial	11.4
Miscellaneous	18.3
Total	149.0

From *Compilation of Air Pollution Emission Factors, Second Edition*. U.S. EPA Publication AP-42. 1973.

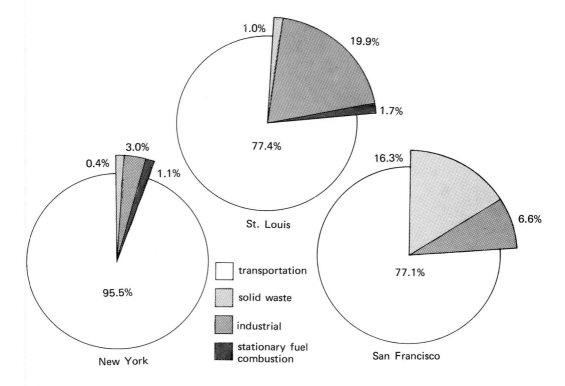

FIGURE 6.3 Carbon monoxide emissions by source for three U.S. metropolitan areas.

been found to contain 100 to 200 ppm carbon monoxide in air. (See Figure 6.4.)

While the carbon monoxide concentration in cities has been rising, the criteria for safety have become more rigid as a result of increased awareness of the hazards on the part of air pollution control workers and public health officials. The criteria are based on the period of time during which the concentration remains at a given level, because the longer the exposure, the more serious the effect. A concentration of 30 ppm for 8 hours was set as the "serious" level by Los Angeles County in 1969. This level had been exceeded 19 times in 1964, 58 times in 1965, 23 times in 1966, 26 times in 1967, and 6 times in 1968. The State of California Air Resources Board set a limit of 20 ppm for 8 hours. The Board stated:

> Prolonged exposures to such ambient concentrations can produce an increase of more than 2% COHb in non-smokers and higher levels in smokers. These amounts may produce impaired CNS[3] function and interfere with oxygen transport by blood.

Following this frank admission of the serious effects to be expected from exposure to a concentration meeting its own standards, the Board added:

> There is some evidence suggesting a more stringent standard and current work is likely to require an early reevaluation.

---

[3]Central Nervous System.

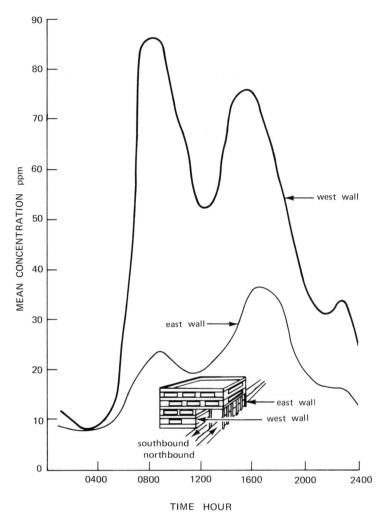

From P. C. Wolf, redrawn from *Environmental Science and Technology*, vol. 5, March 1971, p. 217, copyright 1971 by the American Chemical Society. Reprinted by permission of the copyright owner.

FIGURE 6.4 Carbon monoxide concentrations at peak rush hours, showing the difference between an enclosed area (west wall) and a partially open area (east wall).

The limit was lowered in November 1970 to 10 ppm average for 12 hours. This closely approximates the federal standard.

In the 1950s, Los Angeles had established three alert levels at 100, 200, and 300 ppm. Alerts were to be announced if the instantaneous concentration reached these levels. Prolonged exposure to the lowest alert level causes most people extreme discomfort, and many sufferers have symptoms of poisoning such as headaches, dizziness, lassitude, and more serious manifestations of carbon monoxide intoxication. In December 1970, Los Angeles changed the carbon monoxide alert levels to 50, 100, and 150 ppm.

Effect on plants   Carbon monoxide has the effect of inhibiting nitrogen fixation. At concentrations of 0.01 to 1 percent (100 to 10,000 ppm), visible effects are abscission (leaf drop), premature aging, and the initiation of roots on stems. These effects resemble those of ethylene, a gas produced naturally by ripening fruit and some pathogenic fungi. Carbon monoxide is believed to inhibit cellular respiration in plants by reacting with the enzyme system cytochrome-oxidase which is vital to the utilization of sugars as metabolic fuel. In view of this, there has also been interest in the possible effects of carbon monoxide on the cytochrome-oxidase system of animals. Cytochrome oxidase is the principal oxygen-reducing enzyme of the body. It is a heme protein that will combine with carbon monoxide, reducing the enzyme activity. However, it appears unlikely that the carbon monoxide concentration becomes high enough to inhibit cytochrome activity except under extreme conditions.

Carbon monoxide from natural sources

Higher animals produce some carbon monoxide from the breakdown of hemoglobin. In humans, an average person produces about 0.4 milliliters per hour. The output is increased in hemolytic disorders. Some carbon monoxide is probably liberated also from nonhemoglobin hemes such as bile pigments.

Some bacteria can produce carbon monoxide, but the amount is small and probably of little importance. Kelp *(Nereocyctis)* contains in its bladders up to 800 ppm carbon monoxide. Some members of the marine *Siphonophora,* an order of the class Hydrozoa, emit gas bubbles with as much as 80 percent carbon monoxide. The Portuguese man-of-war, a poisonous, stinging jellyfish, is a releaser of carbon monoxide. The significance of this biological production of the gas is not known.

The gas is produced naturally from a number of other noncombustion sources. Coal gas, from the distillation of coal, contains up to 10 percent carbon monoxide. Water gas, made by passing steam over incandescent coke or a mixture of coke and coal, contains 30 to 40 percent carbon monoxide. Water gas, known as illuminating gas, is used in many large cities and is generally enriched with natural gas or petroleum gas. Natural gas usually contains no carbon monoxide.

The surface waters of the western Atlantic were found to be supersaturated with carbon monoxide. The ocean appears, therefore, to act as a source of carbon monoxide and may be the most important natural source of carbon monoxide in the atmosphere, possibly contributing as much as 5 percent of the amount emitted by the burning of fuels.

OZONE

Ozone is an important component of so-called oxidant smog.[4] There is very little emission of ozone from automobile exhaust. However, ozone plays a key role in the photochemical formation of atmospheric pollutants as well as being an end product in the intricate complex of reactions that take place (see Photochemical Smog). Ozone is the most reactive form of molecular oxygen and the fourth most powerful oxidizing agent known, exceeded only by fluorine (F_2), oxygen difluoride (OF_2), and atomic oxygen ($O°$). Ozone has a distinctively pleasant odor at concentrations of about 2 ppm or slightly less, but it is irritating at higher concentrations. It

---

[4]The term *smog* is a contraction of *smoke-fog*. Its composition is highly variable.

is used as a disinfectant for air and water, and industrially for bleaching waxes and oils, as well as for organic synthesis. The use of ozone as an air disinfectant for closed spaces is dangerous.

Effect on animals

Ozone is extremely toxic. Rats exposed to 1 ppm for an 8-hour day develop bronchitis, fibrosis (formation of fibrous tissue), and bronchiolitis. A concentration of 1.25 ppm for 1 hour causes an increase in the residual lung volume (reduced expiration) and a decrease in the breathing capacity. Higher levels cause pulmonary edema (accumulation of fluid in the lungs), hemorrhaging, and impairment of gas exchange through the alveolar membrane. The maximum allowable concentration for occupational exposure for an 8-hour work day has been set at 0.1 ppm.

A curious characteristic of ozone toxicity is that tolerance develops in laboratory animals from exposure to subinjurious concentrations. The tolerance develops quickly, within 24 hours. It may persist for 4 to 6 weeks in rats and 100 days in mice, as shown by their survival from exposure to concentrations that would ordinarily be lethal. After tolerance develops, the pulmonary edema and hemorrhaging that usually follow challenging exposures are absent. Moreover, cross-tolerance can be demonstrated between ozone and the lethal gases phosgene and nitrogen oxides.

Ozone increases susceptibility to infection. Experiments with mice showed that ozone increases the survival of bacteria inside the lungs. In rabbits, cells washed from the lungs after exposure to ozone showed that there was a reduction in the ability of the alveolar macrophages to engulf bacteria. Also, laboratory animals that have been subjected to prolonged exposure to ozone show signs of premature aging.

The injurious effect of ozone may be partly due to its ability to oxidize lipids (fats). A deficiency of tocopherol (member of the vitamin E complex containing several tocopherols) is known to make tissues more susceptible to oxidation. Rats with a deficiency of vitamin E are more susceptible to lethal concentrations of ozone than those not deficient.

Effect on plants

Ozone has long been known to be toxic to vegetation, but injury to crops from atmospheric pollution has only recently been recognized. In 1958 grape leaves in California vineyards were described as having the upper surfaces stippled from ozone toxicity. Shortly afterward similar injury to tobacco leaves, called "weather fleck," was identified as ozone injury. Since then, ozone toxicity has appeared as stipple, flecks, spots, streaks, tipburn, or premature yellowing on at least twenty crop plants. Ozone affects the mature leaves first and spots are usually visible only on the upper surface. White pine needles are also affected with tipburn caused by ozone.

Plants are extremely sensitive to ozone. Concentrations as low as 0.08 ppm for periods of exposure as short as 1 hour cause injury to sensitive plant tissues. The sensitivity of plants to ozone is dependent upon the species, variety, and environmental conditions both before and during exposure. Among the edible plants that are susceptible to ozone, some varieties of tomatoes and radishes are injured by 0.15 to 0.25 ppm when exposed for 1 to 2 hours.

Certain tobacco varieties are the most sensitive organisms known for indicating the presence of ozone and will show symptoms when the atmospheric concentration is above 0.05 ppm. The presence of sulfur

USDA photo.

Ozone damage on tobacco leaves. Air pollutants cause great damage to crops and ornamental plants each year.

dioxide enhances the effect of ozone. For example, symptoms appeared after 2 hours of exposure of 0.037 ppm ozone plus 0.24 ppm sulfur dioxide, but when the plants were exposed to the same concentrations of the gases separately, they did not develop visible signs of injury. This effect is described technically as reducing the threshold concentration by synergistic action.

Ozone is normally present in the atmosphere at about 0.05 ppm at sea level. However, the concentration is variable from an average of about 0.02 ppm in winter to about 0.07 ppm in summer. It is produced naturally in the atmosphere by the action of electric discharges on oxygen. The reaction probably involves the combination of atomic oxygen ($O°$) with oxygen at high temperatures or by the energy provided by ultraviolet radiation:

$$3O_2 \longrightarrow 2O_3$$

Ultraviolet radiation in the wave-length range of 260 nanometers (nm) is absorbed by oxygen molecules, causing them to react.[5] In this manner, radiation from the sun probably causes high concentrations of ozone in the outer portions of the earth's atmosphere.

The role of ozone as an atmospheric pollutant is further discussed under the section Photochemical Smog.

---

[5] 1 nm = 1 billionth of a meter.

NITROGEN OXIDES

Nitrogen combines with oxygen to form a family of oxides, collectively called nitrogen oxides (indicated by the symbol NO_x), which play a multiple role in air pollution. They contribute substantially to visible pollution due to the color of the common nitrogen dioxide. They also enter into chemical reactions with other substances in the atmosphere with the production of additional irritating and toxic ingredients of smog.

We said in Chapter 3 that the fixation of nitrogen is a vital biological process in certain plants and microorganisms and that industrial nitrogen fixation by the chemical processing of air is an important commercial enterprise. The automobile and other combustion-power devices are also nitrogen-fixing machines, although in combustion the immediate end product is a contaminant rather than a useful product (see Figure 6.5). In gasoline and diesel engines, the high temperature and pressure in the combustion chamber provide suitable conditions for the production of nitrogen oxides, which, unfortunately, enter into a number of reactions in the atmosphere responsible for a panoply of air pollutants.

The lethal family   The nitrogen oxides are among the most toxic substances found in the atmosphere. One of the lesser toxic members, *nitrous oxide,* called "laughing gas," is useful in medicine as a mild anaesthetic. Nitric oxide and nitrogen dioxide, because of their abundance and toxicity, are the most dangerous in terms of public health. A third highly toxic member, *nitrogen pentoxide,* is a potential threat.

Nitric oxide (NO), a colorless gas, is the primary nitrogen product of combustion. It is produced when the temperature is high enough to cause a reaction to take place between the nitrogen and oxygen in the air that is sucked into the cylinder with the air-fuel mixture. Thus the form of the nitrogen oxide that is emitted to the atmosphere in the exhaust is almost all nitric oxide. However, a large portion of it is converted to the more toxic nitrogen dioxide by a complex series of reactions in the atmosphere (see Figure 6.6).

On contact with air, nitric oxide combines with oxygen, or even more rapidly with ozone, to form the more poisonous nitrogen dioxide:

$$2NO \;+\; O_2 \;\rightleftharpoons\; 2NO_2$$

nitric   oxygen   nitrogen
oxide   dioxide

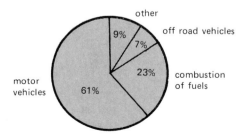

Redrawn from *Air Pollution in California,* Annual Report, 1973.

FIGURE 6.5 Sources of nitrogen oxides in California, 3,800 tons per day.

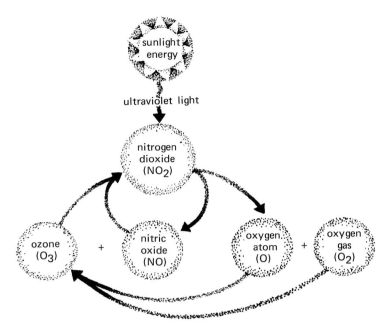

Adapted from National Air Pollution Control Administration Publication No. AP-63, 1970.

FIGURE 6.6 Atmospheric nitrogen dioxide photolytic cycle. The ultraviolet light acts upon nitrogen dioxide, breaking its bond and yielding two substances, nitric oxide (NO) and oxygen atom (O). The oxygen atom then reacts with oxygen (O_2) in the atmosphere, forming ozone (O_3). The ozone reacts upon the nitric oxide to reform nitrogen dioxide. Thus the energy of ultraviolet light acts as a "pump" in the rapid destruction and reformation of nitrogen dioxide.

At high temperatures the equilibrium is reversed, converting nitrogen dioxide back to nitric oxide.

Nitrogen dioxide (NO_2) is the only widely prevalent pollutant gas that is colored. Pure nitrogen dioxide is a deep reddish brown and causes much of the atmospheric discoloration on bad smog days in metropolitan areas. Power plants and other energy conversion systems as well as a variety of chemical process industries emit nitrogen dioxide into the atmosphere, but in most metropolitan areas the most important source is vehicular exhaust.

Nitrogen dioxide is responsible for much of the ozone in polluted atmospheres due to its decomposition by the action of ultraviolet light:

$$1) \quad NO_2 \quad + \quad UV \quad \rightarrow \quad NO \quad + \quad O°$$

nitrogen   ultraviolet   nitric   atomic
dioxide   light   oxide   oxygen

$$2) \quad O° \quad + \quad O_2 \quad \rightarrow \quad O_3$$

atomic   ordinary   ozone
oxygen   oxygen
     gas

These reactions, in conjunction with others of a complex nature, account for daily cyclical fluctuations in the concentration of ozone in polluted areas of high sunlight intensity such as the Los Angeles basin where the amount of ozone is characteristically low in the morning but rises sharply by afternoon.

These reactions involving nitrogen oxides may cause a problem in the upper atmosphere if SSTs come into widespread use because changes in the ozone concentration are apt to affect the amount of ultraviolet light that reaches the surface of the earth, causing unpredictable biological effects.

Nitrogen dioxide itself is a deadly poison and is one of the most treacherous gases known. Exposure may go unnoticed until it is too late for the victim to recover from the delusion of well-being. Inflammation of the lungs is immediate, but it may not cause enough pain to be noticed. Edema (accumulation of fluid) may occur several days later, resulting in death. In some cases the effects may be prolonged for weeks or even months. Apollo astronauts Thomas P. Stafford, Donald K. Slayton, and Vance D. Brand narrowly escaped death while making a landing after their historic rendezvous in space with a Soviet spacecraft in July 1975. They were hospitalized after inhaling a poisonous gas, believed to be nitrogen tetroxide, that leaked into the spacecraft from the fuel system. (Much of the nitrogen dioxide in the atmosphere exists in the form of nitrogen tetroxide, called a polymer, with which it is in equilibrium.)

Even short exposure to nitrogen dioxide is dangerous. Breathing air containing 20 ppm of the gas for a brief time may be fatal. With prolonged exposure, much smaller amounts in the air are dangerous. The MAC (maximum allowable concentration) for occupational exposure has been set at 5 ppm (9 milligrams per cubic meter) for an 8-hour period.

Experiments with rats showed that the acute toxicity of this gas is less than that of ozone, but greater than that of carbon monoxide. However, delayed, rather than immediate, death is the characteristic response of nitrogen dioxide exposure.

The State of California Air Resources Board established the "adverse" level of nitrogen dioxide at 0.25 ppm for 1 hour. Yet this level is consistently exceeded during smoggy periods in the Los Angeles basin, and it is predicted that the nitrogen dioxide concentration will continue to get worse for several years until adequate controls can be established. Nitrogen dioxide remains one of the most potentially dangerous of modern-day pollutants.

Tobacco smoke

Cigarette and cigar smoke contains 330 to 1,500 ppm nitrogen oxides which are removed completely by inhalation, presumably by absorption in the lungs. In view of the extremely toxic properties of nitrogen oxides, knowledge of the effects on smokers is greatly needed.

Plants

Nitrogen dioxide is also highly injurious to plants. It suppresses growth when plants are exposed to 0.3 to 0.5 ppm for 10 to 22 days. A concentration of 1 ppm for 8 hours produces significant growth reduction with no visible damage. Exposure to 4 to 8 ppm for 1 to 4 hours causes visible leaf injury to sensitive plants. Low-light intensity increases the plants' sensitivity.

Nitrogen dioxide reacts with water to form nitric acid and nitric oxide:

$$3NO_2 \; + \; H_2O \; \rightarrow \; 2HNO_3 \; + \; NO$$

nitrogen   water   nitric   nitric
dioxide   acid   oxide

Much of the nitric acid in the atmosphere ultimately combines with ammonia to form ammonium nitrate, and probably all of the nitrogen oxides that do not react photochemically become nitrate salts that form aerosols (suspensions of small particles) in the atmosphere in sizes larger than 1 micron.

HYDROCARBONS

Hydrocarbons are organic compounds consisting of hydrogen and carbon. They comprise the major portions of coal, petroleum, petroleum derivatives, and similar products. The smallest hydrocarbon is methane, a molecule consisting of 1 carbon atom and 4 hydrogen atoms.

Hydrocarbon gases

Methane is a colorless, odorless, tasteless, nearly nonpoisonous, flammable gas. It is a dangerous asphyxiant because it is heavier than air and displaces oxygen at or near ground level. At high concentrations in the absence of oxygen, it is a narcotic (producing sleep or stupor). The gas is widely distributed in nature. The earth's atmosphere contains about 0.00022 percent by volume, making it the seventh most abundant gas, nearly 5 times as much as hydrogen. Methane is sometimes produced by the decay of organic material in swampy locations, hence the common designation "marsh gas." It is a constituent of natural gas, comprising an average of about 85 percent of the product. Methane is highly flammable, making it one of the more useful fuels, but it is explosive in concentrations above 5.53 percent in mixture with air.

Methane burns with a pale, faintly luminous flame. It can be used as a substitute for gasoline in internal combustion engines, and if it were widely adopted, the major exhaust pollutants would be greatly reduced. However, problems of transport, storage, supply, and mileage performance would greatly increase the cost. Known reserves of natural gas are extremely limited and there is increasing industrial demand for methane in the chemical processing industries. There is little likelihood that it can be used widely as a replacement for gasoline by the general public in the near future.

The second smallest hydrocarbon, *ethylene,* occurs in illuminating gas in concentrations up to 4 percent. It is also produced naturally by ripening fruit and by certain microorganisms, and in the absence of ventilation it is highly toxic to plants, causing defoliation at concentrations of a few parts per billion. Ethylene has been used as an inhalation anaesthetic but is dangerous because of its high flammability and explosiveness in mixture with air. It is only slightly toxic to higher animals, the lethal concentration for mice being 950,000 ppm.

$$
\begin{array}{c}
H \\
| \\
H - C - H \\
| \\
H
\end{array}
\qquad \text{or} \qquad CH_4
$$

methane

$$
\begin{array}{c}
H \quad H \\
| \quad | \\
H - C = C - H
\end{array}
\qquad \text{or} \qquad CH_2 = CH_2
$$

ethylene

There is little promise that pure ethylene, or its relative, *ethane* (CH_3CH_3), can be used extensively for fuel. However, the higher molecular weight members of the family, *propane* and *butane*, with 3 and 4 carbon atoms respectively, are well known as "bottled gas," LP gas, or liquified petroleum. Internal combustion engines operate efficiently with these gases used as fuel, but availability and problems of supply are limitations to their wide-spread adoption.

$$
\begin{array}{c}
H \quad H \quad H \quad H \\
| \quad | \quad | \quad | \\
H - C - C - C - C - H \\
| \quad | \quad | \quad | \\
H \quad H \quad H \quad H
\end{array}
\qquad \text{or} \qquad CH_3CH_2CH_2CH_3
$$

normal butane

Gasoline   Gasoline is a complex mixture of hydrocarbons containing from 4 to about 12 carbon atoms. The components differ in boiling point from about 100° to 400°F, referred to as the product's boiling range. The composition of gasolines varies widely. "Straight-run" gasoline is the product from the first distillation of petroleum, but this accounts for a very small portion of the gasoline consumed. In order to meet the demand, refiners augment the supply by converting the higher hydrocarbon consituents of crude oil into the desired lower molecular weight compounds.[6]

The phenomenal success of gasoline as fuel for the internal combustion engine is based on the fact that at the very high temperatures that occur in the engine cylinder, oxygen combines violently with the hydrocarbon vapors, causing rapid expansion while converting the reactants to carbon dioxide, carbon monoxide, and water, as well as other end products that are discussed elsewhere in this chapter.

Sources of hydrocarbon pollution   More than half of the atmospheric hydrocarbon pollution comes from automobiles (see Figure 6.7). Much of it consists of unburned gasoline in the exhaust from incomplete combustion, although some of

---

[6]This is done by means of chemical processing procedures. Methods include thermal decomposition (cracking), catalytic cracking, polymerization, alkylation, isomerization, reforming, dehydrogenation, and aromatization. *Cracking* is a highly descriptive term used in the petroleum industry; in chemical terminology, the word for thermal decomposition is *pyrolysis*.

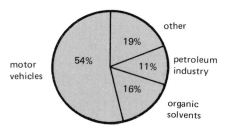

Redrawn from *Air Pollution in California*, Annual Report, 1973.

FIGURE 6.7 Sources of hydrocarbons (organic gases) in California, 5,900 tons per day.

the hydrocarbons in automobile exhaust are reaction products that are not found in gasoline. The second largest source of hydrocarbons is from industry, accounting for about one-fourth of the hydrocarbon pollution. Evaporation from refineries and storage tanks is an important source. Hydrocarbons derived from petroleum are used extensively in industry as solvents and cleaners. Paint thinner is an example of a mixture of volatile hydrocarbons whose ultimate repository is the air. The total hydrocarbon emission of about 35 million tons per year contributes to pollution directly and to smog indirectly by the reactions between hydrocarbons and other constituents in the atmosphere (see Photochemical Smog).

CHLOROFLUOROCARBONS

Gases called chlorofluorocarbons are widely used to pressurize aerosol spray cans such as those containing household insecticides and hair sprays. One of the trade names under which they are manufactured is *Freon*®. The Freons are similar to the light hydrocarbon gases methane and propane but with fluorine and chlorine substituted for the hydrogen atoms in the molecules. These substitutions make the gases more stable; for example, unlike methane and propane, chlorofluorocarbons are practically nonflammable. By 1975, more than 2 million tons of these gases were used annually in refrigeration systems and aerosol sprays.

Though chemically inert under normal circumstances, with an atmospheric lifetime of longer than 10 years, the chlorofluorocarbons can be broken down by conditions in the upper reaches of the atmosphere. It has been theorized that the gases eventually diffuse into the stratosphere where they are attacked by the high intensity of ultraviolet light and excited oxygen atoms, causing the release of free chlorine atoms which, in turn, attack ozone, changing it into oxygen. Thus, it is believed that the ozone layer may be slowly depleted.

The ozone layer, which surrounds the earth at a height of about 15 miles, is important to nearly all forms of life on earth because it shields the lower atmosphere and the earth from the killing effects of high-energy radiation. The ozone layer also serves to absorb infrared heat rays and plays a role in warming the upper atmosphere. Its depletion might have far-reaching effects on the earth's climate and weather patterns.

SULFUR OXIDES

Sulfur pollutants are belched into the air in enormous quantities as part of the industrial effluent. Atmospheric sulfur comes mainly from the sulfur content of fuels. Nearly 80 percent of the sulfur in sulfur dioxide is initially emitted as hydrogen sulfide and is converted to sulfur dioxide in the atmosphere. Sulfur dioxide and the products of its reactions are devastatingly damaging to structural and other materials, as well as being physiologically harmful and injurious to health. Sulfur pollutants were probably the predominant pollutants in most of the "killer smogs."

Sulfur dioxide is a colorless gas with a penetrating and pungent odor. It produces respiratory irritation at low concentrations. As an industrial air pollutant, it is formed by the oxidation of the sulfur compounds in fuel. The gas is moderately soluble, and in the presence of moisture it forms sulfurous acid, a fair oxidizing agent:

$$SO_2 \ + \ H_2O \ \rightarrow \ H_2SO_3$$

sulfur   water   sulfurous
dioxide   acid

Sulfurous acid is converted slowly to sulfuric acid (H_2SO_4). Sulfur dioxide also reacts very slowly in oxygen to form sulfur trioxide gas (SO_3), which in turn reacts quickly with water, forming sulfuric acid. Aerosols (fogs) of sulfuric acid contribute to the haze from the combustion of high sulfur fuels in industrial areas. The bluish white plume often seen coming from the stacks of electric power plants is often largely a sulfuric acid aerosol.

Sources of sulfur oxides

The major natural sources of sulfur dioxide are volcanic in origin. However, this contribution to pollution is believed to be small. Most of the total emission of sulfur oxides is from burning of fossil fuel by public utilities, industries, and furnaces used for heating commercial and residential buildings (see Figure 6.8). Public utilities account for over one-half of the total. Fossil-fuel power plants, for generating electricity, are the major source, and they are being built more rapidly than anticipated because nuclear power plants are being installed more slowly than planned. Most of the remaining 25 percent of the sulfur oxide emissions is accounted for by nonferrous smelters and petroleum refineries. The major sources are highly varied in size and design, and reduction of their emissions to acceptable levels requires a diversity of highly complex control systems. Emissions are increasing at an alarming rate. The estimated ouput in 1970 was 37 million tons in the United States, and by 1990, the potential to emit sulfur oxides will be nearly 95 million tons.

Physical damage

Sulfur dioxide and its end products have a severe corrosive and deteriorating effect on many materials. Sulfur dioxide and sulfur trioxide are especially damaging to building materials that contain carbonates, such as limestone, marble, mortar, and roofing slate. The reaction forms soluble calcium salts that are washed away by condensation and rainwater. Works of art in or near industrial areas and cities are in serious jeopardy. One of the two specimens of Cleopatra's Needle was moved to New York's Central Park in 1880. The obelisk has deteriorated more since its arrival than during its previous 3,000 years in Egypt where it originally

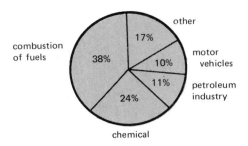

Redrawn from *Air Pollution in California*, Annual Report, 1973.

FIGURE 6.8 Sources of sulfur dioxide in California, 990 tons per day.

stood before the Temple of the Sun at Heliopolis. Some works of art are being moved indoors to preserve them.

The atmospheric deterioration of marble and limestone structures in Italy is referred to as "stone cancer." The art works and irreplaceable structures of Venice are decaying from the emissions of nearby industries. The smokestacks of Mestre, Marghera, and surrounding areas on the mainland pour the equivalent of 15,000 tons of sulfuric acid into the atmosphere every year from the burning of high-sulfur fuel. This, added to the municipal *vaporetto* from oil heating, motor boats, buses, and automobiles, brings a blight on the city. Nearly all of the thousand or so *palazzi* and churches contain material of marble having a high carbonate content, highly susceptible to sulfur oxide corrosion. Venice has other environmental problems including floods, humidity, salinity, industrial water pollution, changes in the populations of plants and animals in the lagoon, impaired tidal flow for removal of sewage, a rising water level,

Power plants are a major source of sulfur dioxide. The problem can be partially alleviated by using low-sulfur fuels.

Photo courtesy of National Air Pollution Control, Environmental Protection Agency.

Cleopatra's Needle standing in Central Park has deteriorated more since it was moved to New York in 1881 than during more than 3,000 years in Egypt. What is the effect of the air on the lungs of the two small boys?

and sinkage of the foundations. These, combined with corrosive air pollution, make Venice appear to be a dying city.

Sulfur oxides even cause the destruction of extremely durable structural materials. Steel corrodes up to four times as fast in cities as in rural areas. Particulate matter accelerates the attack of sulfur oxides on metals, especially damaging when moisture is present. The oxide and the acid attack iron, steel, copper, nickel, zinc, and at high concentrations, even aluminum. However, the corrosive action of sulfur oxides contributes to the attractively greenish *patina* seen on copper or bronze ornamental and structural materials in cities. The film forms a protective coating against further deterioration. But similar action causes serious damage to electrical equipment by increasing the resistance at copper contacts.

Sulfur oxides attack some paint pigments and modify fresh paint films in a way that delays drying and increases the susceptibility to moisture. Sulfur pollutants even attack nylon, to the discomfort of women in cities with heavily polluted atmosphere. It makes paper brittle due to the presence of metallic impurities that catalyze (accelerate) the conversion of sulfur dioxide to sulfuric acid. Sulfur dioxide also attacks leather and textiles, causing slow deterioration. Synthetic fibers, but particularly cotton and wool, are susceptible. Laboratory experiments under controlled conditions show that photo-chemically produced compounds in the atmosphere cause fading of dyes in fabrics, and the addition of sulfur dioxide causes a synergistic effect, greatly increasing the damage.

Human health effects

Sulfur dioxide is intensely irritating to the eyes and the respiratory tract. A concentration of 0.3 to 1 ppm can be detected by the average person, and this may be by taste as well as smell. Most people will show a response from exposure to concentrations in the range of 1 to 5 ppm. Exposure to the higher concentration for 1 hour causes a pronounced choking sensation, and 10 ppm for the same period produces severe distress. Exposure can cause serious inflammation of conjunctiva of the eyes. Concentrations of 6 to 12 ppm cause immediate irritation of the nose and throat. A few minutes of exposure to 1 percent (10,000 ppm) produces irritation, even to moist areas of the skin.

Sulfur dioxide is so irritating that at acutely dangerous concentrations, it serves as its own warning agent. However, this safety factor does not always exist at low concentrations. Moderate resistance develops from continuous exposure so that concentrations as high as 5 ppm can no longer be smelled. Concentrations of 400 to 500 ppm may be lethal and are considered to be immediately dangerous to life, while 50 to 100 ppm is regarded as the maximum permissible range for 30 to 60 minutes exposure. The accepted maximum allowable concentration for an exposure of 8 hours is 10 ppm (26 milligrams per cubic meter of air).

Moist air and fogs probably increase the danger of sulfur dioxide greatly, partly due to the formation of sulfurous acid and sulfite ions. A further explanation may be that sulfur dioxide becomes adsorbed on finely divided aerosol particles, resulting in deposition deep in the air spaces of the lungs. The toxic end product of sulfur dioxide is probably sulfuric acid, which is a stronger irritant than sulfur dioxide. Sulfuric acid causes 4 to 20 times the effect on an equimolar basis (molecule for molecule).

Dr. Sidney Laskin of the New York University Medical Center reported that the susceptibility of mice to the carcinogen benzpyrene is increased by breathing air containing sulfur dioxide. A curious finding was that guinea pigs exposed to sulfur dioxide at 5 ppm survived longer and had less lung disease than those exposed to lower concentrations. In contrast to the adverse effects from lower concentrations, the 5-ppm group of animals did as well as the control animals that were breathing pure air. Apparently a protective mechanism is involved, either by the inducement of alveolar microphages[7] or by the water solubility of sulfur dioxide limiting its attack to the superficial air passageways.

The California Standards for ambient air quality specify the "adverse" level of sulfur dioxide as 0.04 ppm for 24 hours, or 0.5 ppm for 1 hour. These levels were reduced from earlier standards of 1 ppm for 1 hour or 0.3 ppm for 8 hours because these exposures were known to cause damage to vegetation.

---

[7]Alveolar microphages are phagocyte cells of unknown origin. *Phagein* (Gr), "to eat."

Injury to plants   Sulfur dioxide has been known for many years to be a major cause of injury to higher plants. Some species are so sensitive that its effects are sometimes the first indication that the air is polluted. The cause of the symptoms, however, is not always easy to identify due to the fact that the injury often resembles other afflictions such as sunscald, frost, aging, or other disorders having obscure environmental or cultural causes.

Plants with high physiological activity are among the most sensitive species. Alfalfa, grains, squash, cotton, grapes, apple, endive, and white pine are susceptible. Species with thick, waxy leaves, such as citrus and privet, are more resistant. Trees with needles, such as most pines, are generally resistant except for those with newly formed needles. Conditions that are ordinarily beneficial to plant growth, such as high morning light intensity, high humidity, good soil moisture, and moderate temperature, are apt to enhance injury.

Prolonged exposure of plants to sublethal levels of sulfur dioxide causes them to develop a chronic type of injury which first appears between the veins, leaving the vein areas green. Older leaves gradually become yellow, then white, and eventually die. However, in leaves that are only partially injured, the unaffected parts regain their function upon removal from exposure, and new foliage grows normally. The symptoms of such chronic exposure are sometimes described as "early aging" and are not always easy to identify. Concentrations that cause chronic injury are usually less than 0.4 ppm.

The most serious cases of plant damage have been from high concentrations in the vicinity of smelters. A pollution espisode occurred at Anaconda, Montana, during 1910 to 1911. Large amounts of sulfur oxides were emitted, causing nearly all the important species of trees to die within a radius of 5 to 8 miles. During another incident in 1929 in the area of a smelter at Trail, British Columbia, damage to plants occurred as far as 52 miles from the smelter. Douglas fir, ponderosa pine, and other forest vegetation were 60 to 100 percent damaged for a distance of 30 miles. The emission rate from the smelter was as high as 20,000 tons and averaged 18,000 tons of sulfur dioxide per month. Inhibition of cone production by pine trees was still evident two years after installation of a control system. Acute injury to broadleaved plants causes the leaves, or areas of the leaves, to dry out and become bleached.[8] Injured grasses show a similar pattern except that the tips are killed and the injury generally has a more streaked appearance. The response in plants is clearly the result of the oxidizing or reducing properties of sulfur oxide per se and not due to the action of acid. Sulfuric acid aerosols are not generally toxic to plants, although emissions of large droplets from factories have been known to damage vegetation.

Sulfur dioxide has a powerful toxic action on microorganisms. It is used as a fungicidal fumigant, as a preservative for food and wine, and as a disinfectant in breweries. Its toxic action is believed to depend on the formation of sulfurous acid or sulfite ions in the tissues of the organism.

Acid rain   The exact fate of sulfur pollutants depends on the atmospheric condi-

---

[8]Acute effects are those which show up quickly, usually within 24 hours.

tions. A first step is the oxidation to sulfur trioxide. This dissolves in the water droplets of mist and forms sulfuric acid, which can react further to form sulfate salts, such as ammonium sulfate. Another route that sulfur dioxide can take in polluted atmosphere is to undergo photochemical (light-stimulated) oxidation to form sulfuric acid. This oxidation of sulfur dioxide is catalyzed (speeded) by the presence of metal ions, such as iron or magnesium, and by ammonia, also an emission product of the combustion of coal.

The sulfuric acid or sulfate salt aerosols eventually settle out, much of it in precipitation. The time required for fallout is estimated at 5 to 14 days. The rain in some parts of Sweden has been measured at pH 2.5, about as acid as Coca Cola. Some people attribute this to the sulfur oxides blown from the industries of England and other areas of northern Europe.

Prevention of sulfur oxides

Since the sulfur oxide comes from fuels with high sulfur content, the most immediately effective control measure is the use of low-sulfur fuel. Coal is the main source of sulfur pollution; dramatic improvements have been made by reducing the use of high-sulfur coal. Recourse to low-sulfur fuels, however, is at the present time limited. Low-sulfur natural gas is a beneficial substitute for coal and fuel oil, but it is limited in supply. The magnitude of the problem is evident from a report issued by a panel of the National Academy of Engineering and National Research Council stating that satisfactory methods of controlling sulfur oxide emissions have not been accomplished on a large commercial scale.

HYDROGEN SULFIDE

During the nights of November 23 and 24, 1950, in the small town of Poza Rica, Mexico, a poisonous gas drifted into homes and struck down people of all ages. The toxic gas, hydrogen sulfide, came from an industrial plant built to manufacture sulfur from natural gas. Hydrogen sulfide is ordinarily removed from natural gas as one of the steps in the process of recovering by-product sulfur, but in the early morning of November 24, the hydrogen sulfide was inadvertently released and caught many of the residents in bed. Hydrogen sulfide causes rapid respiratory irritation and central nervous system impairment, including vertigo, unconsciousness, and death. In all, 320 persons were hospitalized and 22 died.

Hydrogen sulfide is a colorless gas, having about the same toxicity as cyanide. The gas has a penetrating odor resembling rotten eggs, and is therefore detectable by the human nose at very low concentrations, about 2 ppb (parts per billion). However, the olfactory sensory organs become fatigued quickly and can no longer respond. The sense of smell for hydrogen sulfide is lost within 2 to 15 minutes at 100 ppm or higher. This characteristic makes hydrogen sulfide extremely dangerous in cases of excessive or prolonged exposure.

Low concentrations of hydrogen sulfide cause headache, nausea, lassitude, collapse, coma, and death. Hydrogen sulfide readily passes through the alveolar membrane of the lungs and penetrates to the blood stream. Death may come within a few seconds from even one or two inhalations. At 1,000 ppm it is almost instantaneously fatal. Death is from respiratory failure. The California standards for ambient air quality specify an "adverse" level of 0.1 ppm for 1 hour as sufficient to produce sensory irritation. The maximum allowable concentration for an 8-hour day is 20 ppm.

Natural sources

The principal natural sources of hydrogen sulfide are decaying vegetation and animal material, especially in shallow aquatic and marine environments. Anaerobic bacteria attack sulfates in deep sediments, reducing the sulfate to sulfide. Anaerobic decomposition sometimes takes place in underground organic deposits, causing hydrogen sulfide to be present in domestic water supplies. Water from such sources has an objectionable odor, but in some parts of Europe water containing hydrogen sulfide is considered to have medicinal properties. A concentration of 0.05 ppm is detectable by taste. Hydrogen sulfide is given off by sulfur springs, volcanic discharges, gas wells, coal pits, and sewers. Early French writers referred to *plomb des fosses* (lead of the pits) because the colic and diarrhea that afflicted sewer workers who were exposed to hydrogen sulfide resembled the well-known symptoms of lead poisoning. It has been estimated that about 30 million tons of hydrogen sulfide are released annually by ocean areas and 60 to 80 million tons by land areas.

Industrial sources

Industrial pollution may not exceed 3 million tons annually and is probably not often a significant factor in public health, although it can be a threat in localized areas and from exposure during industrial operations. The important industrial sources of hydrogen sulfide are those which also give rise to sulfur oxides and sulfuric acid, the users of high-sulfur fuels. About 80 percent of the sulfur compounds in fuels are converted to hydrogen sulfide, which is soon oxidized in the atmosphere to other sulfur compounds. Notable emitters of hydrogen sulfide are petroleum refineries, kraft paper mills, tannery wastes, and mining operations.

Most of the hydrogen sulfide released to the atmosphere does not remain in that form for long. It is oxidized to elemental sulfur and sulfur dioxide. The latter is further oxidized to sulfur trioxide and subsequently to sulfuric acid and sulfates. (These related sulfur compounds are discussed in the section on Sulfur Dioxide, this chapter.) An important reaction in the atmosphere is the oxidation of hydrogen sulfide with ozone:

$$H_2S + O_3 \rightarrow SO_2 + H_2O$$

hydrogen sulfide $\quad$ ozone $\quad$ sulfur dioxide $\quad$ water

The reaction takes place fairly rapidly, with a half-life of atmospheric hydrogen sulfide of 2 to 28 hours, depending on location and conditions. Thus hydrogen sulfide does not remain as a pollutant for a long period of time, but is a transient toxicant. It can be highly injurious, however, if concentrations are high or if exposure is prolonged.

AEROSOLS

An aerosol is a suspension of finely divided solid particles or droplets of liquid in the air. Examples are smoke, fog, dust, and the condensation plume from steam (see Figure 6.9). The term *particulate* is often used to

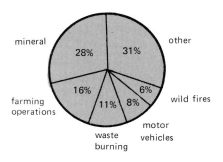

Redrawn from *Air Pollution in California,* Annual Report, 1973.

FIGURE 6.9 Particulate matter in aerosol-polluted air comes from various sources. These comparisons were based on analyses in California, where the average output was 1,900 tons of suspended particulate material per day.

refer to both solid and liquid aerosol particles in the air. The amount and nature of suspended particulates in the atmosphere and their chemical and physiological effects are among the least understood aspects of air pollution.

Source of particulates

Particulates are not necessarily visible. Suspended particles of microscopic dust, chemicals, liquids, sea-salt, pollens, aeroallergens, bacteria, and viruses are abundantly present in the atmosphere, and some of them are the cause of much human discomfort and disease. Other injurious aerosol constituents include soot, lead particles from exhaust, asbestos, fly ash, volcanic emissions, pesticides, sulfuric acid mists, and metallic dusts. Dozens of substances have been identified as minor particulate components of air, possibly including many toxic materials that have not yet been determined to be injurious.

A surprising amount of dust settles on the earth from outer space. Every day 8 billion solid particles penetrate the atmosphere. The average size is less than a grain of sand. Only about 2,000 per day of the larger ones, in the form of meteorites, ever reach the earth's surface. Those that do average about 50 pounds of material, so dense that it would be about the size of two building bricks. The rest of the particles burn up or disintegrate in their swift passage through the atmosphere. There is enough dust and ash left over from the attrition of the larger bodies and the remains of the disintegrated material to leave an accumulation of about 2 million tons of debris that settle on the surface of the earth annually. It drifts down and is found everywhere from the high mountain peaks to the sediment at the bottom of the oceans.

Microscopic study of air samples taken in Austria showed that an enormous number of particles are dispersed in the air. The number varied widely from 5,000 per cubic centimeter (about 82,000 per cubic inch) in downtown Vienna to 800 particles per cubic centimeter (about 13,000 per cubic inch) in the residential area, as compared to 300 per cubic centimeter (5,000 per cubic inch) in natural aerosols. The size frequency was the same both in the city and in the mountains.

Physical and physiological effects

Particulate matter has damaging physical effects, partly due to the acceleration of the corrosive effects of other pollutants such as sulfur oxides and sulfuric acid and partly due to the formation of grime. Dirt and soot are damaging to fabrics and other materials, which are then costly to clean, requiring the use of large quantities of solvents and detergents which further add to the pollution cycle.

The physiologically injurious effects of pollutants such as sulfur dioxide and its reaction products are made more severe by the presence of particulate matter. The synergistic effect was noted in the case of the London "killer smog" of 1952; the evidence indicated that it was the combination of particulates and sulfur oxides that caused 4,000 deaths during that episode.

Cancer-inducing substances have long been known to be present in particulate pollutants. They induce cancer when painted on the backs of mice or when injected under the skin but do not consistently do so when injected into the lungs. However, lung tumors, resembling human lung cancer, are produced in experimental animals when benzpyrene is mixed with iron oxide or hematite and injected into the trachea. These materials are commonly found in mixture with other atmospheric pollutants. Toxicity experiments involving injections of extracts from polluted air into newly born mice show that there is a broad spectrum of carcinogenic materials in the atmosphere.

Soots are physically complex structures, mostly highly porous agglomerates of great absorptive power. Since soots are produced in conjunction with other combustion products, many of them potent toxicants, the potential for synergistic action is inherent. The first clinical evidence for a cause of cancer was the observation in 1775 that scrotal cancer affecting London chimneysweeps was associated with the accumulation of soot in the crotch (see Cancer and Causes, Chapter 13). Polycyclic hydrocarbons, to which group the potent carcinogen benzpyrene belongs, are produced generally in combustion processes. In humans certain "inert" particulates, including those in cigarette smoke, cause resistance to the passage of air in the respiratory system.

Size of particulates

Aerosol particulates in the atmosphere range in diameter from about 0.001 micron[9] to more than 100 microns (see Figure 6.10). Most of the particulates are in the range of 0.1 to 10 microns in diameter. Since the rate of fallout by gravity is determined by the density times the cube of the diameter of the particle (assuming a sphere), the larger particles settle out more rapidly than the smaller ones. However, there are many other factors involved, including the shape of the particle, its velocity, the air density, wind movement, turbulence, thermal conditions, and the electrostatics at the surface of the particle. Particles larger than 10 microns settle rapidly, although they may remain air-borne for moderately long periods under turbulent conditions. Particles of 0.1 to 1.0 micron would normally remain air-borne for extended periods, except that collisions and electrostatic attractions, as well as other forces, result in coagulation and the formation of effectively larger particles. Particles of less than

---

[9]A *micron*, often designated by the Greek letter mu (written μ) is 0.001 millimeter (about $1/_{25,000}$ inch.

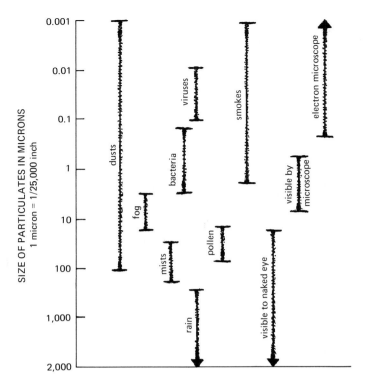

FIGURE 6.10 Relative size of atmospheric particles.

0.1 micron can form extremely stable aerosols, but by random movement in the air, eventually collide with other particles and form larger particles by coagulation.

In most cases particulate dispersions contain agglomerates rather than discrete particles. Most of the particles injected into the atmosphere fall out within minutes, hours, or days. But the smaller particles may remain air-borne for years. Many of the aerosol particles originating from human-made sources in cities and industrial areas, as well as from volcanoes, forest fires, and nuclear explosions, notoriously travel across international borders.

Some of the aerosol particles, mainly sizes above 2 microns, are removed from the air by rainfall. Particles also disappear from the air by impinging on objects such as trees and buildings. Extremely small particles do not readily impinge on objects because of their aerodynamic properties. The size judged to be most liable to enter the respiratory tract and become lodged in the air spaces of the lungs is about 1 micron.

Particles emitted from automobile exhaust are mostly less than 5 microns in diameter. They consist of carbon particles, lead compounds, crankcase oil, and reaction products from the combustion of motor oil. Other particulates originating indirectly from automobile pollution of aerosols are formed as reaction products from nitrogen oxides and hydrocarbons in the atmosphere. These products are discussed under Vehicular Exhaust. Diesel engines emit more particulate material than gasoline engines. Diesel exhaust is estimated to contain 62.5 percent par-

TABLE 6.3   Relation Between Volume and Surface
of Spherical Droplets Having a Total Volume of
523,600 Cubic Microns*

Microns diameter	Number of particles	Surface area, square microns	
		Total	Ratio
100	1	31,416	1×
10	1,000	314,160	10×
1	1,000,000	3,141,600	100×

*Volume of a sphere = $\frac{1}{6} \pi d^3$; surface of a sphere = πd^2

ticles less than 5 microns in size and 37.5 percent in the 5- to 20-micron range.

Surface area   The effects of small particles are best understood by considering the relationship between the size of the particles and their surface area. Assuming the shape of particles in an aerosol to be perfect spheres, the total surface area is inversely proportional to the diameter of the particles. In order to visualize this, let us take a droplet having a diameter of 100 microns and a volume of 523,600 cubic microns. Then split this large droplet into smaller droplets, each having a diameter of 10 microns. The volume will be the same, but instead of 1 droplet there will now be 1,000 droplets, and they will have a total of 10 times the surface area of the original droplet. Split these droplets further into smaller ones each having a diameter of 1 micron, and there will now be 1,000,000 droplets with a surface area 100 times the original droplet (see Table 6.3).

The enormous surface created when chemicals are divided into small droplets greatly increases the possibility that chemical reactions will take place. Large surfaces also bring about physical changes such as increased absorption of heat and light scattering. The effects of small particles in carrying toxicants and depositing them in the air spaces of the respiratory system, as well as their effect on reactions with tissues, need further study.

VEHICULAR EXHAUST

The automobile   The automobile is man's greatest achievement in mass-produced machines. To many people its operation and display represent the pinnacle of prestige, pleasure, and dominance over nature. For what other creature can take control of a package of power equivalent to 400 prancing horses, and at the slightest touch, command it to propel him at speeds beyond the most uninhibited imaginations of men who lived a mere century ago? But the metal and plastic successor to the horse does not have the intelligence to correct human mistakes, and its exhaust is not only more obnoxious, but poisonous as well.

Through use and misuse, automobiles and other highway vehicles killed more than 60,000 people during 1973 and more than 50,000[10] during

[10]This drop in vehicular fatalities was due largely to a reduction in the use of gasoline.

1974 in the United States alone. This is more human life than the fatalities from wars, murders, drugs, and pesticides rolled into one. Moreover, the crushing physical impact of man's gleaming technological pride leaves an annual toll of several hundred thousand injured and maimed men, women, and children, each a grim statistic of undocumented misery. Despite the colossal enormity of this human sacrifice to transient pleasure and noisy convenience, a more damaging by-product of the machine may be the silently insidious effect on human health from the poisonous exhaust of its internal combustion engine, making the automobile probably the most serious pollution problem of the technological age. In 1974, a panel of the National Academy of Sciences submitted a report on air quality standards to the U. S. Senate Subcommittee on Environmental Pollution. The Panel attributed as many as 4,000 deaths and 4 million illnesses per year to automotive exhaust emissions.

Breath of death   Two of the primary pollutants, carbon monoxide and nitrogen dioxide, are extremely poisonous gases frequently present at toxic levels in the atmosphere of metropolitan areas. A third primary pollutant, unburned hydrocarbons of gasoline, is probably less injurious but indirectly gives rise to the more toxic *peroxyacetyl nitrate* (PAN) and its relatives, collectively called peroxyacyl nitrates or PANs. The PANs and another dangerous toxicant, ozone, are usually referred to collectively as *oxidant* by pollution control officials. The PANs are not only extremely irritating to mucous membranes and the respiratory tract, but are highly toxic as well and may be carcinogenic.

Lead, a poison of interminable persistence, has periodically afflicted mankind for 2,000 years. It is spewed into the environment at 5 times the rate of DDT at its peak. Many other by-products of gasoline exhaust have been identified, including small amounts of *benzpyrene,* one of the most potent cancer-causing agents known. Finally, the long-term effect of carbon dioxide emission may be potentially catastrophic; the possibility requires attention and serious study. The best that can be said of the exhaust from the gasoline-fed internal combustion engine is that the water vapor formed as a by-product is innocuous.

The toxic nature of the gaseous and particulate excreta of the internal combustion engine is not easy to correct. The gasoline engine is designed for presently available fuels within rigid limits; beyond those mechanical limitations, it is balky and uneconomical. The changes that have been made in internal combustion engines to reduce carbon monoxide output have been directed toward bringing about more complete combustion— to carbon dioxide. This is determined primarily by the air-fuel ratio. The best results are obtained with a high proportion of air in the vapor of the fuel-air mixture fed to the cylinders. If the mixture is made leaner (more air and less fuel), the carbon monoxide emission is reduced. In practice this has limitations, because a large reduction in carbon monoxide requires a mixture so lean that the engine does not run well. The most difficult problem is that changes in the setting and operation of the engine to reduce carbon monoxide output cause an increase in the output of nitrogen oxides, compounds of even more serious import because of their contribution to visible smog, irritation, and toxicity.

The dilemma   Carbon monoxide emission is reduced by the higher engine temperatures used for the most efficient operation, but particularly if high temperatures can be maintained in the exhaust manifold long enough to achieve complete oxidation of carbon monoxide. A dilemma is caused by the fact that the reverse is true for nitrogen oxides (NO_x), the formation of which is increased by high-combustion temperatures calculated at 4,000 to 5,000°F. Such temperatures are highly favorable for the reaction between nitrogen and oxygen in the air of the fuel mixture to form nitric oxide.

$$N_2 \quad + \quad O_2 \quad \rightleftharpoons \quad 2NO$$
nitrogen   oxygen   nitric
                              oxide

Also, production of nitric oxide by this reaction is favored by the high ratio of air to fuel, which is otherwise desirable because it cuts down on the output of carbon monoxide.

Hydrocarbon (Hc) emissions are dependent on several factors. In general, operating conditions that reduce carbon monoxide emissions also cut down on hydrocarbon output because conditions favoring complete oxidation are desired for both results. One would expect, theoretically, that all the fuel would burn up in the cylinder if there were enough air present. Such is not the case. The gasoline engine is not that efficient under operating conditions. About 40 percent of the exhaust hydrocarbons are unburned fuel components; the remainder are the products of combustion—low molecular weight combustion products not found in gasoline.

In summary, carbon monoxide production results from *low* air content of the fuel mixture and *low* temperatures, whereas nitric oxide production is promoted by *high* air content and *high* combustion temperatures. Hydrocarbon production does not adhere exactly to the pattern for either carbon monoxide or nitric oxide, but follows more closely that of carbon monoxide. The engineering obstacles to achieving reasonably complete elimination of all three pollutants without major redesign or conversion to other means of power is a formidable one. It is evident that either major changes in engine design are needed or radically different devices for improving combustion will have to be installed.

DOCTORING THE COMBUSTION

Most of the steps taken to eliminate exhaust pollutants have been directed toward the oxidation of carbon monoxide and hydrocarbons to the nontoxic products of complete combustion, carbon dioxide and water. Changes have been made based on the knowledge that carbon monoxide and hydrocarbons can be converted to harmless forms if kept sufficiently hot and exposed to oxygen. Numerous modifications have been tried, including better air-fuel control, adjustments to spark timing, and containment of vapors from the crankcase, fuel tank, and carburetor (see Figure 6.11). Mechanical design changes have also been made, such as combustion chamber shape, gasketing, piston and ring configuration, and compression ratio. Other proposals include devices attached to the exhaust manifold to conserve the heat, and an engine-driven pump to supply additional air.

An obvious method of reducing nitrogen oxide output is to reduce the peak temperature occurring in the combustion chamber. This is done by retarding the spark or lowering the compression ratio. Another very effective way to reduce the combustion temperature is to dilute the air-fuel mixture before it enters the combustion chamber with an inert material to act as a "heat sink." The most readily available diluent is the exhaust gas. Recycling the exhaust gas is fairly simple and inexpensive, but there are adverse effects. One of these is that deposits build up in the engine intake system; another is that vehicle "drivability" may be impaired. In view of the fact that changes which bring about improvements in carbon monoxide and hydrocarbon emission increase the output of nitrogen oxides, adjustments for nitrogen oxide emission control must always be made in combination with adjustments for carbon monoxide and hydrocarbon control. The required accommodations, for the most part, leave the output of nitrogen oxides at a distressingly high level (Table 6.4).

Engines are highly variable in the efficiency with which they use fuel. The average engine probably wastes 5 to 7 percent of the gasoline fed to the cylinders. About one-third of the wasted fuel blows by the piston rings into the crankcase. "Blow-by" devices to return the crankcase fumes to the intake manifold are installed at the factory on all cars sold in the United States. A device under study is one that will provide air-bleed into the intake manifold during deceleration.

Several additional systems of exhaust pollution control for the internal combustion engine have been tried. The most promising is a device called the *catalytic converter*, placed in the exhaust line in front of the muffler. An obstacle is that the lead content of most gasoline quickly "poisons"

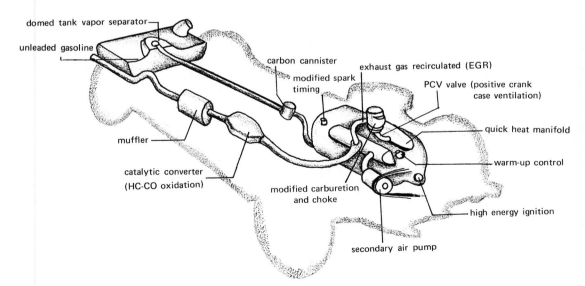

FIGURE 6.11 Modifications in engine, carburetion, and exhaust design are needed to produce a low-emission car.

TABLE 6.4   Light Duty Vehicles: Emissions of Carbon Monoxide, Hydrocarbons, and Nitrogen Oxides at Low Altitudes for 1970 and 1975 Model Years, Based on Low Mileage.

	Grams per mile	
	1970 model	1975 model*
Carbon monoxide	36	1.8
Hydrocarbons	3.6	0.23
Nitrogen oxides	5.1	2.3

From *Compilation of Air Pollution Emission Factors, Second Edition.* EPA Publ. AP-42. 1973.
*EPA estimates.

the catalyst (platinum and/or palladium) and renders the expensive device inoperative. This problem has been solved by requiring the use of unleaded gasoline in cars equipped with the catalytic converter. The catalytic device is intended to reduce the output of hydrocarbons (Hc) and carbon monoxide (CO), while the emission of nitrogen oxide, mostly nitric oxide (NO), is to be handled by the "exhaust gas recirculation" (EGR) system, a more sophisticated version of exhaust recirculating systems that have been used for several years. The idea here is to use the exhaust itself as a source of inert gases to dilute the charge in the combustion chamber, which has the effect of slowing down the combustion process, thereby reducing the peak combustion temperature and the output of nitrogen oxide. The catalytic device also permits advancing the spark timing to a more normal setting that consumes less fuel and improves performance.

An unexpected side effect of the catalytic converter is the finding that the device emits large amounts of sulfates and sulfuric acid. Thus, it is possible that the catalysts, which were ordered to be installed on 1975 cars to meet stringent standards for hydrocarbon and carbon monoxide emissions, may themselves create new health hazards. For that reason, the Environmental Protection Agency recommended freezing the 1975 nationwide standards for auto emissions of hydrocarbons and carbon monoxide through the 1979 model year. Meanwhile, research is continuing to determine just how serious the sulfuric acid threat is going to be.

The manipulations required to reduce the three primary emission pollutants, carbon monoxide, hydrocarbon, and nitrogen oxides, to acceptable levels requires a degree of mechanical legerdemain that is difficult to achieve within the limitations of the internal combustion engine. Since the life of an automobile is estimated to be an average of 11 years, engineering improvements require a long "lead time."

Alternate types of engines that could replace the conventional internal combustion engine are being studied. They include the rotary internal combustion engine, the steam engine, the gas turbine, and the stratified charge engine. The rotary design is adaptable to small engines and is proving to be successful, but it is disappointing in fuel economy. The original stratified charge engine had a combustion prechamber, but a new design called the Ford Proco, for programmed combustion, will have the stratification of combustion created by a fuel injector that shoots a cloud of fuel directly into the cylinder, the cloud being rich in the center and lean on the outside. The steam engine, utilizing external combustion, is efficient and reduces the output of all three primary pollutants to extremely small amounts, but manufacturers face obstacles of design and practicability for engines of automobile size. The gas turbine, an outgrowth of aircraft engine design, is a relatively low polluter and has been put into operational use on trucks and experimentally on passenger cars. It is well adapted for use where high-power sources are required, such as on large trucks. Acceptability of the noise level and high volume of exhaust from passenger cars is questionable.

Electric power is an intriguing possibility, but no one has been able to develop a battery light enough or long-lasting enough to be widely used. Rechargeable batteries would not be pollution-free because power would have to be generated at some central source.

Recent tests on an early, and nearly forgotten, invention for an unconventional "hot air" engine show promise for the development of a new type low-pollution, low-fuel consumption automobile engine. Called the Stirling engine, the concept was developed in the early nineteenth century by a Scottish clergyman. It uses alternately heated and cooled gases to push the pistons back and forth, and is theoretically more efficient than any other type of engine. At present it appears too big and heavy for automobile use, so the problem now being worked on by one of the major auto makers is to reduce the size, complexity of manufacture, and cost.

Diesel engines   Diesel engines differ greatly from gasoline engines in their exhaust emissions. Diesel-powered trucks emit about half as much carbon monoxide and about the same amount of unburned hydrocarbons, on a mileage basis, as 1970 automobiles. But nitrogen oxides from diesel trucks are 6 or 7 times as high as from 1970 automobiles (see Tables 6.4 and 6.5). If the EPA's estimates of emissions from automobiles of 1975 and later models are correct, this situation is changing in favor of passenger cars with respect to all three of the major components of vehicle exhaust. Heavy-duty diesel emissions of carbon monoxide and hydrocarbons will be approximately 10 times higher, not less, than those from gasoline powered automobiles, and nitrogen oxide emissions will be 15 times higher! However, tests on light-duty diesel engines used in passenger cars show that they rate favorably with the estimates for 1975 gasoline powered cars, and may be even somewhat lower in nitrogen oxide emissions.

TABLE 6.5   Heavy-duty Diesel-Powered Vehicles:
Major Exhaust Emissions

Pollutant	Grams per mile
Carbon monoxide	20.4
Hydrocarbons	3.4
Nitrogen oxides*	34.0
Sulfur oxides†	2.4
Particulates	1.2

From *Compilation of Air Pollution Emission Factors, Second Edition.* EPA Publ. AP-42. 1973.
*As NO_2.
†Based on fuel with average sulfur content of 0.2 percent.

The diesel engine is of a type called *compression-ignition*. It is designed with extremely high compression ratios, from 13:1 to as much as 20:1, depending on cylinder size. The fuel is not ignited with a spark but by the heat of compression. When the air in the diesel engine cylinder is suddenly compressed, the temperature rises to as high as 4,500°F. The fuel is injected and is spontaneously ignited, creating a pressure as high as 1,500 pounds per square inch. The extremely high temperature and pressure of the fuel-air mixture favor combustion of the hydrocarbons to carbon dioxide and water. However, as with the gasoline engine, the same conditions that reduce the emission of carbon monoxide and unburned hydrocarbons also favor the production of nitrogen oxides and the other undesirable reaction products. Therefore, the nitrogen oxide content of emissions from heavy duty diesel engines is exceedingly high, although small automobile diesel engines are in use that satisfactorily meet air emission standards.

No method has been developed for the control of diesel exhaust pollutants from large diesel engines. The need for an effective control system for trucks, buses, and other diesel-powered vehicles and machines is one of the most critically urgent and difficult problems in the air pollution control program.

FUEL ADDITIVES

Tetraethyl lead

Numerous substances can be added to gasoline to inhibit the explosion-type combustion that causes knocking. Iodine and aniline were among the earliest antiknock substances to be so recognized. But the metal alkyls (a class of hydrocarbon combined with a metal) are the most effective. *Tetraethyl lead* is the best. But alone, tetraethyl lead has a serious disadvantage. A decomposition product of combustion is lead oxide, which is deposited in the combustion chamber and further reduced to metallic lead, which causes pitting. To correct this, gasoline manufacturers add *ethylene dibromide* or a mixture of ethylene dibromide and ethylene dichloride, which form lead halides (lead bromide and lead chloride) instead of the oxide. The lead halides are more resistant to reduction than the oxide and are therefore blown out with the exhaust in the form of particulate aerosols.

The amount of lead antiknock compound in gasoline when leaded gasolines were at the peak of their popularity was equivalent to an average of 2.4 grams metallic lead per gallon. The most economical way to produce gasoline of high-octane quality without the use of lead alkyl additive is to use more severe refining methods, which reduces the yield and increases the cost by 2 to 6 cents per gallon (see Hydrocarbons for further discussion of octane ratings; see Chapter 8 for further discussion of lead compounds).

An alternate solution is to reduce the compression ratio of gasoline engines to adapt them to fuel of lower-octane rating. This would allow the cost of gasoline to remain about the same, but would result in a loss of fuel economy and performance. The most appropriate answer will be determined by public preference. Barring unforeseen developments, it will be necessary to "get the lead out" because the most effective exhaust smog control devices are catalytic burners, which are quickly rendered ineffective by the deposition of lead on the catalyst. Conversion, of course, to other types of engines such as steam or the gas turbine would obviate both the catalytic smog device and lead additives.

Lead pollution of air

Lead is the most important particulate matter resulting from automotive pollution. There is a direct correlation between gasoline consumption and particulate lead in the air. Levels of 1.0 and 2.0 micrograms per cubic meter are typical averages for the larger cities. The lead content in the air of nonurban areas is much lower. Among 117 composite samples of air from nonurban areas, only 100 of the 117 samples had lead in detectable amounts (above 0.01 micrograms per cubic meter). The danger to human health and the environment is discussed elsewhere (see Lead in the Atmosphere, Chapter 8).

Detergents

With most untreated gasolines, gum tends to form in the carburetor jets and carbon accumulates in the cylinder, eventually causing various malfunctions. Several compounds have been used to keep carburetor and cylinder parts clean. The most commonly used detergents are tricresyl phosphate (TCP), certain boron compounds, and polybutene amine. TCP has been used for many years, primarily for its effectiveness as a lead scavenger. Polybutene amine is said by the manufacturer to be a major breakthrough in reducing unburned hydrocarbon and carbon monoxide emissions from dirty engines. However, the exhaust emission products from the detergent additives, their biological activity, and possible toxicity remain subjects for investigation.

SMOG

The word *smog* is supposed to have originated in Great Britain as a contraction of "smoke-fog." The term may have been suggested by a report on smoke-fog deaths by H. A. Des Voeux in 1911. The report dealt with two occasions during the autumn of 1909 in Glasgow, Scotland, when 1,063 deaths were attributed to smoke and fog. In recent years, atmospheric pollution has become increasingly critical. There have been several catastrophic instances of smog resulting in serious injury to health and multiple deaths.

KILLER SMOGS

From time to time a combination of high emission of air pollutants and adverse atmospheric conditions combine to cause the formation of "killer smogs" (see Figure 6.12). During a disastrous week in the winter of 1930 in the Meuse Valley of Belgium, a large number of people were made ill and many were killed by smog. The Meuse Valley is a narrow river valley, 15 miles long with hills on each side reaching to about 300 feet. The area is highly industrialized with coke ovens and blast furnaces of steel mills, zinc smelting plants, glass factories, and sulfuric acid plants. On December 1, 1930, and for the remainder of the week, a thermal inversion layer confined the increasingly heavy concentration of industrial emissions to the valley. By the third day many people complained of respiratory ailments, and before the atmospheric conditions changed to clear the smog from the valley, 60 people died. There were also fatalities in cattle. It could not be determined with certainty what toxicants caused the difficulty, but probably a combination of several pollutants was responsible. It was estimated that the sulfur dioxide content of the air was 25 to 100 milligrams per cubic meter (9.6 to 38.4 ppm) but no actual measurements were made. If the sulfur dioxide had been completely oxidized to sulfuric acid (although this is improbable), the concentration of sulfuric acid mist theoretically could have reached 38 to 152 milligrams per cubic meter (14.6 to 58.5 ppm).

Donora, Pennsylvania, was the scene of a smog condition in 1948 that made 43 percent of the population ill and killed 20 people. Donora nestles within a horseshoe-shaped valley of the Monongahela River. There are steep hills on both sides of the valley rising to several hundred feet. In the town there is a large steel mill along with a sulfuric acid plant, a zinc plant, and other industries. There were 14,000 people living in the valley. The weather was "raw, cloudy and dead calm" on the morning of October 26, 1948, when a heavy fog moved in over Donora. According to one graphic report, "the fog piled up all that day and the next" and by the third day "it had stiffened adhesively into a motionless clot of smoke." It was then barely possible to see across the street. The air had the sickening smell and bittersweet taste of sulfur dioxide, a normal part of the environment in Donora but on this occasion more penetrating than usual. By the second day many people complained of cough and irritation of the respiratory

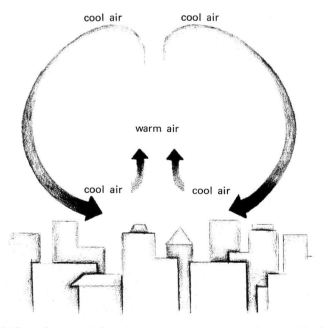

FIGURE 6.12 Flow of warm air from heat produced by city is obstructed by buildings, causing what is called the *heat island effect,* and holding high concentrations of toxins.

tract, eyes, nose, and throat. Some of the afflicted people suffered from constriction of the chest, headache, nausea, and vomiting. Complaints ceased abruptly on the evening of the fifth day in spite of the fact that the fog remained. The fog droplets may have increased in size so that they did not penetrate deeply into the respiratory tract. Pre-existing cardiac or respiratory disease was present among those who died. Again, the specific cause was not determined. No measurements of sulfur dioxide were made during the incident, but the concentration was estimated to be between 0.5 and 2.0 ppm. Other substances, including aerosol particulate matter, may have contributed to the problem. A survey of the mortality and health records for the area suggested that a similar episode may have occurred in 1945.

The "black fog" that settled on London in December of 1952 caused 3,500 to 4,000 deaths. From December 5 to December 9 there was fog accompanied by thermal inversion over most of the British Isles. London, which lies in the broad valley of the Thames River, was in the heart of an area where an unusually large number of people became seriously ill, most of them in the older age groups. In most cases, the onset of the illness was on the third or fourth day of the smog. Symptoms were shortness of breath, cyanosis (bluish discoloration due to poor oxygenation of the blood), fever, and rales (an abnormal sound produced by breathing). Deaths were due to chronic bronchitis, bronchopneumonia, and heart disease. The death rate remained high for several weeks after the adverse climatic conditions changed. During the height of the smog, measurements over a two-day period showed that the average concentration of sulfur dioxide was 1.34 ppm. Another episode occurred in 1959, but it was not as severe as the 1952 smog.

The residents of the Kanto Plains area of Yokohama, Japan, were afflicted with an accumulation of smog during the winters of 1945 and 1946. There was an unusually high rate of respiratory tract irritation and asthma. Smog trouble in the area continues to recur. A similar attack occurred in New Orleans, Louisiana, in 1958, when a series of smog episodes was associated with high frequency rates of asthma.

The "killer smogs" were spectacular instances of multiple deaths caused by atmospheric pollution. More often, however, the effects are more subtle, less immediately lethal, of a more chronic nature, and afflict a larger number of people.

CHRONIC HEALTH EFFECTS

Eye irritation is often the first symptom of smog injury to human tissues. The effect on eyes appears to be temporary and there is little, if any, evidence that such injury results in permanent damage. Respiratory irritation and difficulty in breathing are also common, and the correlation of the oxidizing power of smog with its irritating action has implications beyond the simple annoyance factor. The air passageways to the lungs do more than carry the air and waste gases to and from the air sacs. The passageways are lined with mucus-secreting epithelium and serve to condition the air before it reaches the alveoli (see Figures 6.13 and 6.14). The

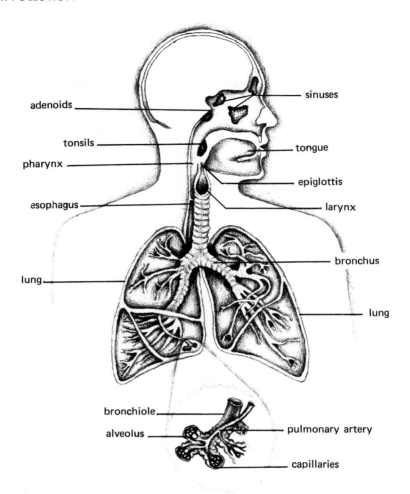

FIGURE 6.13 The respiratory system contains sensitive tissues that are vulnerable to toxic gases and atmospheric particles.

air is changed in three important ways: it is warmed to body temperature, it is saturated with moisture, and it is filtered free from most of the dust particles and other foreign substances. The cells that form the epithelial lining are ciliated, that is, they have hair-like protoplasmic appendages that project into the passageways and, by continuous, coordinated, rhythmic action, work the mucus and entrapped foreign material upward to the throat where it is either swallowed or expelled. Both the mucus-secreting ability and the ciliary action of the epithelium are sensitive to adversities, and excessive exposure to foreign material such as smoke, dust, and toxic substances impair their functioning, causing an increase in susceptibility to various respiratory disorders.

The major constituents of smog, with the exception of carbon dioxide, are powerful poisons. Many aspects of their toxic effects, both alone and in combination with each other, are not completely known. The danger, however, to people with susceptibility to respiratory and cardiac difficulties is widely recognized. Those with incipient or advanced bronchial

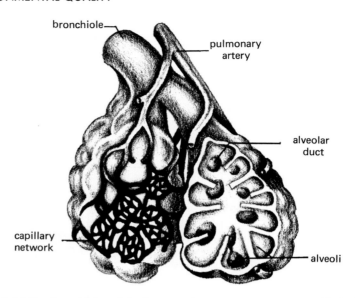

FIGURE 6.14 The bronchioles of the lung terminate at minute air sacs called *alveoli,* separated from a network of capillaries by a thin membrane through which gas exchange takes place. The alveolar membranes are easily damaged by toxins and atmospheric particulates. Pulmonary emphysema, caused by a breakdown in the alveolar walls, may be the result.

asthma, chronic bronchitis, and pulmonary emphysema are apt to be hard hit. The various components of smog may affect people with different susceptibilities in different way.

Emphysema   The most widespread effects of air pollution are those involving respiratory diseases. Emphysema is a progressive deterioration of the alveoli (air sacs) in the lungs (see Figure 6.14). It is the fastest-growing cause of death in the United States. During the years from 1950 to 1959, the death rate of males from emphysema rose from 1.5 per 100,000 to 8 per 100,000. Each month, 1,000 more people are forced to quit work and go into premature retirement because of the disease. In Great Britain nearly 10 percent of all deaths and more than 10 percent of all industrial absences are caused by emphysema, characterized as *chronic productive cough.* Using the same criteria, public health investigators found that in the United States 21 percent of men in the 40 to 59 age bracket had chronic bronchitis. Other findings were that 13 percent to 20 percent of all males had the disease.

A related condition is *chronic constrictive ventilatory disease,* characterized by constriction of the air passages, which makes breathing difficult. One of the known causes is sulfur oxides. *Bronchial asthma* also is often aggravated by air pollution.

A curious disease described as *Tokyo-Yokohama asthma,* or *T-Y asthma* as it came to be known, afflicted the American troops in Tokyo in 1946. The patients did not respond to usual treatments, but most of them recovered upon leaving the area. It was concluded that the attacks were related to the level of air pollution.

Some public health investigators think that there is a connection between air pollution and the *common cold.* The cause of colds may derive largely from the presence of particulate matter in smog. The common cold is one of the major causes of industrial absenteeism.

Deaths from *lung cancer* have increased. While the causes are manifold, evidence points increasingly toward air pollution as one of the contributors. There are twice as many cases of lung cancer per capita in the larger cities as in rural areas, and the death rate from lung cancer appears to be proportional to city size. The rate of lung cancer in Norway is half that of the United States.

EFFECTS ON LABORATORY ANIMALS

When guinea pigs are exposed to artificial smog, the result is reduced physical activity, diminished fertility, and an increase in their susceptibility to spontaneous pneumonia. Mice exposed to synthetic smog also suffered reduced fertility and reduced infant survival. In mice, there is fragmentation of the cells of the alveolar lining of the lungs, breakdown of the endothelial lining of the blood vessels, and rupturing of the membranes of the red blood cells. Older animals are especially vulnerable. One of the effects of artificial smog on mice is to increase their susceptibility to bacterial infection from streptococci. Synthetic smog containing 25 ppm of carbon monoxide and 0.15 ppm total "oxidant" (typical concentrations found in metropolitan air) increased the number of deaths from pneumonia.

Smog can damage the lungs by the formation of compounds that harden the tissue. Peter K. Mueller, in the sanitation laboratory of the California State Department of Public Health, found that in rabbits and rats two of the ingredients of photochemical smog, nitrogen dioxide and ozone, form what are called *carbonyl* compounds. The carbonyl group has an oxygen attached through a double bond to a carbon atom:

The group is present in all aldehydes, ketones, and organic acids. The structure greatly increases the reactivity of adjacent carbon-hydrogen radicals. Mueller found that these compounds caused a stiffening of the delicate molecular structure of the proteins *collagen* and *elastin*. These are the materials in the lung tissue that give it elasticity. Thus the effect was to harden the lungs. He also found that inhalation of smog for 4 hours caused changes in the fatty acids in lung lipids (fats).

EFFECT OF SMOG ON PLANTS

Before the biological importance of air pollution was fully realized, it was noted that plants grown in bad smog areas suffered injury from what often appeared to be obscure causes. It is now apparent that many plants are more sensitive than humans to the acute effect of smog and, in fact, can be used as indicators of harmful levels of air contaminants.

Two ingredients of photochemical smog are recognized as being particularly injurious: ozone and PAN (peroxyacetyl nitrate). With PAN, injury appears on some plants as a characteristic silvering on the bottoms of the leaves. Ozone, on the other hand, kills cells on the tops of the leaves and appears as flecking or stippling of the upper surfaces.

Injury to plants from photochemical air pollution was noted in the mid-1940s when losses occurred to spinach and other sensitive plants in California. Since then, crop damage has occurred in many parts of the United States, particularly California, New Jersey, and Florida, and in many foreign countries. The damage to agriculture in the United States in 1966 was estimated at 500 million dollars. In California alone, the damage to agricultural crops during 1969 was estimated at 44.5 million dollars.

In California, it is now impossible to raise orchids in metropolitan areas. Vegetable and citrus growing have been hit hard by injury and reduced yields. In New Jersey, injury was reported to thirty-six commercial crops. Orange trees in central and southern Florida have been severely damaged, causing some growers, as in California, to relocate.

Ornamental plants have not escaped the onslaught of air pollution. City parks are becoming increasingly limited in the plants that can be successfully grown, and park superintendents find that they must be more restrictive in selecting types of plants that can tolerate smog. Home gardeners are often puzzled by the failure of their favorite specimens to flourish. Injury is common to petunia, snapdragon, chrysanthemum, larkspur, carnation, orchid, pansy, rose, and zinnia.

A disturbing aspect of air pollution injury to plants is that many of the pine trees in the mountainous areas of southern California are dying. Symptoms are characteristic of smog injury, and it is recognized that nitrogen oxides and PANs contribute to the death of the forest pines.

PHOTOCHEMICAL SMOG

The most irritating, and some of the most injurious, components of smog are the products of reactions in the atmosphere between oxygen, ozone, and emission pollutants. The mixture of undesirable products is sometimes called *photochemical smog* because some of the chemical reactions are initiated by the energy from ultraviolet light. The cycle of reactions is not completely understood; much of the work on air pollution is done in *smog chambers,* which may differ in their characteristics from conditions in the atmosphere. However, the important steps in the atmospheric production of photochemical smog have been determined. They consist of a series of reactions in which oxygen, ozone, nitrogen oxides, and hydrocarbons participate to produce compounds that are both irritating and toxic. Some of the principal reactions are as follows:

Step 1:

One of the most important steps in the formation of photochemical smog is the action of ultraviolet light on nitrogen dioxide, producing nitric oxide and atomic oxygen:

$$NO_2 \xrightarrow{\text{UV light}} NO + O°$$

nitrogen dioxide   nitric oxide   atomic oxygen

Step 2:

The atomic oxygen further reacts with the usual form of oxygen in the atmosphere, molecular oxygen (oxygen gas), with the formation of ozone:

$$O° \quad + \quad O_2 \quad \rightarrow \quad O_3$$

atomic   molecular   ozone
oxygen   oxygen

Some of the ozone is used up in oxidizing nitric oxide to nitrogen dioxide. Ozone is also involved in the production of aldehydes. Additional ozone is produced, however, by the action of oxygen on oxidized hydrocarbons called *radicals*.

Step 3:

Part of the atomic oxygen formed in Step 1 combines with hydrocarbons, to form hydrocarbon (acyl) radicals:

$$O° \quad + \quad Hc \quad \rightarrow \quad HcO^-$$

atomic oxygen   hydrocarbon   radical

The acyl radicals undergo a complex series of reactions, some of which consist of a further reaction with oxygen to form more highly reactive substances called *peroxy radicals,* which in turn react with hydrocarbons to form aldehydes, ketones, and other compounds.

Step 4:

A crucial stage in the photochemical smog complex is the reaction between peroxy radicals and nitrogen dioxide to form the highly toxic PANs (peroxyacyl nitrates):

$$HcOO^- \quad + \quad NO_2 \quad \rightarrow \quad HcOONO_2$$

peroxy radical   nitrogen dioxide   PAN
                                      (peroxyacyl nitrate)

The foregoing is a greatly abbreviated and simplified version of the scheme for photochemical smog.

PANs   The PANs (peroxyacyl nitrates) are believed to be responsible for most of the eye irritation in photochemical smog, although other compounds are known to contribute. Not all of the substances responsible for irritation have been identified. Formaldehyde and acrolein, both highly toxic substances, are known to be present along with various types of peroxides and free radicals.

The first known member of the group is *peroxyacetyl nitrate,* or simply PAN.

$$CH_3-\overset{\overset{\displaystyle O}{\displaystyle \|}}{C}-OONO$$

It is a potent eye irritant at concentrations of about 1 ppm or less. PANs and other eye irritants are known to persist for more than 24 hours in contrast to ozone, which is believed to have, in photochemical smog, an average half-life of about 1 hour. One distant cousin of the PAN family, PBzN (peroxybenzoyl nitrate), causes eye irritation at a concentration only $\frac{1}{200}$ that of formaldehyde. Since it takes approximately twice as much formaldehyde as PAN to produce irritation, the results indicate that PBzN is 100 times more powerful than the familiar PAN.

In view of the potency of the PAN family of irritants and the probability of their having incompletely determined biological properties, including carcinogenic potentialities, more work is needed to determine their toxic activity to animals and humans as well as to make records of concentrations in various areas at regular intervals. Investigators conducting a regular measurement program at the Air Pollution Research Center at Riverside, about 60 miles inland from the center of Los Angeles, found as much as 58 ppb (parts per billion) although the concentration was usually below 50 ppb by volume. Concentrations up to 214 ppb were found in Los Angeles County.

PAN has long been known to be highly toxic to plants, and the extent of characteristic injury is indicative of widespread occurrence of the toxicant. It causes injury to many plants when they are exposed to less than

Photo by Clayton Knipper, *The Cleveland Press,* courtesy of Environmental Protection Agency.

Cleveland, Ohio, under smog.

10 ppm for 2 to 4 hours. Many crop plants are affected, including spinach, beets, celery, tobacco, pepper, endive, romaine lettuce, swiss chard, and alfalfa. Among the highly susceptible ornamental plants are petunia, snapdragon, primrose, aster, and fuchsia. Most woody shrubs are comparatively resistant, as are grapes, cabbage, and onion. PAN typically causes a glazing or silvering of the undersides. The leaves take on a silver or bronze color, depending on the pigmentation of the leaves. The cells collapse, causing the epidermis to separate slightly from the tissue underneath. Finally, the tissue dries and blotches of dead areas appear on the undersides or through the entire leaf.

Oxidant   It is the practice to express the amount of irritating air pollution as parts per million of *oxidants*. The most potent oxidants in photochemical smog include nitrogen dioxide, PANs (peroxyacyl nitrates), and ozone. These were formerly referred to as a group and called *oxidant*. More recently it is customary to report nitrogen dioxide separately (see Figure 6.15). In Los Angeles the effect of various oxidant concentrations are given as follows:

Oxidant Concentration	*Effect*
0.08–0.1 ppm	Light smog
0.13–0.15 ppm	Slight eye irritation
0.15 ppm	California standard at which eye irritation, reduced visibility and plant damage occur

Ozone may comprise as much as three-fourths of the total oxidant in irritating smog. Ozone, itself, does not account for the degree of eye irritation experienced. But the concentration of more highly irritating substances is sometimes expressed in terms of the equivalent of ozone in oxidizing power. The first, second, and third alert stages for ozone are set at 0.5, 1.0, and 1.5 ppm, respectively. A maximum of 0.9 ppm was reached in Los Angeles in 1955. In 1974, during the heavy smog months of July, August, and September, oxidant levels of 0.20 ppm were equaled or exceeded on 84 days, and the average oxidant peaks during those three months was 0.51 ppm.

Many people experience discomfort even at concentrations of ozone below the limits set by official standards. For example, in San Francisco, 26 percent of the population were reported to have been adversely affected when the ozone concentration reached 0.05 to 0.10 ppm.

Forest haze   Ozone is a contributor to the hazy atmosphere frequently seen in forested mountain areas. Terpene hydrocarbons (a class of unsaturated organic compounds) are exuded from vegetation. The volatility and concentration of these compounds is such that in calm weather there may be significant concentrations in the atmosphere in and above wooded areas. F. W. Went and R. Rasmussen made a detailed study of the emanations from vegetation. They believe that the terpenes undergo photochemical polymerization to an atmospheric aerosol (finely divided droplets of particulate matter), resulting in the "blue haze" seen in many forested areas. Went estimated that the total release of terpene-type hydrocarbons from vegetation is about 170 million tons per year. Ozone in the atmosphere is

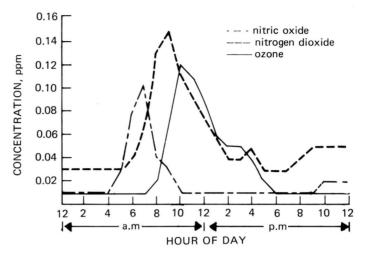

Redrawn from National Air Pollution Control Administration Publication No. AP-63, 1970.

FIGURE 6.15 Concentrations of nitric oxide, nitrogen dioxide, and ozone at different times of the day in Los Angeles. Nitric oxide, emitted in the exhaust, is converted in the atmosphere to nitrogen dioxide, which is acted upon by ultraviolet light and, in a stepwise reaction with atmospheric oxygen, forms ozone. The steps in this photochemical reaction are shown in Figure 6.6

probably responsible for most of the terpene reactions that cause the bluish smoke. Thus, a natural haze or "smog" is formed that may be seen especially on warm, calm days in forested or other heavily vegetated areas.

SOUND AND NOISE

Noise is any sound that is not wanted. It is one of the more common forms of atmospheric pollution. Because individual perception and interpretation of sounds differ, what is music to one person may be noise to another. Family disagreements over phonograph and radio selections are only too familiar. The city dweller who visits his relatives in the country cannot go to sleep in the evening because it is too quiet; and he loses sleep in the morning because the rooster crows and the dogs bark. By contrast, the suburban or country dweller is apt to experience sleepless nights in the city due to the annoyance of loud music in a neighbor's apartment, incessant traffic noises, and the clatter of early-morning delivery trucks. Daytime noises are even more intense. While it is possible for most people to become inured to the din of machines and partially lose consciousness of its existence, there is mounting evidence that psychological and physiological damage is inescapable.

INJURIOUS EFFECTS OF NOISE

Violent noises can cause temporary or permanent impairment of hearing, thus the expression "deafening." Continual noise can lead to gradual decline in auditory acuity and eventual deafness (see Figure 6.16). Prolonged exposure to less than ear-splitting noises may have other effects that are more subtle but possibly as serious.

Noise causes several undesirable effects. Damage to hearing and loss of sleep are only two of the more obvious insults to the relationship between man and his environment. Noise interferes with speech, sometimes making it unintelligible; sounds of warning are misunderstood or not heard. Even low-level noises impair verbal communication because they require more attention and effort and cause misinterpretations. Noises produce irritability and a feeling of fatigue and may reduce a worker's efficiency. There is evidence that noise is one of the major causes of stress and many of the other human afflictions associated with tension— anxiety, insomnia, accident-proneness, high blood pressure, and other cardiovascular diseases. Even in its less serious forms, noise is commonly a source of annoyance and cause for complaint. All these are of personal, social, and economic importance.

The effects of noise on the fetus are not fully known. Medical scientists have noted that an unborn child will move and kick when there is a loud noise. It also responds with a sudden increase in the heartbeat as though it were disturbed or frightened.

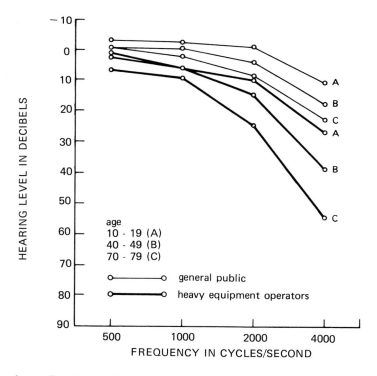

Redrawn from William Bronson, "10,000,000,000,000,000,000 . . . Baby Mice . . . Ear Pollution," *Cry California, The Journal of California Tomorrow,* Fall 1967.

FIGURE 6.16 Hearing loss by age group and by general public versus those exposed to excessive noise.

NATURE OF SOUND AND NOISE

Sound is produced when an object vibrates, alternately compressing and expandinng the air. The compressions and expansions travel wavelike from the source. They are called *sound waves* or simply vibrations. The two most basic characteristics of sound are *pitch* and *amplitude,* and these are related to *loudness*.

Pitch   When an object serves as a source of sound, the pitch is determined by how rapidly the object vibrates. The rate of vibration is called the *frequency*. The higher the frequency, the higher the pitch, and the lower the frequency, the lower the pitch. This can be easily demonstrated by playing a phonograph record at a slower speed than intended. The words or tones will not only emerge more slowly, but will have a throaty sound and a distinctly lower pitch. The reverse effect is produced by playing the record at an excessively high speed. Similar results are produced as the speed of an electric fan or that of a siren increases.

Technically, the *frequency* of an object, or of its sound waves, is defined as the number of vibrations per second. People can hear sounds with frequencies from about 20 to 20,000 vibrations per second, but there is a great deal of individual variation (see Figure 6.17). Some persons can hear frequencies that others are not able to detect, and sometimes there is poor hearing in part of the frequency range due to disease or injury.

The notes on a piano range from a frequency of 27 vibrations per second for the lowest to about 4,000 for the highest. The quality of a stereo system is determined partly by the range of frequencies that it can reproduce with a high degree of integrity. There are also differences between humans and animals in the ability to detect different frequencies. Commercial "dog whistles" produce sounds with frequencies above those that can be detected by humans. Bats produce and hear sounds

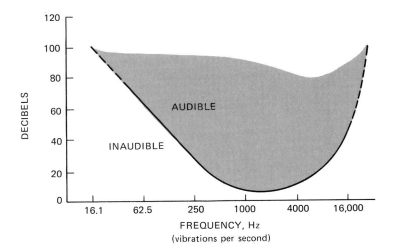

Redrawn from *The Noise Around Us Including Technical Backup,* U.S. Dept. of Commerce, 1970.

FIGURE 6.17 The threshold hearing for young, healthy ears depends on frequency.

having frequencies far above the range of the human ear. Nature is full of sounds to which humans are insensitive.

A change in pitch occurs as one moves rapidly toward or away from a source of sound, as when a train-rider approaches or leaves a railway crossing signal. The lower pitch caused by "stretching out" the sound waves, or the reverse effect produced by "squeezing together" the sound waves, is called the *Doppler effect*.

Amplitude

Sounds having the same frequency, and therefore the same pitch, may differ in the amount of energy involved. The energy that goes into producing a sound determines the intensity of the sound. But intensity and loudness differ; loudness is the strength of the *sensation* of sound perceived by the individual (see Figure 6.18).

The distance that a vibrating object, such as a tuning fork or a guitar string, moves as it vibrates is called the *amplitude of vibration*. The greater the energy that goes into producing the sound, the greater the amplitude of the sound that is produced. Also, a large vibrating object will produce a sound of greater intensity and loudness than a small vibrating object. This can be demonstrated by making the prongs of a table fork vibrate, then placing the handle of the fork on the top of the table. The sound becomes instantly louder because the table top increases the area of vibration. The intensity and loudness of sound are also affected by the density of the medium. The sound produced by striking two rocks together under water is louder than when they are struck together in air because the denser the material, the louder the sound. There is no sound in a vacuum because there is nothing there to form sound waves.

Loudness

The loudness of sound (the strength of the sensation) does not depend entirely on the energy or amplitude of the sound wave. Loudness also depends on the frequency or pitch. The human ear has a low sensitivity for sounds of low pitch and for those of extremely high pitch (Figure 6.17). For example, to produce the same sensation of loudness, a low pitch that has 64 vibrations per second must be 10,000 times as powerful as a high pitch of 3,000 vibrations per second. The maximum sensitivity comes at about 1,000 to 3,000 vibrations per second. (The frequency range of speech is from about 200 to about 1,000, while middle C on the piano has a frequency of about 256.)

For any given frequency, the energy of a sound must be increased about 100 times to make the sound 2 times as loud, and must be increased about a million times to make the sound 6 times as loud. If we notice that $100 = 10^2$ and $1,000,000 = 10^6$, it can be seen that loudness increases only by the exponent of 10 (2 and 6) and not by the energy required. The unit of loudness expressed by the exponent of 10 is called a *bel*. In the first example the loudness is 2 bels and in the second example the loudness is 6 bels.[11]

---

[11]Another way to look at this is that the number 2 is the logarithm of 100 and the number 6 is the logarithm of 1,000,000. (The logarithm to the base 10 of a number is the exponent of 10—that is, the number of times 10 must be multiplied by itself—to give the number.) Therefore, the loudness expressed as bels is equal to the logarithm of the sound energy.

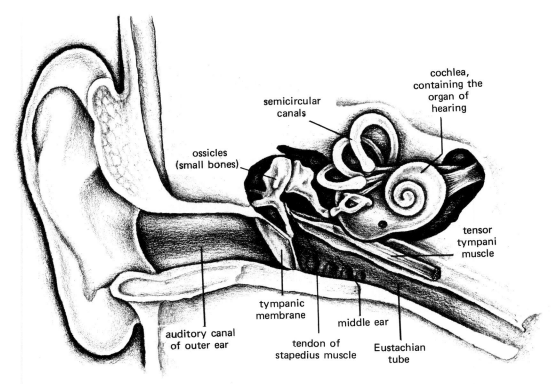

FIGURE 6.18 Sound waves coming in through the outer ear impinge against the ear-drum, called the *tympanic membrane*. Next, inside the air-filled middle ear, three tiny bones transmit the vibrations to the inner ear, which contains the organ of hearing. The ability to detect delicate sounds, as well as to avoid damage from loud, low-pitched sounds, is controlled by two small muscles in the inner ear that cooperate to control the tension of the tympanic membrane.

The term *decibel* is often used to indicate one-tenth bel.[12] A sound of between 0 and 1 decibel is about the weakest that the average human can hear. For testing purposes, 0 decibel is considered to be the threshold of hearing. A whisper is about 20 to 30 decibels and an average speaking voice is about 60 decibels. Note that according to the logarithmic relationship, 20 decibels is 100 times as intense as the threshold and 60 decibels is 1,000,000 times as intense. The loudest sound that a person can stand without discomfort is about 80 decibels (see Figure 6.19). Automobile horns may reach 90 decibels and a jet airplane at a distance of 100 feet may have an intensity of about 140 decibels.[13]

---

[12]Technically, a decibel is the amount of sound wave pressure that equals 0.0002 dyne per square centimeter. A dyne is a force which if exerted for 1 second will move 1 gram (about $1/_{28}$ ounce) a distance of 1 centimeter.

[13]Because different frequencies of sound have different degrees of loudness, the most precise unit of measurement to express loudness is the *phon*. The number of phons is equated with the number of decibels produced at a frequency of 1,000 vibrations per second. For example, a sound of 60 decibels at 1,000 vibrations per second has a loudness of 60 phons. Above and below that frequency, phons of loudness drop off because the ear is less sensitive to vibrations above or below 1,000 vibrations per second.

Kinds of sound   The kind of sound—whether noisy, musical, abrasive, soothing, displeasing, or pleasant—depends on a combination of characteristics in addition to pitch, amplitude, and loudness. A vibrating object produces many frequencies even though, as with a musical instrument, one tone may predominate. A blend of many frequencies produces a characteristic sound, which in music is called *quality*. It distinguishes a guitar, for example, from a mandolin. The vibrations that occur in addition to the fundamental frequency are called *overtones* or *harmonics*.

If two objects vibrate at different frequencies, there will result pulsations of loudness called *beats*. Since the number of beats per second is equal to the difference in frequencies of the two objects, the closer the tones, the faster the beats. When the number of beats reaches about 30 per second from tones that are very close but not quite the same, the sound is very unpleasant. Musicians call this a form of *dissonance*. At beats above 30 per second, the frequencies are so close that the human ear cannot detect the beats. Since the ear cannot tell that anything is wrong, the disagreeable sensation disappears.

Whether sound is pleasant or unpleasant is determined partly by *resonance*, produced when repeated vibrations of the same frequency create a strong response. A sound box of the right size and shape increases resonance. *Sympathetic vibrations* are a form a resonance produced in an unattached object having the same natural resonance as the object that produces the sound. For example, dishes in a cupboard or a picture on a wall may rattle when a certain note is sounded on the piano. A comparable situation is when a strong wind creates resonant vibrations in a suspension bridge. The bridge may fall if the vibration becomes intense. Soldiers are always ordered to break step when crossing a bridge because when marching in unison, they can set up vibrations that may cause the bridge to break apart.

Speed of sound   Sound travels at different speeds, depending on the density and elasticity of the medium through which it travels. The denser and less elastic the substance, the slower the speed of the sound waves. Sound travels at a velocity of 1,130 feet per second in air when at a temperature of 20°C (68°F). This is equal to about 344 meters per second or about 770 miles per hour. When the temperature drops to 0°C (32°F), the air is denser and the sound waves travel at a slightly slower speed, about 1,090 feet per second. The velocity increases about 2 feet (0.6 meter) per second for each centigrade degree rise in temperature, equal to an increase of about 1 foot per second for each fahrenheit degree. Water is denser than air but also more elastic. Sound travels through water at a velocity of about 4,800 feet (1,440 meters) per second. Steel is 6,000 times denser than air and this quality retards the movement of sound waves, but steel is also much more elastic than air by a magnitude of 2 million times; therefore, sound travels more rapidly through steel, at a velocity of 16,400 feet (5,000 meters) per second, or about 11,000 miles per hour.

Thus the speed of sound, regardless of the substance through which it travels, is much slower than the speed of light, which is 186,282 miles per second. It is easy to estimate the distance from the source of a far-off sound if there is some other signal to identify it, such as a puff of smoke or a flash of light. Five seconds between a lightning flash and a clap of thunder equals a distance of about 1 mile.

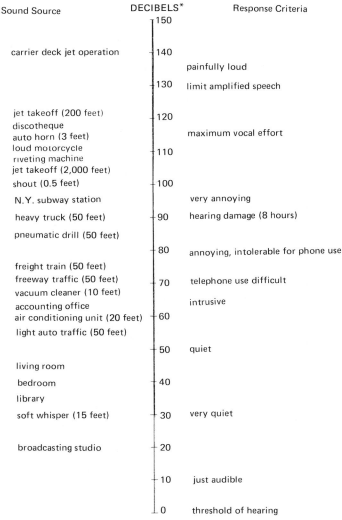

Sound Source	DECIBELS*	Response Criteria
	┌150	
carrier deck jet operation	┤140	
		painfully loud
	┤130	limit amplified speech
jet takeoff (200 feet)	┤120	
discotheque		
auto horn (3 feet)		maximum vocal effort
loud motorcycle	┤110	
riveting machine		
jet takeoff (2,000 feet)		
shout (0.5 feet)	┤100	
N.Y. subway station		very annoying
heavy truck (50 feet)	┤90	hearing damage (8 hours)
pneumatic drill (50 feet)		
	┤80	annoying, intolerable for phone use
freight train (50 feet)		
freeway traffic (50 feet)	┤70	telephone use difficult
vacuum cleaner (10 feet)		
accounting office		intrusive
air conditioning unit (20 feet)	┤60	
light auto traffic (50 feet)		
	┤50	quiet
living room		
bedroom	┤40	
library		
soft whisper (15 feet)	┤30	very quiet
broadcasting studio	┤20	
	┤10	just audible
	└0	threshold of hearing

*Typical A—Weighted sound levels taken with a sound-level meter and expressed as decibels on the scale. The "A" scale approximates the frequency response of the human ear.

Adapted from *Emerging Problems*, The President's 1971 Environmental Program, Book 3 (Washington, D.C.: The Domestic Council Executive Office of the President, 1971), p.9.

FIGURE 6.19 Sound levels and human response.

SOURCES OF NOISE

The sources of noise pollution in an industrial society are manifold. The main source in cities and other highly developed areas are noises associated with transportation. Automobiles, trucks, motorcycles, buses, fire engines, police cars, ambulances, airplanes, freight trains, and accessory noisemakers such as horns, sirens, and other raucous devices produce a cacophony of noise that has few equals on earth for intensity and annoyance. Much of it is unnecessary. The value of horns and sirens, except in extreme emergencies, is questionable. Cities that have passed laws against horn-tooting have reported no increases in accidents. In New York, city-operated ambulances, which are prohibited from sounding their sirens, have had fewer accidents than before.

Noises associated with manufacturing, building construction, and street work such as tearing out pavement for trenching and road repair are second in intensity to transportation noises, perhaps only because these sources tend to be more sporadic. Jack hammers, compressors, and pile drivers can out-decibel trucks and sirens and even make the din of traffic sound subdued. Because noise suppression is often costly, voluntary corrective measures cannot be depended on. Legislation to enforce reasonable standards would probably be needed to satisfactorily alleviate the problem. (See Figure 6.20.)

In outlying areas, garbage trucks are a source of community noise. But traffic is also a major contributor, especially near highways, freeways, and principal arteries. Motorcycle buffs inhabit most of the residential areas. Loud radios, stereos, and television sets and boisterous parties are pollution problems in apartment sections. Even inside the house, noise from washing machines, diswashers, food mixers, fans, furnaces, blowers, and air conditioners are a source of annoyance. Improvements could be made if more attention were given to design, placement, and acoustical protection.

Aircraft, due to their increase in size and enormous increase in numbers, have become one of the most troublesome sources of noise pollution. For example, O'Hare International Airport in Chicago, the busiest airport in the world, has more than a thousand operational jet flights on an average day. This means a flight in or out every 40 seconds. Within a 15-mile radius of the airport, there are several hundred thousand people clustered in homes, apartment houses, businesses, schools, hospitals, and nursing homes. The scream of jet engines is not conducive to tranquility and repose. In some locations, conversations cease every few seconds, television viewing and listening becomes impossible, and sleep and relax-

Left photo courtesy of Tim Welch. Right photo courtesy of General Dynamics.

Left photo: Motorcycle noise can be reduced by improved engine and muffler deisgn.
Right photo: Aircraft noise is an increasing annoyance.

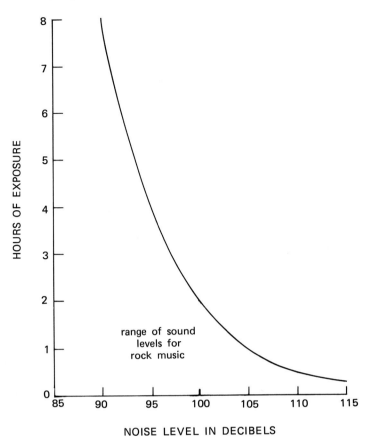

range of sound
levels for
rock music

NOISE LEVEL IN DECIBELS

Redrawn from *The Noise Around Us Including Technical Backup,* U.S. Dept. of Commerce, 1970.

FIGURE 6.20 Daily exposure to noise allowable under Federal regulations.

ation are interrupted. This condition is multiplied many times throughout the country and the world.

Control of aircraft noise requires several changes. Procedures already in use call for prescribed courses and reduction of power at certain altitudes. Other methods being tested include a study of climb-out profiles, glide slopes for landing, and the design and development of quieter engines and airplanes. A basic problem is that the safest jet approach to a landing is a long, low approach; but this is also the noisiest. One solution is to rezone the area under the landing and takeoff pattern, but this is costly and seldom solves the problem. Relocation of airports is another solution.

The supersonic transport (SST) brought noise pollution problems. A supersonic airplane is one that travels faster than sound. When an aircraft travels at a velocity greater than the speed of sound, there is an enormous increase in air resistance. A *shock wave* is created that can be extremely energetic, depending on the size, speed, and route of the aircraft (see Figure 6.21). When a shock wave strikes things below, it can have a jolting and devastating effect. The shock wave, or *sonic boom,* may spread over an area of 10 to 80 miles (see Figure 6.22). Although many

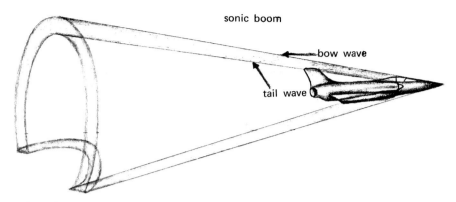

Redrawn from *Second Federal Aircraft Noise Abatement Plan,* 1971, U.S. Dept. of Transportation.

FIGURE 6.21 Sonic boom. When an airplane flies faster than the speed of sound, it compresses the air, pushing a shock wave before it—much like when a boat creates a spreading bow wave when cruising on a lake. This bow wave, or cone of increased air pressure, spreads out behind the airplane. At an altitude of 60,000 to 70,000 feet, the edge of the cone could cover an area 60 miles across.

cases of damage to structures caused by sonic booms from relatively small military aircraft have been reported and numerous damage claims have been filed, the destructiveness and nuisance effects of large commercial supersonic aircraft have not been fully evaluated. Objections to the development and commercialization of an SST are based on the possibility that their shock waves might become intolerable.

NOISE CONTROL

Noise control depends on the importance that the community attaches to the noise level and whether recreational or economic interests are associated with the source. These include widely diverse activities, such as highway driving, air travel, motorcycling, and garbage collecting. Agreement can seldom be reached on the utility or necessity for any particular noisemaking device. One thing that is needed to evaluate the damaging effects is a standard method for measuring the noise output of the various sources.

In some industries, noise control does not necessarily mean noise reduction. Earplugs worn by shipyard workers may bring about satisfactory control without changing the noise intensity. One way to reduce airport noise is to limit activities to daytime and early-evening flights. Although there is no reduction in total noise, the annoyance level is improved. Another way to cut down noise is by the construction of barriers or insulation; this does not put an end to the noise, but results in the harmless absorption of a portion of the sound waves.

Thus, there are four basic ways to reduce noise and its damaging effects. The first and most satisfactory way is to *reduce the noise level* at its source, for example, by designing and developing quieter machines, trucks, airplane engines, automobiles, and household appliances. The second method is to use *insulation* or absorbing materials, such as the

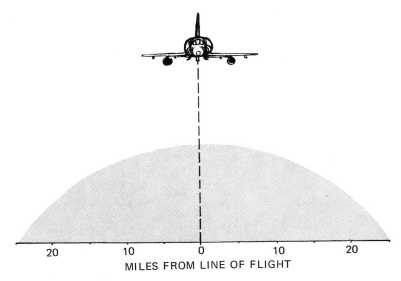

Adapted from the U.S. Dept. of Commerce.

FIGURE 6.22 The region on the ground over which the sonic boom will be detected is dependent on atmospheric conditions and airplane characteristics. The volume, weight, length, lift characteristics, altitude, and Mach number of the airplane affect both the amplitude and duration of the boom. Many miles are required on each side of the flight path before the sonic boom deteriorates to a low rumble.

installation of acoustic tile and the construction of enclosures around industrial machinery. The third method may be described as *operational,* such as rescheduling airline flights. Finally, the fourth, and least satisfactory method from the long-range view, is to *protect personnel* by the use of such devices as earplugs or ear muffs.

A quiet atmosphere in which to relax or work is rapidly becoming one of the most urgent environmental requirements. In the end, the solution must depend on a combination of methods. However, more emphasis is needed on reducing the noise at its source, a method that in many cases lacks only the will and the effort.

PROBLEM AREAS

Few environmental problems have been fraught with so many complexities as those involving air pollution control. While some air pollution abatement efforts have been eminently successful, for example, the reduction of particulate emissions—mainly soot and certain other substances—from coal- and oil-fired furnaces, many of the more difficult problems have not yet succumbed to control despite the best efforts of air pollution workers. We have permitted machines, factories, and other human activities of various kinds to pour hundreds of foreign substances into one of the most fragile segments of the environment while having only vague notions about their total effects. Air pollutants tend to be complex mixtures of chemical substances, and they often undergo chemical changes in anywhere from microseconds to months after release into the atmosphere. The toxic hazards of many of them have not been completely studied and are not well understood.

Automobiles, though only one part of the overall problem, illustrate the frustrations that can be encountered in cleaning up a bad offender—auto exhaust. Tinkering with the internal combustion engine to reduce one kind of emission led to an increase in another kind of emission almost as obnoxious. The addition of emission control devices, and the adjustments needed to make them work, made the engines balky and inefficient, increased gasoline consumption, ran up the cost of repairs, and required a degree of maintenance alertness that seemed to be beyond the reach of the average driver. Advanced systems, such as the catalytic converter, added greatly to the cost of the automboile at a time when the total cost of a new car—presumably smog-free—was out of reach for many people, thereby keeping outworn, high polluting automobiles on the highways long after they should have been relegated to the junk heap.

Air quality standards set by the Environmental Protection Agency under the Clean Air Act of 1970 do not purport to *eliminate* air pollutants. Their goal is to bring the concentrations of emissions down to tolerable levels. Deciding tolerable levels, in itself, causes contention. Some auto manufacturers claim the standards are too tough, while some authorities contend they are not tough enough. Because attainable standards are tied to the practicalities of engineering design, lead time for the industry, costs, and consumer acceptability, the restoration of air quality that is both attainable and healthful is still hampered by contention and dissatisfaction.

The inherent limitations of the internal combustion engine may force drastic changes in the type of engine used for most vehicles in the near future. Although it is not yet clear which of several types now under study will eventually be adopted, the change will inevitably be costly in engineering, retooling, and restocking, and there will remain billions of dollars worth of outdated and unusable spare parts in the inventories of dealers, parts stores, and repair shops. Thus, higher costs for private transportation may hasten other innovations including smaller vehicles and more effective public transportation systems.

Electric power plants are among the most troublesome air polluters. Nuclear power plants, low in air pollution emissions, are being built less rapidly than had been anticipated by planners due to public fear of other nuclear hazards from their operation. Thus, fossil fuel emissions from power plant stacks are increasing at a faster rate than would otherwise be the case. Part of the power plant emission problem can be alleviated by burning low-sulfur fuel oil or natural gas and, where these measures are not feasible, by the installation of equipment to remove sulfur oxides and sulfuric acids. This is an expensive procedure, especially where high-sulfur coal is used. In general, nitrogen oxides remain unabated. Another major solution, but one of dubious long-range value, is to build power plants far removed from urban areas. For example, a total of nine power plants are planned for the Four Corners region of the southwest, where in the midst of the wild beauty of the area there are large deposits of strippable coal available for fuel. Major environmental problems are involved, however, and these are discussed in Chapters 18 and 19.

READINGS

Air Contaminants

Air Conservation. The Report of the Air Conservation Commission of the American Association for the Advancement of Science. Publ. No. 80, Washington, D.C., American Assoc. Advancement Sci., 1965.

CORMAN, R. *Air Pollution Primer*. New York, National Tuberculosis and Respiratory Disease Association, 1969, pp. 1—4.

GILPIN, A. *Air Pollution*. University of Queensland Press, 1971. 67 pages.

GOLDSMITH J. R. and LANDAW, S. A., "Carbon Monoxide and Human Helath," *Science* 162:1352—59 (1968).

HINDAWI, I. J., *Air Pollution Injury to Vegetation*. National Air Pollution Control Administration Publ. AP-71. Washington, D.C. U.S. Dept. Health, Education and Welfare, 1970.

Profile of Air Pollution Control. Air Pollution Control District, County of Los Angeles, 1971.

REVELLE, R., "Atomspheric Carbon Dioxide," *Restoring the Quality of our Environment*. Report on the Environment. Pollution Panel, President's Science Adv. Comm., U.S. Govt. Printing office, 1965.

SCHRENK, H. H., HEINMANN, H. CLAYTON, G. D., and GAFAFER, W. M., *Air Pollution in Donora, Pa*. Public Health Bulletin No. 306. Washington, D.C., Public Health Service, 1949.

Vehicular Exhaust

Automotive Fuels and Air Pollution. Report of the Panel on Automotive Fuels and Air Pollution. U.S. Dept. of Commerce, March, 1971.

Smog

Automotive Air Pollution. Sixth Report of the Secretary of Health, Education and Welfare to the Congress of the United States. Washington, D.C., U.S. Government Printing Office, 1967.

RASMUSSEN, R. and WENT, F. W., in *Science* 144:566 (1964).

STEPHENS, E. R., "The Formation, Reactions, and Properties of Peroxyacyl Nitrates (PANs) in Photochemical Air Pollution," *Advances in Environmental Sciences* 1:119—46 (1969). New York, Wiley-Interscience.

WENT, F. W., *Proc. Nat. Acad. Sci.* 46:212 (1960).

Sound and Noise

Anon., "Noise—Fourth Form of Pollution," *Environmental Science and Technology* 4(9):720—22 (1970).

BARON, R. A., *The Tyranny of Noise*. New York, St. Martin's Press, 1970.

BERLAND, T., "Silencing Invisible Pollution," *Today's Health,* July 1970. Chicago, Ill., American Medical Association.

BRONSON, W. "10,000,000,000,000,000,000 Mice . . . Ear Pollution," *Cry California* 2(4):28—30 (Fall 1967).

HARRIS, C. M., "Noise," *Environmental Science and Technology* 1(4):293—296 (1967).

The Noise Around Us. Findings and Recommendations. Washington, D.C., U.S. Dept. Commerce, 1970.

The Noise Around Us Including Technical Backup. U.S. Dept. Commerce Publ. COM 71—00147, 1970.

CHAPTER 7

WATER POLLUTION

An old saying was that "the solution to pollution is dilution." Even today, we depend mainly on dilution for disposing of wastes in air, water, and soil, for some of the waste problems in nature are taken care of by the large amount of living space. Unfortunately, when living space in any part of the biosphere becomes limited, as, for example, by congestion or industrial expansion, dilution is no longer a satisfactory solution and other means must be taken to reduce the ratio of waste to space.

Most of the wastes of civilization are drained into streams and rivers and ultimately into lakes or oceans. In many cases the waste is dumped into the same bodies of water from which drinking water is withdrawn. The effect of sewage, industrial waste, and agricultural drainage on plant and animal life in closed bodies of water is sometimes catastrophic. The accumulation of excessive plant nutrients, called *eutrophication* (which we will consider in detail later in this chapter), has occurred in Lake Erie, Lake Washington at Seattle, the lakes in the Madison, Wisconsin, area, and Lake Zurich in Switzerland.

Coastal waters are also changing due to organic material from sewage outfalls. Marine algae are disappearing from sections of the California coastline. Kelp, an important source of algin used in food and medicines, is diminishing. Heavy metals and pesticide residues are found in fish in amounts above those declared by the Food and Drug Administration to be safe for human consumption. Some species of fish-eating birds accumulate DDT in such large quantities that the pesticide is suspected of being the cause of thin egg shells and the decline of certain species, such as the brown pelican, in some of their breeding grounds. Other species of commercial fish and marine birds are decreasing in numbers at an alarming rate. The specific cause is unknown in most cases.

Early Hebrew literature tells about the pollution of water throughout all of Egypt during the struggle for freedom against the Egyptian conquerors. Yahweh gave Moses and his brother Aaron the power to cause rivers, streams, ponds, and pools to become "blood red" and to "stink." The fish died and the water could not be used for drinking. The people had to dig wells near the river's edge to obtain usable water.

The Romans understood the importance of clean, fresh water. In the first century A.D., according to Frontinus,[1] eight aqueducts carried more than 200,000,000 gallons of water daily into Rome. Most of it went to public fountains, from where it had to be carried to private homes. The ruins of Roman aqueducts can be seen in many places throughout Europe, and some of them are still in use. The sewage of Rome, however, flowed into the Tiber, and some people think that the pollution partly accounted for the disease and unsanitary conditions of the Pontine swamps and adjacent agricultural areas and villages, eventually contributing to the decline of the Empire.

The vastness of the earth's water had led many people to think of the rivers and eventually the oceans as the ideal receptacle for waste—a garbage dump of seemingly unlimited dimensions. But the delicate balance of many organisms in their environment is often overlooked. Extraneous materials deposited along or near the shores may remain in concentrated form for long periods of time. Some of the pollutants can disrupt the life cycle of plants and animals or readily gain entry into the food chain with potentially serious or even disastrous results.

The oceans, which cover 70 percent of the earth's surface, receive an estimated 4 billion tons of dissolved matter annually. Some of it is man-made. It is essential to find out what it contains and the effect of each constituent on water, wildlife, and people.

EUTROPHICATION

Lakes, large and small, are constantly being enriched by organic matter from decomposing plant and animal remains, providing nutrition for algae and larger aquatic plants. This fertilization, whether from natural sources or man-made, is called *eutrophication,* from the Greek meaning "well fed." This condition occurs often in nature and is not necessarily bad, for the organic material means food for fish and other aquatic life.

But too much mineral nutrient in the water can be more serious than too little. Many people have noticed the reddish or greenish cast of lakes, bays, or coastal areas in summer or fall caused by enormous numbers of algae and other phytoplankton. Their presence is often followed by the death of fish and other water life, sometimes on a massive scale. Such occurrences are the result of microscopic life having become overabundant. When the algae and other plankton die, decomposition of the superabundance of organic material causes the oxygen in the water to become suddenly depleted. Water cannot absorb replacement oxygen fast enough to take care of the needs of the living organisms. They suffocate. Also, toxins are sometimes produced.

The "red tide" that is frequently seen along the coasts of California, the Gulf of Mexico, and India is caused by the buildup of microscopic *dinoflagellates*. These are single-cell swimming organisms that have some of the attributes of both animals and plants. They impart a red or dirty brown coloration to the water and at night create a spectacular display of

---

[1] *On Aqueducts,* by Frontinus (first century A.D.).

luminescence in the froth of waves or in the wake of a boat. Their most injurious effect is the production of a lethal poison called *saxitoxin,* which accumulates in the bodies of clams and mussels. People who eat them are liable to be stricken with "paralytic shellfish poisoning," a dangerous form of food poisoning having a 21 percent mortality rate. The red tide along the coast of India is thought to be caused by an overabundance of mineral nutrients deposited by the heavily mineral-loaded run-off of flood water during the monsoon rains. Along the California coast, it is suspected that the nutrients from sewage outfalls contribute to excessive growth of the organisms when suitably high temperatures and other favorable conditions prevail.

Nutrient of the sea

In contrast to many of the land areas where the scarcity of water is the chief limitation to the growth of vegetation, the oceans are afflicted with a lack of nutrients. This is the most critical factor, but not the only one, in determining the growth of the small plant and animal life of the sea, called plankton.

Most of the open sea is a biological desert, an essentially barren area comprising 90 percent of the ocean or nearly three-fourths of the earth's surface. Half the world's fish supply is produced in coastal waters and in a few offshore areas of comparably high fertility. The other half is produced in regions of upwelling water that total no more than 0.1 percent of the ocean's surface. Thus, the most productive portions of the ocean—the coastal waters—are those which are the most polluted and where the marine life is the most susceptible to disturbances in the environment.

ALGAL BLOOMS

Every lake or pond goes through a life cycle that may take from a few days up to thousands of years for the larger bodies of water. Each lake or pond faces the prospect of becoming more shallow with time as silt and organic matter are swept in by streams and run-off and deposited in the water. Shallow ends, often near the outlets of streams, tend to become marshes. Algae and other plants proliferate under these conditions and often appear as a green scum (Figure 7.1). We can see various stages in the life history of lakes, from those with fresh, clear, sparkling water to the shallow, growth-choked, scum-surfaced swamps on the verge of becoming dead lakes.

In lakes and slowly moving waters that have been enriched, algal blooms or dense plant growth accumulate along the shores and in shallow portions where they may form decaying masses and foul-smelling scums. Blue-green algae of several types are the most troublesome in water warmer than 70°F. Sometimes a toxin, as yet unidentified, is poisonous to livestock and wild animals. (See Figure 7.2.)

Algal blooms are sometimes caused by conditions that are not related to added nutrients. Artificial warming of the water, such as that from cooling towers and condensers, may stimulate the growth of blue-green algae. The elimination of other aquatic flora, either mechanically or by the presence of some toxic substance, promotes the growth of bloom-forming algae by removing competition for the available nutrient supply. The release of organic nutrients by the decomposition of aquatic plants killed

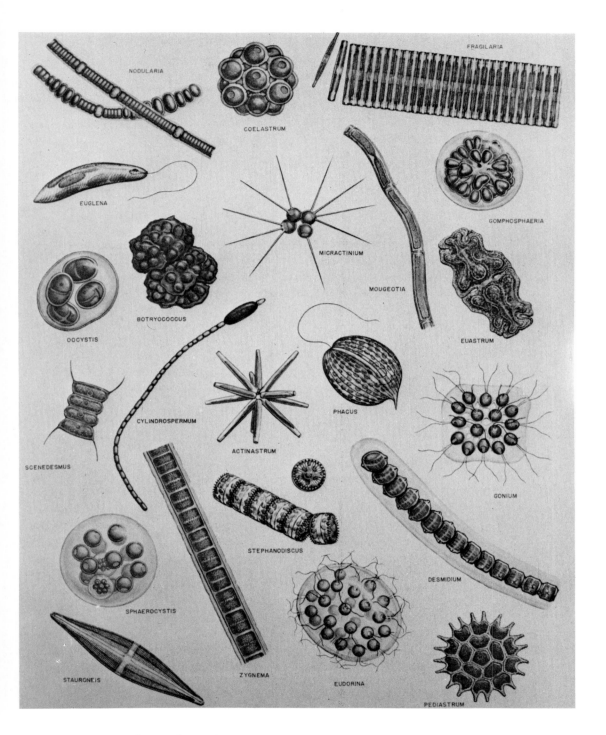

Courtesy C. M. Palmer, from *Algae in Water Supplies*, 1959, U.S. Public Health Service.

FIGURE 7.1 Plankton and other surface-water algae.

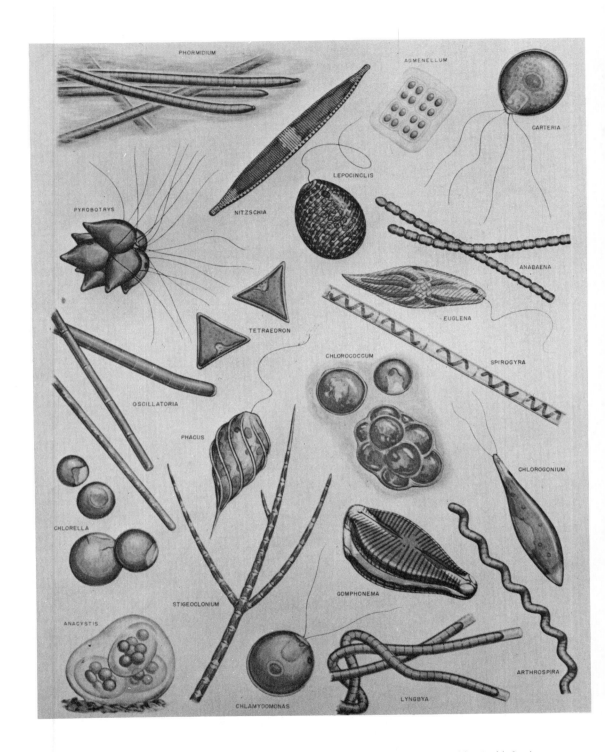

Courtesy C. M. Palmer, from *Algae in Water Supplies*, 1959, U.S. Public Health Service.

FIGURE 7.2 Polluted water algae.

Left photo courtesy of Frank E. Clarke, U.S. Geological Survey. Right photo courtesy of *Minneapolis* (Minn.) *Star*, and Environmental Protection Agency.

Left photo: Lake in the remote interior of Alaska is in an advanced stage of natural eutrophication, nearly a "dead lake." *Right photo:* Mass of dead and dying algae prevents any life in these lake waters. Algae feed off nutrients poured into the lake from cesspools, sewage treatment plants, and run-off from agricultural lands. The algae grows excessively, consuming all available oxygen, eventually causing even its own death. This is along the bank of Lake Minnetonka, Minnesota.

by herbicides may cause an increase in algal growth, and the absence of plankton-feeding organisms may favor excessive production of bloom.

Sources of nutrients

Algae, which are the principal growth products of eutrophication, need at least fifteen elements to sustain growth. If any one of them is in short supply, it will be a limiting factor in the growth and development of the algae. However, many of the essential nutrients, particularly the so-called micronutrients, are nearly always present in abundance in lake waters. Also, algae normally have access to an abundant supply of carbon, especially from carbon dioxide in the air and from carbonates dissolved in the water. (Some scientific investigators believe that a deficiency of carbon dioxide can become a limiting factor.) Nutrients that are utilized by the algal plants in large amounts are apt to be the most important in determining the rate and degree of growth. The primary plant nutrients are nitrogen, phosphorus, and potassium. Of these three, nitrogen and phosphorus are the ones most apt to be in short supply naturally, and since some of the blue-green algae have nitrogen-fixing capabilities, phosphorus is in many cases the most important added nutrient in bringing about excessive growth.

There are several man-made sources of the nutrients that cause eutrophication. They include domestic, industrial, and agricultural wastes. Sewage is an important source of nitrogen and phosphorus as well as organic matter. Detergents (discussed later) are important contributors of phosphates. Agricultural fertilizers and livestock and poultry wastes are also major sources. Cattle, horses, hogs, sheep, chickens, turkeys, and ducks in the United States produce more than a billion tons of solid wastes and nearly one-half billion tons of liquid waste annually. Cattle, with a national population nearly half that of the human population, produce nearly 20 times as much waste.

Wastes from slaughter houses and food processing plants often end up in the waterways. Boats discharge a variety of pollutants including sanitary wastes. More than 8 million watercraft ply the navigable waters of the United States. Laws governing the discharge of sewage differ among the states, and the laws in existence are often of negligible effectiveness.

Therefore, several points of attack are needed to solve the problems. Better sewage treatment to remove nitrogen and phosphorus and more effective disposal of the organic material as sludge are needed on a large scale. Control of industrial and agricultural wastes would mitigate the problems. Since phosphate-containing detergents are a major contributor of phosphorus, the elimination of phosphates from detergents would go far toward retarding eutrophication. Because food and other agricultural products come from the land, the return of agricultural wastes to the land would complete their cycle. More emphasis is needed on research directed toward that end.

Several large bodies of inland water in the United States are noted for their algal development due to eutrophication over a period of many years. These include Lake Zoar in Connecticut, Lake Sebasticook in Maine, the Madison Lakes in Wisconsin, Lake Erie, the Detroit Lakes in Minnesota, Green Lake and Lake Washington in the state of Washington, and Klamath Lake in Oregon.

There are many examples of lakes that are in an early stage of their natural life history yet are in danger of eutrophication from man-made pollution. Lake Tahoe, nestled 6,225 feet high in the Sierra Nevada Mountains of California and Nevada, is at present one of the clearest lakes in the world. The lake is 21.6 miles long and 12 miles wide. The bottom is 1,645 feet beneath the surface at the lake's deepest point, with an average depth of 990 feet. The waters of Lake Tahoe do not, as yet, support dense aquatic growth nor are they appreciably enriched. Blue-green and filamentous algae and other types that cause objectionable surface growth are now found near the shoreline, but their presence is not critical because of the low nutrient content. Phosphorus concentrations have reached critical levels, but nitrogen levels are still low and this is believed to be the limiting factor in the further development of algal growth. However, additional residential and recreational development on the perimeter of the lake and further waste disposal in the lake may have serious consequences.

LAKE ERIE

The Great Lakes are one of the most magnificent natural wonders of the world. Gouged out by glaciation during the Ice Age, they contain about 20 percent of the fresh water in all the lakes and rivers on earth. They provide important fisheries with more than 1,200 miles of water transportation, generation of hydroelectric power for far-flung cities and industries, and natural recreational areas for millions of people. However, the Great Lakes are also a sink for the wastes of many of the 35 million people living in the adjacent regions of Canada and the United States. The Lakes are the repository for massive quantities of municipal, industrial, and agricultural wastes, all of which have an important impact on the biological balance of the lake waters.

Lake Erie is perhaps the largest body of fresh water to have been seriously enriched with nutrients by human activities. It is inaccurate to say that Lake Erie is dead. In the words of Dr. William T. Pecora, former director of the U. S. Geological Survey, "Lake Erie, if anything, is too alive." The western part of the lake is a shallow shelf that is the recipient of large amounts of natural organic material deposited by rivers from the surrounding area. The algal growth that is typical of such waters is food for fish. The fish harvest of Lake Erie is equal to the rest of the Great Lakes combined. Lake Michigan's Green Bay, named for its natural growth of phytoplankton, is also productive of fish. However, when eutrophication takes place to excess as it has in Lake Erie, the results can be catastrophic. When algal blooms decay, they deplete the oxygen from the bottom layers of the lake. The scum that washes ashore and the stench that arises make the beaches unusable. Sometimes the decaying algae cause discoloration of drinking water. The beaches have sometimes been fouled with 6 inches to 3 feet of decaying algae. While the overall take of fish has soared, the catch is medium-quality fish. The major species of more desirable commercial fish have nearly disappeared. These changes have been taking place for more than half a century.

According to a five-year study by the Federal Water Pollution Control Administration, the chief source of pollution of Lake Erie is municipal waste. Detroit heads the list of contributors providing more wastes than all of the other cities combined. Two other major contributors are Cleveland and Toledo. Steps have now been taken to build new sewage treatment plants, improve existing plants, construct new sewers, and initiate other water pollution control programs.

Industrial wastes are probably the major source of pollution in the tributaries and harbors of Lake Erie. The lake receives an estimated 9.6 billion gallons of industrial waste water per day. There are at least 360 known industry sources responsible for 87 percent of the industrial waste that flows into the lake. Electric power plants account for 72 percent and steel production accounts for 19 percent of the industrial discharges.

DETERGENTS

About 5 billion pounds of detergents are used annually in the United States. Most of it is flushed into the sewer systems from bathrooms, kitchens, laundries, factories, and other industrial establishments. Detergents consist of *surfactants* mixed with conditioning and water-softening agents, mainly polyphosphates.[2]

---

[2]*Surfactants* are surface active agents, substances having properties similar to soap.

Detergents that contain water softeners do a better job of cleaning. They help reduce bacteria and other microbes and materially contribute to minimizing cross-infection. If water softeners were eliminated from detergents and not replaced with other conditioners, cleanliness in the home would be reduced. Automatic dishwashers would clean so poorly that they would be almost useless. There would be serious impairment of sanitation in schools, restaurants, hotels, and hospitals. Food-processing industries such as dairies, poultry farms, meat-packing houses, and canneries would be required to modify their cleaning practices or else suffer from lowered sanitation standards.

PHOSPHATES

Phosphate action

The major components of synthetic detergents, known in the trade as "syndets," are one or more inorganic phosphate compounds that are added to make the products resist hard water and clean more efficiently. Syndets contain anywhere from less than 1 percent to more than 70 percent phosphate. The actual phosphorus content of most household laundry detergents formerly averaged between 10 and 12½ percent, but the amount used is steadily decreasing as substitute preparations are placed on the market. Sodium tripolyphosphate, the most widely used phosphate "builder" in detergents, is one of a family of phosphates used extensively as water softeners.

$$
\begin{array}{ccccccc}
 & O & & O & & O & \\
 & \parallel & & \parallel & & \parallel & \\
Na-O-P & -O- & P & -O- & P & -O-Na \\
 & | & & | & & | & \\
 & O & & O & & O & \\
 & | & & | & & | & \\
 & Na & & Na & & Na & \\
\end{array}
$$

sodium tripolyphosphate

The value of the polyphosphates is in their ability to form soluble complexes with alkaline earth metal ions, mainly calcuim (Ca^{++}) and magnesium (Mg^{++}) found in hard waters. Compounds having this action are called *sequestering agents*. The calcium ion in water is captured by *coordination sites* within the molecule of the tripolyphosphate. In the absence of water softeners, the metal ions react with soap and other surface-active cleaning agents with the formation of insoluble precipitates. These precipitates are responsible for the "curds" formed in hard water. They aggravate the "ring" of dirt that forms in the bathtub.

Another property of the polyphosphates is the ability to break up curd-like particles, an action called *deflocculation*, and to hold the water-insoluble materials in the form of a *colloidal* suspension. Colloidal particles are those that are suspended in very finely divided form and often form extremely stable suspensions.

Phosphate pollution

Detergents have been named as the principal contributor of phosphorus to surface waters afflicted with man-made eutrophication. According to a 1970 congressional report, detergents were the largest single source of phosphorus pollution in Lake Erie and Lake Ontario. It was estimated that detergents accounted for more phosphorus in Lake Erie than all other sources combined. Seventy percent of the phosphorus contributed to Lake Erie came from municipal and industrial wastes, and about 50 to 70 percent of the phosphorus content of these wastes was from detergents.

Phosphates are not necessarily contaminants. Phosphates are widely recognized and needed as valuable fertilizers in agriculture and horticulture. They are important, therefore, in human food production as well as in plant and animal nutrition. Phosphates also have a wide variety of domestic and industrial uses. It is only when phosphates accumulate in excessive concentrations in surface waters that they become pollutants. Even then, high concentrations will not cause trouble if one or more of the other nutrients are inadequate for the development of phytoplankton. Too often, however, our penchant for perfection has the unintentional effect of setting the stage for eutrophication: enrichment of the water with nutrients, accelerated growth of algae, oxygen depletion of the surface water, and impaired fish production. The ecological balance is destroyed.

Phosphate removal

Methods for removing phosphates from sewage are being studied in an effort to devise procedures that would be both effective and economical. The removal of the phosphates from detergents would eliminate about 280 million pounds of phosphorus per year from surface waters, but at least 680 million pounds from other sources would remain. Even if all man-made sources were eliminated, at least 250 million pounds of phosphorus would be deposited in surface waters from natural sources.

Costs

As much as 90 to 95 percent of the phosphates can be removed from municipal sewage with present technology. The cost is estimated at less than 5 cents per 1,000 gallons of waste water. Procedures that presently appear to offer the greatest promise involve chemical treatment of the waste water. One method consists of treatment with lime and the removal of the phosphorus as insoluble calcium phosphate. Another method brings about coagulation by treatment with an alum.[3] A chemical-biological process involves chemical treatment at the primary sedimentation stage. Other processes and refinements to improve efficiency and reduce costs are now being developed.

The most desirable procedure would be to eliminate phosphates at their source. However, if using substitutes in detergents were to add several cents a box to the cost, the added premium might exceed the expense of removing algal nutrients from wastes.

It is not always possible to predict that reducing phosphorus or nitrogen will result in a proportional decrease in algal growth. There are too many other factors that determine the growth of algae:

- In general, the rate of algal growth doubles with every 20°F rise in temperature between 32°F and 90°F.
- The amount of each nutrient in the water is an important factor in the environment of the algae.

---

[3]Alums are a class of double sulfates. Commercial alum or potash alum is $KAl(SO_4)_2 \cdot 12 H_2O$. Aluminum sulfate, $Al_2(SO_4)_3 \cdot 18 H_2O$, is sometimes incorrectly called alum or *filter alum*.

- It was concluded from one set of calculations that if 99 percent of the phosphate coming into Lake Erie were cut off, the remaining 1 percent would be more than enough to support the growth of algae over the entire lake.
- Excessive growths are apt to occur if the average concentration of inorganic nitrogen exceeds 0.3 ppm and the average concentration of phosphorus exceeds 0.1 ppm.
- There is wide variation in the critical concentration of phosphorus. Excessive growth can occur in water containing less than 0.001 ppm, and algae can be absent in some waters with fifty times as much phosphorus.
- In some cases there may be better ways of controlling algae than by making drastic reductions in phosphorus.
- Further research is needed to determine the other basic requirements for algal growth, such as the presence of certain trace elements.

The substitute:
NTA

Phosphates are the most economical materials available for softening water and improving the efficiency of cleansing agents. But phosphates are not the only substances that can do the job. There are a large number of materials that resemble the polyphosphates in their ability to form soluble complexes with metals, including the main culprits in hard water, calcium and magnesium. Such materials have been used for many years for a wide variety of applications. They are called *chelating* agents, from the Greek word *chela,* meaning "claw." Many of the chelating agents are synthetic amino acids. One of the simplest and cheapest is nitrilotriacetic acid, or NTA (see Figure 7.3).

NTA was found to be effective as a water softener, to aid in the dispersion of dirt, to help maintain the proper alkalinity of the water, and

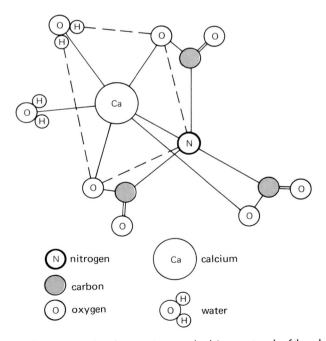

FIGURE 7.3 A calcium ion in hard water is enmeshed in a network of bonds by the chelating agent NTA (nitrilotriacetic acid).

to aid the surfactant in loosening the dirt. NTA was also found to be 70 percent degradable during the biological activity phase of sewage treatment. The water-softening action of NTA resembles that of the tripolyphosphate. A calcium or other metal ion is captured (chelated) in a network of bonds where it is held so tightly that it cannot cause trouble.

Early experiments with animals gave no evidence of high acute toxicity from NTA, and there had been no indication of genetic effects. However, new information from toxicological studies became available shortly after producers of NTA had expanded production capacity and the major detergent manufacturers had changed their formulations to contain the chelating agent. Experiments on rats and mice with NTA in combination with cadmium or methyl mercury showed an increase in congenital abnormalities as compared to the same dosages of metals when administered alone. NTA triggered a tenfold increase in fetal abnormalities and fatalities.

There were other objections. NTA is believed to be converted mainly to nitrite and nitrate, which are toxicants under some circumstances and also furnish nutrients to microorganisms and higher plants. Replacement of one nutrient for another is of questionable value. But a more serious possibility is the metabolic formation of *nitrosamines,* a group of compounds which includes some highly carcinogenic (cancer-causing) substances. NTA was removed from the market pending further investigations of its biological effects.

Other substitutes   Several other materials can be used in place of phosphates as detergent "builders." Sodium carbonate, known also as *soda ash*, is a mildly alkaline material commonly added to detergent formulas, but it can be injurious when ingested, especially by children. One of the borates known as *borax* is widely used as a cleansing agent. But sensitive trees and shrubs have been damaged by waste water from industrial cleaning operations in which borates were used.

One substitute for phosphates is a material called *sodium metasilicate*. The silicates are effective water softeners, but they have a major drawback. They are strongly alkaline materials which irritate the skin and eyes. Sodium metasilicate was fatal to laboratory rats when fed to them in small quantities. About 3,000 cases of accidental detergent ingestion are reported to occur in the United States each year. While no permanent injuries have been recorded from "heavy duty" laundry detergents containing phosphates, those made with the highly alkaline metasilicate can injure the esophagus and stomach if ingested. The Food and Drug Administration now requires warning statements on packages containing phosphate-free detergents with metasilicate water softeners. Thus, substitutions to eliminate a pollution problem can result in the creation of a household hazard.

MORE DETERGENT PROBLEMS

Arsenic   Purified phosphates are among the safest compounds used in consumer products, being incorporated in foods, beverages, and medicines. However, one of the impurities in mineral phosphate deposits, and hence, in commercial phosphate water softeners, is arsenic. This contaminant has been detected at concentrations of 10 to 70 ppm in several commonly marketed household detergents. In view of the fact that much of the sewage is dumped into waterways, it is not surprising that the Kansas River has arsenic values of 2 to 8 ppb (parts per billion), close to the limit of 10 ppb in drinking water recommended by the U.S. Public Health Service, although much of the arsenic may be from natural sources.

There is danger that arsenic in laundry water may be absorbed through unbroken skin. Skin eruptions and other types of dermatitis allergies have been reported in sensitive people. Baby diaper rash and hand rashes are associated with arsenic in detergents. When present at 50 ppm, arsenic inhibits the healing of wounds. The accumulation of arsenic in human hair following the use of arsenic-containing detergents was reported as early as 1958. There is also evidence that arsenic accumulates in the livers of mammals.

Enzymes

Enzyme detergents were introduced in the United States in 1968 after having been used for several years in Europe. The enzymes are intended to attack stains that other detergents cannot remove, especially stains from protein-containing materials: blood, urine, milk, meat juice, and chocolate. The enzymes are produced by cultures of the microorganism *Bacillus subtilis,* which secretes the enzymes during its digestive process. Unfortunately, the enzyme dusts in manufacturing plants also attack the skin of workers' hands. When inhaled, the enzymes sensitize the lungs, causing allergic reactions resembling hay fever or asthma. The enzymes were found to contain several potentially toxic substances that caused disintegration of blood elements and agglutination of red blood cells when injected into the body cavity of mice. Whether or not the smaller concentrations of enzymes to which detergent users are normally exposed comprise a widespread allergic hazard is under intensive investigation. The potential effects of detergent enzymes on aquatic organisms are also unknown.

Foamy water

The appearance of bubbles and foam on a glass of water drawn from the kitchen tap was the first indication to many people that the domestic water in some areas was not very pure. The frothy "head" on a glass of drinking water was caused by the presence of *surfactants*. The term is a contraction of "surface active agent." Surfactants increase the wetting and cleaning power of water. They are the principal active ingredients in bath soap, toothpaste, shaving cream, shampoo, dishwashing compounds, detergents, and cleansing agents of various kinds. The largest use for surfactants is in household detergents, but there are also many industrial applications.

From the ground up

How do surfactants get into the water supply? They enter the ground from cesspools, sewage plant outlets, and waste water spreading beds and will readily percolate into the underground water supply. If water wells are in the vicinity or within range of the underground flow, the water that is pumped from the wells will be contaminated with surfactant.

When foaming of domestic and irrigation well water was first noticed, the most commonly used surfactant in detergents was ABS, an abbreviation of *alkyl benzene sulfonate*. ABS was cheap and effective, but highly resistant to microbial degradation. Foaming caused by ABS was severe in some irrigation waters and in lakes and streams used as recipients for municipal and industrial effluent. Extreme foaming occurred in many

USDA photo.

Foam caused by detergents in a stream.

municipal sewage plants where foam was blown about by the wind, causing a nuisance and unsanitary conditions in surrounding neighborhoods.

Biodegradable detergents

People who worked on the foamy-water problem reasoned that if something could be done to make ABS attractive to bacteria, the molecule would be so impaired that its foaming properties would be destroyed. This was accomplished by a modification in the chemical structure of ABS that produced biodegradable detergents called LAS (linear alkylate sulfonates).[4] Why this makes a difference to the microbes no one knows for sure, but it is a happy fact that the microorganisms can easily digest the linear chain and use it as a source of carbon for food. However, phenolic decomposition products of largely unknown biological activity remain. But the foam, odor, and taste formerly imparted to water by ABS have been virtually eliminated.

INDUSTRIAL POLLUTION

The contribution of the chemical industry to human welfare is evident in the wide range of chemicals used in drugs, cosmetics, plastics, synthetic fibers, paints, cleansing agents, and many other kinds of consumer products. Unfortunately, most of the chemicals end up in the environment. A high proportion is sluiced down the sewers and ultimately into rivers, lakes, and oceans. Many of the chemicals are highly toxic to living organisms. Some of them find their way into the food chain, disrupting or impairing the natural biotic cycle, threatening populations of organisms with extinction, and jeopardizing the quality of man's food supply. Natural waters from some rivers and lakes have become unfit for domestic use and some streams have become biologically barren.

---

[4]Materials that are decomposed by microorganisms were referred to as *biodegradable*.

Photo courtesy of Environmental Protection Agency.

Oil refineries below Charleston on the Kanawha River. Note the nine outfalls on the near side of the plants.

Chemical pollutants

The dependence of present-day society on chemical technology is almost beyond comprehension. Synthetic chemicals of every conceivable kind, from medicines to miniskirts, are part of our everyday life. Nearly two million chemical compounds are known and several thousand new ones are discovered each year. More than 9,000 synthetic organic compounds are used commercially, with 300 to 500 new ones being added annually. Production of the top 50 chemicals during 1973 was in excess of 400 billion pounds in the United States alone.

When the explorer Thor Heyerdahl sailed his papyrus raft westward across the Atlantic in 1969, he was dismayed at the abundance of flotsam that he encountered. In some areas the floating debris carried foul-smelling oily material. About twenty years earlier, he had sailed across the Pacific in a similar craft, the *Kon Tiki,* and had experienced a relatively unpolluted Pacific Ocean. During the intervening period, relentless pollution with garbage, sewage, industrial wastes, and similar debris had established a markedly deteriorating trend in the quality of the oceans.

Oceans do not have the rapid turnover that occurs in rivers, springs, and lakes. The oceans are more stable and maintain their chemical composition over periods of thousands of years. Injurious chemicals can build up in concentration over long periods of time unless they are decomposed. Poisonous metals are indestructible. A classic example is the disastrous result of discharging industrial effluent containing mercury, such as that which occurred in Minamata Bay of Japan. Even some synthetic organic chemicals have remarkable stability, as, for example, DDT and many of the plastics.

Under the anaerobic conditions of silt and mud, chemical substances may be relatively free of oxidation and other chemical degradation as well as resistant to bacterial decomposition. But it cannot be assumed that chemical pollutants will remain indefinitely at the point of deposition. Some of the pollutants may become buried in the sediment near the coastline where they can be consumed by microorganisms and mud-inhabiting worms. Thus, chemicals and their degradation products are taken up in the food chain and passed from one species of organism to others that prey upon them. Some of the organisms may be carried by ocean currents; fish and other organisms may travel under their own power for great distances.

A large number of organic and inorganic substances have been identified in domestic sewage. Materials that are known to be potentially serious pollutants include sundry chemicals from domestic disposal, such as from household cleaners, medicines, and chemicals in food wastes; a wide variety of industrial chemical wastes, including petroleum products, phenols, solvents, chemical intermediates, metallic wastes, and by-products which may be highly toxic to living organisms; pesticides from agricultural use or industrial waste disposal; and radioactive chemicals, some of which have toxic properties lasting for thousands of years and whose fate in the food chain is largely speculative, though they are among the most toxic substances known. Many of these substances are discussed elsewhere in this book.

Sometimes contamination of the underground water supply, as well as surface waters, comes from unsuspected sources, as in cold climates where roads are commonly salted during the coldest winter months to lower the freezing point of water and prevent the formation of a hard, slippery coating of ice. In porous soils, large amounts of salt leach into and through the soil where it can make water unusable.

Industrial disposal   Industry is the largest single user of water, accounting for 50 percent of the daily water requirement of the United States, and this is expected to climb to 65 percent by 1989. Treatment of industrial waste waters is complicated by the presence of a wide variety of both inorganic and synthetic organic pollutants, many of which are not readily susceptible to biodegradation. Solvents, oils, plastics, plasticizers, metallic wastes, suspended solids, phenols, and various chemical derivatives of manufacturing processes are apt to be difficult to identify and impossible to remove without more advanced technology than we now possess. Some of the substances are known to be highly toxic to living organisms but are of unknown effect as environmental pollutants.

One method used to dispose of industrial toxins is to inject the waste water deep underground. A large petrochemical plant pumps wastes containing phenolic compounds into a well 6,000 feet deep and wastes containing a class of chemicals called nitriles into a separate well 7,000 feet deep. Disposal of toxic substances deep underground requires careful study of the possibility of irretrievably contaminating the underground water supplies because subsurface water may move for considerable distances underground. Another danger of underground disposal is that the pressure might cause slippage of deep layers of rock, resulting in earthquakes. A series of earth tremors in the Denver area were apparently associated with deep underground disposal of wastes.

Pulp and paper is the fifth largest industry in the United States and is the third highest in the total industrial use of water. Strenuous efforts are being made to reduce the amount of water used per ton of product. The rate of reuse is now about 216 percent, among the highest in industry. Pollution stems from suspended matter and large amounts of dissolved organic substances, the removal of which is generally difficult and expensive. More than 7,000 gallons of water are needed for each ton of wood pulp produced. Treatment costs come to about 30 to 36 million dollars a year. This would increase to about one billion dollars if the entire industry were to achieve 85 percent B.O.D. (biochemical oxygen demand) reduction and nearly complete removal of suspended solids. (B.O.D. is discussed later in this chapter.)

The iron and steel industries are large users of water. By the nature of their operations, they have critical requirements for specialized pollution control facilities. The amount of water needed usually ranges from 20,000 to 50,000 gallons per ton of steel produced, although one mill, by using advanced water conservation methods, has reduced its requirement to less than 2,000 gallons per ton. Water reuse in the industry is generally about 40 percent of the total throughput. Effluents may run at rates of 10 to 25,000 gallons per minute. The processes of a large steel mill may generate as many as 100 distinct kinds of discharges requiring 5 to 10 separate waste flow systems, two-thirds of which involve chemical, sedimentation, or filter treatment. The principal pollutants are scale, oils and greases, and miscellaneous chemical wastes.

Some industries recycle a large proportion of their water needs, especially water used for cooling. One large petrochemical plant uses 1.5 billion gallons of water per day of which only 10 percent is new water. The industry average is probably 1 gallon of new water in each 3 gallons of water used. There is increasing interest on the part of industry to reuse its water, not only for reasons of expense and pollution control but also because there is an increasingly limited supply of available water.

WATER POLLUTION AND WILDLIFE

FISH KILLS

There are many cases on record of the destruction of wildlife by polluted waters. Mass killing of fish was among the earliest and most dramatic results of indiscriminate pollution of water. The cases are too numerous to list. During the ten-year period from 1960 to 1969, a total of 145 million fish were reported killed in more than 4,200 cases of pollution, the contamination coming from diverse sources which vary from year to year.

A high point in fish kills came in 1971 when an estimated 74 million fish were killed in 46 states. The largest number, about 25 million, were "done in" by municipal wastes of all kinds, especially sewage. About 5 million fish were killed by industrial pollutants, more than one million were killed by agricultural wastes, mostly manure-silage drainage, and 704,000 fish were killed by transportation activites. Pesticides accounted for about 264,000 of the fish killed by agricultural activities. Fish kills during 1972 are shown in major categories in Figure 7.4.

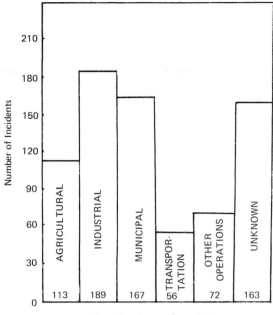

Data from Environmental Protection Agency.

FIGURE 7.4 A total of 760 fish kill incidents were reported in 1972. Industrial operations were the main source, with municipal sources of pollution a close second.

It must not be assumed that all mass die-offs of wildlife are caused by man-made pollution. Large numbers of ducks are sometimes killed by a type of botulism that develops in marshy areas at certain times of the year. For example, 140,000 ducks died of botulism in California during 1970. In the fall of 1964, approximately 50 million dead anchovies piled up in the harbor at Santa Cruz, California. The fish were victims of the "red tide" that periodically assaults areas of the Pacific coast. When conditions are favorable for a buildup in population of the red-tide organism, the single-cell plantlike animals—or animal-like plants—sometimes have disastrous effects on fish, apparently because they become so numerous that they rob the water of oxygen. An infestation was attributed with killing more than 200 tons of commercial fish near the coast of southern California. A number of similar kills of fish and shellfish have been reported. Some of the marine life may have been killed by the poisonous substance *saxitoxin*, produced by the red-tide organisms.

THE BLACK TIDE

Pollution of the aquatic and marine environment by oil can occur almost anywhere at any time. There are many recent examples. A dramatic incident was that of the tanker *Torrey Canyon* when she struck a reef off the southern tip of the British Isles in March of 1967. She was carrying 118,000 tons of crude oil. The tanker began to split apart about a week

Left photo courtesy of W. E. Seibel, County Fish and Game Association, Ohio; USDA photo. Right photo courtesy Environmental Protection Agency.

Left photo: Sugar beet wastes dumped into stream killed these fish. *Right photo:* Outfall carrying wastes from meat-packing plants in Omaha, Nebraska. Dark material is blood and paunch manure. The floating materials are mainly grease.

after going aground, and the effects of oil-polluted beaches on a disastrous scale was realized for the first time.

Oceangoing tankers are a continuing source of oil pollution. About 1 billion tons of petroleum products are shipped worldwide each year. The tonnage is increasing annually with the construction of tankers of greater size for improved efficiency. If spills were only 0.01 percent of the petroleum transported, they would amount to more than 20 million gallons polluting the sea every year.

The *Torrey Canyon* was the largest oil spill up to that time. The pollu-

Photo courtesy of Standard Oil Company of California.

Mammoth tankers are capable of carrying more than 200,000 deadweight tons of cargo.

tion caused widespread destruction of many forms of marine life despite strenuous efforts to clean up the spill. The Nature Conservancy in London reported that 7,000 oil-logged birds had been captured and treated but that only a few hundred lived. The St. Agnes Bird Observatory estimated that 40,000 seabirds died. Within two weeks, one million gallons of twelve different detergents had been used in an effort to clean up the spill. Most of the detergents consisted of 80 percent aromatic solvent (similar to paint thinner) and 20 percent surfactant. But the detergents themsleves were toxic, and the mixture of emulsified oil and detergent was far more toxic than the oil itself. In addition to the catastrophe to marine life and fisheries, the direct monetary cost of the cleanup was about 8 million dollars, according to a British government report. Supertankers twice the size of the *Torrey Canyon* are already in operation.

Refineries and oil drilling operations are other potential sources of oil pollution. The blowout of a well in California's Santa Barbara Channel in January 1969 resulted in a leakage estimated at 500 barrels per day, which amounted to approximately 236,000 gallons over a ten-day period. Damage to sea life including marine birds was extensive and the incident had a shocking impact on the aesthetic qualities of the nearby coastal area. Offshore drilling continues in several areas along the nation's coastlines and is an ever-present threat that requires constant vigilance. Pollution of the seas with oil from all sources is a major global problem.

ATMOSPHERE IN THE SEA

Without dissolved carbon dioxide and oxygen, it would not be possible for the profusion of aquatic (fresh-water) and marine life to exist. The high solubility of carbon dioxide in water is of great importance in promoting photosynthesis by phytoplankton and, consequently, is beneficial to all organisms in the food chain. The relatively low concentration of oxygen in water compared to air is a disadvantage to aquatic and marine animals that depend on gills or similar arrangements for obtaining oxygen from water and accounts for the greater efficiency of air-breathing animals. However, some of the disadvantages are overcome by the intricate structure of gills and mechanisms for gas exchange. The increase in solubility of oxygen and carbon dioxide in water at low temperatures is favorable for the development of marine life and explains why many of the world's most productive fisheries are in cold waters.

The bodies of water are in a state of equilibrium with the atmosphere and the earth because the solubility of the gases increases as the water temperature and salinity drop. However, we do not find the same proportions of the gases in water as in the air, due primarily to differences in solubility. Pure water exposed to the atmosphere contains, *by volume,* about 2 percent nitrogen, 1 percent oxygen, and .05 percent carbon dioxide at 0° and atmospheric pressure.[5] Compared to nitrogen and oxygen, the amount of carbon dioxide is proportionately much higher than in air. However, the concentration of oxygen and carbon dioxide

[5] Air contains about 78 percent nitrogen, 21 percent oxygen, and .03 percent carbon dioxide. The percentage of dissolved gases by weight are much lower than the values calculated on a volume basis.

in natural waters is highly variable. The differences are largely the result of the growth of algae and other aquatic organisms and the stagnation or mobility of the water.

Carbon dioxide

Carbon dioxide in natural water comes primarily from two sources: from dissolved gas from the atmosphere, and from carbon dioxide liberated by organisms during respiration. The concentration varies from a fraction of a part per million to about 20 milligrams per liter (about 20 ppm or .002 percent).

Water is capable of absorbing carbon dioxide by entering into a chemical reaction with it, forming carbonic acid:

$$CO_2 \quad + \quad H_2O \quad \rightleftharpoons \quad H_2CO_3$$

carbon dioxide   water   carbonic acid

Because the reaction is reversible, the carbonic acid serves as a reservoir which makes some carbon dioxide available for plants when the dissolved gas nears depletion. Some of the carbonic acid dissociates (ionizes) with the formation of hydrogen ions and bicarbonate ions:

$$H_2CO_3 \quad \rightleftharpoons \quad H^+ \quad + \quad HCO_3^-$$

carbonic acid   hydrogen ion   bicarbonate ion

And some of the bicarbonate ions dissociate forming more hydrogen ions and carbonate ions:

$$HCO_3^- \quad \rightleftharpoons \quad H^+ \quad + \quad CO_3^=$$

bicarbonate ion   hydrogen ion   carbonate ion

The three reactions are in equilibrium, shifting to the right or left depending on the withdrawal or addition of ions by organisms, the acidity or alkalinity of the water, and the temperature. In the ocean the concentration of bicarbonate ion is an average of about 0.014 percent.

Some investigators suggest that a small increase in air temperature could cause a great increase in atmospheric carbon dioxide, leading to a catastrophic effect on climate if there were no moderating influence. But others say that control mechanisms may exist in the oceans, which are capable of absorbing large amounts of carbon dioxide in chemcial combinations. A "feedback" mechanism may exist in the atmosphere itself, for an increase in temperature is apt to cause an increase in cloudiness, resulting in a smaller amount of solar radiation reaching the surface of the earth. The world's atmospheric content of carbon dioxide and its potential global effects were discussed in Chapters 3 and 6.

Oxygen

In a body of water containing diversified biota, the oxygen content is in a condition of unstable equilibrium. Some of the oxygen is added by direct absorption from the air and some of it comes from the oxygen given off by plants during photosynthesis. Absorption of oxygen from the atmosphere may be aided by surface agitation caused by wind and waves, but such action may also reverse the process of releasing oxygen to the atmosphere when the water becomes supersaturated. Oxygen is removed from the water by respiration of organisms and by the oxidation of dead organic material during its decomposition.

While aquatic plants absorb carbon dioxide from the water for use during photosynthesis, oxygen is released in the dissolved state, except when the water is supersaturated with oxygen, in which case gaseous oxygen is liberated. Oxygen is produced during the first (energy capture) phase of the photosynthetic process called *photolysis,* during which the action of light energy on the chlorophyll indirectly brings about the decomposition of water into its components, hydrogen and oxygen. Since photolysis can occur only in the light, the production of oxygen is limited by the amount of sunlight available and can occur only in the region called the photic zone. The total oxygen productivity of an ecosystem[6] is determined by the amount of solar energy reaching the organisms, by the photosynthetic efficiency of the plants, and by physical factors such as temperature that affect the rate of photosynthesis. A covering of ice and snow reduces both the penetration of light and the absorption of oxygen.

As the temperature goes down, the oxygen capacity of the water increases. Cold water at 39°F can hold 1½ times as much oxygen as warm water at 75°F.[7] The saturation level at 0°C in fresh water is about 15 ppm, but the concentration may increase to two or three times as much when the rate of photosynthesis is high during daylight hours, and may fall far below the saturation level when photosynthesis declines at night.

Depletion of dissolved oxygen

Bacterial decomposition of organic matter may deplete the dissolved oxygen in stagnant deep water during the summer when there is an over-abundance of phytoplankton and other aquatic life. In like manner, the decomposition of organic matter in sewage wastes depletes the oxygen content. Organic pollution from either natural or human sources is sometimes expressed as the amount of oxygen needed by microorganisms to decompose the organic material. An evaluation is made by determining the amount of oxygen used up over a period of 5 days in the laboratory. This is called the B.O.D. *(biochemical oxygen demand).* The B.O.D. is an indication of the amount of organic material present in the water or waste water.

A problem arises if there is a deficiency of oxygen to the extent that anaerobic conditions prevail.[8] Whereas the aerobic oxidation or organic material results in generally harmless end products, anaerobic decomposition produces substances that are both objectionable and toxic.

Oxygen deficiencies are injurious to fish. A reduction in concentration of dissolved oxygen (D.O.) to about 3 milligrams per liter or less causes delayed hatching of eggs, reduced size and vigor of the embryos, deformities in the young, interference with the digestion of food, accelerated clotting of blood, reduced growth rate, poor utilization of food, impairment of swimming speed, and a decreased tolerance to toxicants. The lethal effect of low concentrations of dissolved oxygen appears to be enhanced by excessive carbon dioxide and ammonia as well as by cyanides, zinc, lead, copper, and cresols (cresylic acids).

Global oxygen

Organic pollutants make such demands on the oxygen content of natural waters that serious problems often ensue. In many cases the oxygen level is below that needed for the survival of fish and other aerobic organisms. On a global basis, however, the oxygen reserve is less critical.

---

[6]The ecosystem is the composite family of living organisms and their environment (see Chapter 2).

[7]The air contains 250 milligrams per liter of oxygen at 20°C (68°F) at atmospheric pressure.

[8]*Anaerobic* means not requiring oxygen, or the absence of oxygen.

There is no reason to assume that the earth's oxygen supply will be depleted in the foreseeable future, even though at various times and places organic waste can place an excessive burden on its immediate availability. The world's oxygen supply was discussed in Chapter 3.

WATER AND HEALTH

The quality of domestic water supplies is an old problem. In *De architectura,* a work written between 25 and 23 B.C., Vetruvius Pollio, the great Roman architect and builder, gave numerous examples to show that bad water was injurious to health. Phythios, the Greek builder of the temple of Minerva in Priene on the coast south of Ionia, also dealt with this problem in a textbook on architecture. Clean water was just as difficult to guarantee 2,000 years ago as it is today. The Romans did not have filtration plants, chlorination facilities, or similar modern equipment. They had to rely on local streams and springs or transport the water over long distances. The Roman aqueducts, a number of which are still in use today, provided clean water for cities. Where these sources were not available, wells had to be dug; and to assure good water, they were planned by architects and builders of established reputations.

According to Vetruvius, the water at Susa, the ancient capital of Persia, was unfit for human consumption because its use resulted in tooth decay. It was the earliest reference to the effects of fluorine deficiency. In more recent times, *denti neri,* or black teeth, has been a distinguishing characteristic of the people living in the vicinity of Vesuvius and other areas of volcanic origin in certain parts of Italy. Dark or mottled teeth are characteristic symptoms caused by drinking water that contains an excess of fluorine.

Roman aqueduct still in use at Segovia, Spain, carries water from mountain sources into the city for domestic use.

Vetruvius also cited the case of a spring on the island of Chios that was said to confuse the mind. In the Alps there were springs whose water caused "thick necks," an obvious reference to goiter, caused by an enlarged thyroid gland, which in turn is caused by insufficient iodine in the water or diet. Both goiter and cretinism, a form of mental retardation associated with failure of the thyroid to develop, are still found in some Alpine valleys. Vetruvius warned against water containing too much or too little iodine. He also warned against water from springs in the vicinity of metal mining. He said that the water would be adversely affected by the minerals present in gold, silver, iron, lead, and copper mines.

Vetruvius recommended making chemical tests. One method was to sprinkle water on a bronze vessel. If it did not leave stains, the water was good. Another test was to boil the water in a kettle. If the sediment was sandy after cooling and decanting, the water was good. Vetruvius also suggested that vegetables such as peas and beans be boiled in the water, and if they did not become soft, the water was unfit for drinking and was suitable only for industrial purposes. The simplest and best test of all was to observe the inhabitants in the vicinity of the source. If they were strong and healthy and were not afflicted by diseases of the feet and eyes, this could be taken as proof that the water was healthful and suitable for human consumption.

WATER QUALITY CRITERIA

Drinking-water standards set by the U. S. Public Health Service have been in effect for more than fifty years. However, until recently, the federal government had no direct authority over domestic supplies of drinking water. Responsibility for the quality of local waters resided with each of the state governments. Except for interstate common carriers of water supplies, such as buses and commercial airplanes, the federal standards had no legally binding effect and merely comprised recommendations for performance standards.

The first such federal standards for drinking water were adopted in 1914. They covered only bacteriological properties. Later, inorganic toxic materials such as lead and copper were included. Iron and manganese, though not highly toxic, were also included because they may be aesthetically annoying. They may impart a bad taste, and in the case of iron, a tendency to leave unsightly stains. By 1962, organic chemical pollutants were loosely covered by the standards, and eventually, radioactive toxicants were included.

It was not until 1975 that the federal government was able to establish the first binding national quality standards for drinking water. Acting under the Safe Drinking Water Act signed by the president, December 1974, the Environmental Protection Agency set up interim standards that became effective June 1975, with permanent standards becoming effective December 1976. But the states are still given the primary responsibility for enforcement of national drinking water standards if they elect to do so.

Different water quality standards are required depending on the use for which the water is intended. The major beneficial uses listed below are not necessarily in order of importance or quality:

- Domestic water supply, including drinking water.
- Industrial water supply, including cooling water.
- Agricultural water supply for irrigation.
- Livestock and wildlife water, including refuges for water fowl.
- Shellfish culture.
- Swimming, bathing, water skiing, and other water sports.
- Boating and aesthetic enjoyment.
- Water power and navigation.
- Waste transport, dispersion, and assimilation.

Water quality deficiencies

It should not be assumed that all of the domestic or municipal drinking water supplies are safe (see Figure 7.5). The Bureau of Water Hygiene conducted a survey of 969 public water systems in nine different areas of the country.[9] The findings were:

- 59 percent delivered good water.
- 41 percent of the systems delivered water of inferior quality.
- 36 percent of the tap water samples contained one or more bacterial or chemical contaminants exceeding Public Health Service standards.
- 56 percent of the water supplies had physical deficiencies of various kinds, such as inadequate disinfection capacity.
- 77 percent of the water-plant operators were inadequately trained.
- 79 percent of the systems were not inspected by state or county authorities during an entire year, and in half of these cases the water-plant officials could not remember when state or county health departments last made a survey.

BACTERIAL AND PARASITE POLLUTION

Bacterial contamination is the most common water-borne disease hazard in the United States. In many parts of the world both bacterial and other types of organisms are ever-present threats. Typhoid fever and cholera periodically cause widespread illness and death. Various forms of gastroenteritis are common.[10] The protozoan *Entamoeba hystolitica* causes a type of dysentery that is often serious. Parasitic worms of various kinds are transmitted by water. Eggs of tapeworms and roundworms are found in sewage. Pollution of streams by domestic and wild animals may be a source of contamination. Along the Nile River in Africa, *Schistosoma*—a blood fluke—attacks people who come in contact with the water, especially the farmers who work in the irrigation-flooded fields. The eggs find their way into the kidneys and are discharged in the urine, after which they quickly develop into larvae that must find an alternate host, the water snail, to continue their life cycle. Several species of flukes, all of which are parasitic worms, cause serious diseases in different parts of the world.

An early method of determining the microbial purity of water was to take a reading of its "keeping power." Water that was free of bacteria

---

[9]"Drinking Water: Is It Drinkable?" *Env. Sci. & Tech.* 4(10):811–813. 1970.

[10]Gastroenteritis is inflamation of stomach and intestines.

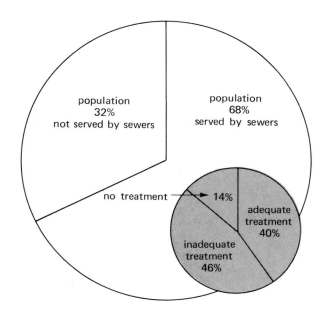

From *Controlling Pollution,* The President's 1971 Environmental Program.

FIGURE 7.5 Municipal wastes: adequacy of treatment.

could be stored for a long time, while contaminated water would tend to develop growths and undergo visible changes. Increasingly widespread use of the microscope made it possible to establish standards based on the numbers and kinds of organisms and organic debris that could be found in the water. Later, bacteriological tests were devised that were reliable and specific for the detection and identification of microorganisms.

Indicators of excrement

Examination of water for each of the various kinds of pathogenic microorganisms is slow, unwieldly, and costly. Therefore, bacteriological tests are used to detect the presence of bacteria that are presumed to be of fecal origin and thus indicative of sewage contamination. When indications of fecal pollution are found, the water is considered to be potentially dangerous to health. The bacteria used as indicators are described as *coliform* organisms, a group which includes diverse microorganisms, the best known of which are strains of *Escherichia coli* and *Enterobacter aerogenes,* which are usually, though not always, of fecal origin. Coliforms may also be nonfecal types that thrive on organic material of vegetable origin.

Coliform bacteria may be removed by chlorination and filtration. Even when this is done, it is important to identify coliform bacteria in the raw water because their presence is an indication of sewage pollution. Sewage may carry other bacteria that are resistant to treatment in addition to viruses, organic matter, and chemical contaminants having uncertain, but potentially dangerous, effects on human health.

Standards for bacterial contamination differ among the various states and countries of the world. The U. S. Public Health Service issued a recommendation in 1946 for water subject to a low degree of contamination and requiring simple chlorination. The standards stipulated that the coliform bacterial content should average not more than 50 bacteria per 100 milliliters of raw water in any month and limited the average monthly coliform content to 1 bacterium per 100 milliliters of water. However, standards are not mandatory and are intended only as a guide for water sanitation.

There is some risk in relying on the level of coliform contamination. The standards have been criticized as arbitrary. However, water-borne diseases in the United States have dropped off dramatically since the beginning of this century. Most of the outbreaks have been attributed to accidents or to breakdowns in sanitation systems. A puzzling case was an outbreak of gastroenteritis in Riverside, California, in 1965 when several thousand people were attacked within a few days by an infection of *Salmonella typhimurium*. The city's water supply was heavily polluted with the organisms, which must have gained entry during a short period. The source of the pollution was never found, or if found, was not reported. How such a large break could occur in a municipal water system remains a mystery.

CHEMICAL QUALITY

Natural chemical content　　Water untouched by man's work is not necessarily clean by nature. The Mississippi River carries more than 2 million tons of mud and silt per day, the equivalent of 40,000 freight carloads, and sweeps the sediment into the Gulf of Mexico. The Paria River of Utah and Arizona carries 500 times as much sediment as the Mississippi. Even water near natural points of origin, such as that issuing from springs, is not necessarily chemically pure and healthful. Dr. William T. Pecora, former director of the U. S. Geological Survey, pointed out that springs feeding the Arkansas River and Red River carry 17 tons of salt per minute. The water of the puckery Lemonade Springs of New Mexico contains 900 pounds of sulfuric acid in each 120,000 gallons of water, equal to 9,000 ppm or 10 times the concentration found in most of the acid-mine streams. Many waters contain harmful quantities of fluorine. The Peace Creek in Florida contains double the amount of fluorine established as the safe limit. The Azure Yampah Spring in Colorado contains 8 times as much radium as the safe limit set by the U. S. Public Health Service. Arsenic, heavy metal ions, and high contentrations of salts of various kinds are naturally present in some waters.

Chemical contaminants　　The effect of chemical pollutants on health is not well understood. Even information on the identity and amounts of the chemicals that are present is incomplete, although the use of chemical products everywhere is on the increase. Some natural waters are known to have injurious chronic effects from prolonged use. Therefore, chemical contaminants that may enter the water supply from sources such as sewage effluents, the recycling of treated wastes, or percolation of wastes into the underground water supply should be under suspicion until proved to be harmless by experiments with laboratory animals.

The appearance of foaming agents in well waters from the use of detergents containing nonbiodegradable surfactants was a warning of the potential dangers. The presence of 0.5 ppm of ABS (alkyl benzene sulfonate) in drinking water would not be acutely toxic, but it would indicate that at least 5 percent of the water is of sewage origin. Moreover, surfactants are capable of solubilizing other compounds that might be harmful. The degradation products of surfactants and organic solvents have largely unknown biological properties.

For a long time, some scientists suspected that much of the drinking water contained organic chemicals other than those of biological origin. But it was not until 1975 when the Environmental Protection Agency reported a comprehensive survey of the water supplies of 79 cities, that the extent of chemical contamination on a national scale became known. Frequent contaminants included two suspected carcinogens, chloroform and carbon tetrachloride, and several other organic chemicals of largely unknown physiological and health effects. The study turned up some surprising contaminants, such as nicotine and vinyl chloride, another carcinogen.

Chlorination

What may turn out to be the most important finding was the strong evidence that chlorination of drinking water contributes to the formation of some of the contaminants, including the carcinogen chloroform. Purification of domestic water by chlorination is not designed to remove chemical contaminants. We do not know whether conventional treatment of sewage effluent leaves most of the toxic components essentially unchanged, degrades and decreases their toxicity, or modifies them in ways that make them more hazardous. The suspicion is growing that the nation's drinking water is not as pure as many people imagine (see Reuse of Waste Water, and Problem Areas).

Nitrates

The danger of drinking water containing high nitrate content has long been recognized. Livestock have been affected by well waters containing 75 to 150 ppm (parts per million) of nitrate nitrogen. Human infants are especially susceptible, not only because of their high gastic pH (low acidity) but also because of their high fluid intake relative to body weight. Nitrate poisoning causes a condition called *methemoglobinemia*, resulting from changes in the hemoglobin of the red blood cells that reduce their capacity to carry oxygen.

Susceptibility of infants to nitrate poisoning is highest during the first few months of life. Although adults are not affected by drinking water that is toxic to infants, breast-fed infants of mothers drinking the water may be poisoned because of high nitrate content of the mother's milk. Cow's milk may also contain sufficient nitrate to cause poisoning of infants. Nitrate poisoning of infants has been reported from drinking water containing nitrate in the range of 15 to 250 ppm of nitrate nitrogen (N), equal to 67 to 1,100 ppm nitrate ion (NO_3^-).

The primary source of excessive nitrates in well water is the leaching of nitrate salts into the underground water supply from agricultural fertilization and the seepage of sewage. A survey of eighteen states during 1951 found 278 cases of nitrate poisoning, and 39 of the children died. In all cases of death, the water contained nitrate in excess of 45 ppm.

It is still not possible to establish an absolutely safe limit for nitrate in drinking water. The U.S. Public Health Service has set a guideline of 10 ppm nitrate nitrogen (45 ppm nitrate ion) on the strength of the fact that no cases of infant poisoning have been reported in the United States from drinking water below that level.

Nitrate poisoning from food additives is further discussed in Chapter 10.

HUMAN WASTE

Part of the living process is to get rid of unwanted material. Otherwise organisms die of the toxins in their own waste. The biological process for ridding the body of wastes is called *excretion* and takes place in some manner in all living creatures. Man and other vertebrates have several mechanisms for excretion. In man, the skin eliminates unwanted water, salts, and carbon dioxide; the lungs expel carbon dioxide and water; the liver, an important detoxifying organ, drains bile constituents into the alimentary canal; the kidneys purify the blood and expel the wastes through the urinary system; the intestinal tract gets rid of salts, minerals, fats, and indigestable material rejected as unsuitable for food or metabolic processes.

The excreta of the alimentary canal are called *feces*. They consist primarily of intestinal bacteria, which comprise the most bulky portion of human waste. The urine, of which about 1½ liters are excreted daily by a human adult,[11] is mostly water with dissolved nitrogenous wastes and salts. The urine also carries away foreign substances such as drugs and other toxins.

Disposal of the wastes of life

When no more than 100,000 people populated the earth during the Ice Age, human excretion was no particular problem. It was accomplished by walking to the woods or the rocks and leaving the waste for disposal by coprophagous (dung-eating) organisms. Nature has several answers to the population problem. One effect of overcongestion is that the waste products eventually become so concentrated that many of the organisms die out from toxins, disease, or abnormal behavior. Sometimes they become cannibalistic or otherwise self-destructive. Some of them may survive by escaping to less congested areas. Civilization, with its concentration of humans living in cities, aggravates the problem of waste disposal beyond that which can be solved by natural means.

Two of the most important discoveries in the history of sanitation were the invention of toilet paper and the water closet. The latter evolved in a primitive form while papyrus was solely an instrument for communicating knowledge.

The people of the Harappan culture of the Indus Valley, circa 2500 B.C., were perhaps the first to become experts in sanitation. Household water supplies, bathing facilities, and drainage systems were widespread. There were bathrooms with waterproof floors. Latrines with seats were built into some houses. There were sloping or stepped channels through the wall to either a pottery receptable or brick drain outside. At Mohenjo-Daro each house had a drain which ran into a central sewer system under the street which fed into cesspools. Manhole covers of brick were installed to allow for repair and cleaning. The citizens of the ancient Indus would no doubt look with disfavor on the chaotic sanitation in some parts of the modern world where sewage is sluiced along gutters and open sewers to spread vermin, bacteria, and pestilence.

The ancient cities of Sumer and Babylon seem to have had similar hygienic devices. In the city of Akhetaton in Egypt, built during the

---

[11]1 liter = 1.06 quart.

Eighteenth Dynasty (1580–1340 B.C.), upper-class houses were equipped with bathrooms, drains, braziers, and stands for holding jars of drinking and washing water. Sanitation facilities in Crete (about 2000 B.C.) were highly developed. Large pits were constructed and lined with stone into which sewage was drained. The palaces were provided with running water by means of tile pipes having finely fitted joints. The early Greeks also made advances in hygiene, but it remained for the engineering genius of the Romans to produce the greatest achievements in public waste disposal in the ancient world.

Excavations of Pompeii reveal that there was a central distributing system that delivered water to each house. Nearly every house in Rome had a cistern, water faucets, and pipes of lead or terra cotta. The *cloacae* of ancient Rome were sewers laid out to drain the marshy grounds between the hills of the city. The most important one, known as the *Cloaca Maxima,* drained the Forum and dated from about the sixth century B.C. The *Cloaca Maxima* would today be called a "combined sewer" inasmuch as it received both sewage and surface drainage water. It discharged into the Tiber. It was 10½ feet wide, 14 feet high, arched in stone, and paved with lava in polygonal blocks. It was so well constructed that it still serves the city. Cities in the provinces also had drainage systems with flush latrines, sewage storage tanks, and sanitary treatment works. Public latrines as they appeared 1,500 years ago may still be seen in the provincial cities and at Timgad in northern Africa. It cannot be assumed, however, that every residence in such well-engineered cities was connected to a sewer.

The use of waste for productive purposes is an ancient solution to the disposal problem. Animal dung has been used as fertilizer for thousands of years. Homer, who portrayed Greek society of about 800 B.C., related how the good King Laertes laid manure on the land with his own hands. Two ancient writers on agriculture, Cato (184 B.C.) and Varro (116–27 B.C.), advocated manuring of crops. In the Orient human dung has been used as fertilizer for centuries and is today a convenient answer to the dual problems of disposal and food production. The practice is not aesthetically or economically satisfactory in an industrial society, although dried sludge is produced by some modern sewage treatment plants as a by-product and sold as horticultural fertilizer.

It was not until 1596 that the water closet, first used by the people of the Indus Valley about 4,000 years earlier, was reinvented in its modern form by the Englishman Sir John Harington, godson of Queen Elizabeth I. It was one of the few advances in hygiene made before the nineteenth century. The rediscovery of the water closet was an important development, for at that time even the finest palaces had nothing more than open privies due to the lack of sewage facilities. Even so, it was slow to be adopted. Harington described his invention in one of the first examples of satire in the English language in a work entitled *A New Discourse on a State Subject, Called the Metamorphosis of Ajax.*[12] The Queen liked the new invention and is said to have had it installed at Richmond Palace.

----

[12]"Jacks" was the slang term for privy. Howard W. Haggard, *Devils, Drugs and Doctors* (New York: Harper & Brothers; Pocket Books, Inc., 1929).

As civilization progressed, the privy and such bathroom facilities as ingenuity could manage began to make living conditions tolerable. But the congestion of people in industrialized areas changed all this. Descriptions of the cities of Great Britain depict a grim picture of conditions during the middle of the nineteenth century.

Typical of the situation in all big cities was the condition of the tenements of Glasgow where there were no drains or even any privies. Dung was left lying in the courtyards next to houses. In another Scottish town there were no private lavatories and only two or three public privies located in the better sections of the city. Sanitation was no better in other European cities and in some it was worse. In those days it was the practice of the citizens to use chamber pots for excreta, and when the pots were filled, to empty the contents out the window and into the street below with a shout, "Gardez l'eau!" for the benefit of the nimble-footed. In London, the sewage of a population of nearly 3 million people was collected in a single, huge cesspit in the center of the city. During the hot summer months the stench was nearly beyond human endurance.

The Great Stench of London that occurred during the summer of 1858 almost brought the deliberations of Parliament to a halt. Steps were taken to overcome the noxious fumes by hanging blankets saturated with chloride of lime over the windows. The courts declared a recess and travelers avoided the city. Surprisingly, there was little increase in disease and people came to realize that diseases were not transmitted through odors and that filth, in itself, did not cause disease. This knowledge, however, was dangerous for it caused the citizenry to ignore the formidable potentiality for disease and epidemics in the slovenly practices.

The terrible epidemic of Asiatic cholera that struck London in 1854

USDA photo.

One of several homes in a community of the eastern United States with no water supply for sanitation or domestic use.

could have been prevented if it had been known that the disease is caused by a germ carried from feces to underground drinking water. John Snow, a London doctor, showed in a classic study that 500 victims who died within ten days had all used the same community water pump. (It is interesting to note that Snow made his study 28 years before Robert Koch, the German scientist-doctor, discovered cholera germs in water contaminated with sewage.) The subject of sanitation became the most popular topic of news and conversation. Corrective measures were taken. As a consequence, the development of drains and sewer systems was probably the most important advance of the Industrial Revolution in the latter half of the nineteenth century.

Development of modern sewer systems

Many of the early systems were of dubious value. Aesthetic reasons, perhaps more than hygienic reasons, stimulated the construction of sanitary facilities. Cesspools were dug everywhere and many of them were prolific breeders of contamination because of seepage into nearby wells. As a result, typhoid fever and cholera epidemics were rampant. In time, the methods were improved. London installed a sewer system in 1865 and many English cities followed the example. Similar advances were made by American cities. By 1900 more than 3,500 United States patents had been granted for improvements on Harington's "Ajax" and nearly 900 patents had been issued for sewage devices.

According to an inventory for 1962, municipal sewage systems in the United States served 125 million people. This still left about 75 million people without sewage facilities. Of those now provided with sewers, 36 million people are served by sewers that are designed to handle both human waste and surface drainage. These are known, technically, as *combined sewers* (see Figure 7.6). They often present formidable problems when storm drainage necessitates the run-off of raw sewage. According to a 1967 survey published by the Federal Water Pollution Control Administration of the U.S. Department of Interior, 31 percent of the cities with combined sewage systems had no treatment plant.

In communities that are served by sewage systems, 60 percent of the waste receives primary and secondary treatments (removal of solids and bacterial breakdown). However, 30 percent of the waste from sewered communities receives only primary treatment (removal of heavier solid material) and 10 percent of the waste is disposed of as raw, untreated sewage. Of the combined sewers (those carrying both storm and waste water) in the United States, the Department of the Interior reported that 22 percent of the cities provided no treatment, 42 percent only primary treatment, while 35 percent provided both primary and secondary treatment. One percent of the cities had some type of third-stage treatment.

The composition of sewage is complex, and it differs depending upon the sources, the type of treatment or lack of it, and whether there is an admixture of storm drainage with industrial waste. Unfortunately, analyses and inventory of waste constituents, a necessary step for rational planning of water pollution control, have been thoroughly done in only a few cases. Inorganic constituents, such as the salts of sodium, potassium, calcium, magnesium, and ammonium, together with chloride, nitrate, bicarbonate, sulfate, and phosphate ions are commonly found in abundance. Organic chemical compounds in waste have been poorly investigated, with the exception of some extensive work on pesticides and detergents and, to a more limited extent, on organic acids and phenols. A large

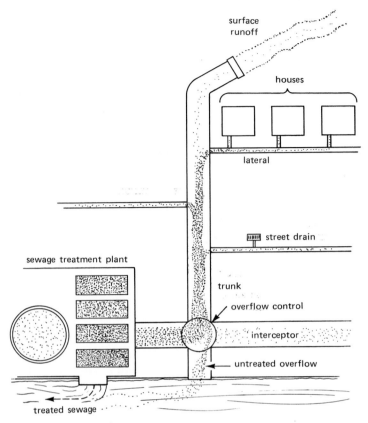

Redrawn from *A Primer on Waste Water Treatment*, Federal Water Pollution Control Administration CWA-12, 1969.

FIGURE 7.6 Combined sewer system.

number of organic and inorganic substances have been identified in domestic sewage. Many of them are probably breakdown products of excreta. More information on the constituents derived from human waste as well as industrial waste and surface drainage into the sewer sinks is urgently needed.

SEWAGE TREATMENT

The methods used in the United States to treat sewage before it is discharged into rivers, lakes, or the ocean are not excessively expensive. The procedures are considered to be effective for the purpose of rendering the waste moderately safe insofar as its bacterial content is concerned. But the virus load and toxic chemical content of sewage discharges have not been as thoroughly evaluated.

The processes of treating municipal sewage are broadly classified as *primary, secondary,* and *tertiary*. The particular process used in a given situation depends on the volume to be treated, the location of the outfall, the dilution factor, the potential hazard to users receiving the water, and in many cases, the cost of the project.

USDA photo.

An example of an early, but mistaken, idea that the solution to pollution is dilution.

Primary treatment   About 40 percent of the sewage in the United States is disposed of as raw sewage or effluent from primary treatment. Primary treatment consists of removing floating and suspended solids by mechanical means (see Figure 7.7). More than one-half of the suspended solids can be removed by primary treatment. First the large solids are screened out and grease and scum are removed. This is followed by sedimentation in a basin called a *primary clarifier,* to remove the remaining solids, called *primary sludge*. The screens, called trash racks, consist of steel bars about 2 to 4 inches apart. In some cases sand and other coarse material is removed by *grit chambers* to further protect pumps and other equipment from damage. Sometimes the waste water is then run through fine screens. Usually, however, after screening and removal of grit, the waste water is run directly into settling tanks. The settling tanks may have skimming devices, or the removal of scum may be done separately.

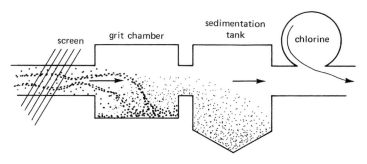

From *A Primer on Waste Water Treatment,* Federal Water Pollution Control Administration CWA-12, 1969.

FIGURE 7.7 Primary sewage treatment.

The primary sludge is a burdensome problem because it is bulky and must be removed. It contains 94 to 99 percent water. Usually the first step is to remove as much of the water as possible. In some cases the sludge is dried in beds with some of the water being removed by filtration. The residue is disposed of on land. Because the sludge itself can comprise a pollution problem, a better method is to bring about microbial decomposition in *sludge digestion tanks* before drying. Sometimes sludge sedimentation and digestion take place in a partitioned structure called an *Imhoff tank*. The microbial action occurs under anaerobic conditions, and because this proceeds slowly at low temperatures, the digestion tanks are usually heated to 80 to 90°F, at which temperatures the sludge decomposes in 20 to 30 days. The digested sludge will have been reduced to about one-third of its original volume and will be relatively inoffensive.

The other products of primary treatment are gases and the fluid or clarified waste water. The gas is mostly methane, which is usually burned as fuel to provide heat for the digesters and other equipment. The clarified waste water has highly objectionable properties and in most cases is put through a secondary treatment.

Secondary treatment   Secondary treatment of waste involves the biological degradation of organic materials by microorganisms under controlled conditions (see Figure 7.8). The usual method is to bring about the biological oxidation of the organic material under aerobic conditions, in which the waste is aerated to supply oxygen for the microorganisms. The degraded material settles out in secondary settling tanks and is therefore described as being removed by sedimentation. The sediment containing the microbial growths and their by-products is called *secondary sludge* or *activated sludge*. The clarified waste water is discharged in the outfalls to rivers, lakes, bays, lagoons, or oceans. Some of the sludge is returned to aeration tanks where it is recycled with the incoming waste. Most of the sludge, however, must be removed, and this operation accounts for one-fourth to one-half of the entire cost of the operation.

Sludge   Both primary and secondary sludges are usually transferred to sludge digestion tanks where decomposition takes place under anaerobic conditions. Various methods for water removal are used. Because the removal of the sludge is a major part of sewage disposal, a great deal of effort has gone into drying and finding uses for the product, or locating disposal sites and transporting the sludge there.

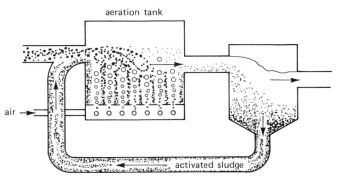

Redrawn from *A Primer on Waste Water Treatment*, Federal Water Pollution Control Administration CWA-12, 1969.

FIGURE 7.8 Secondary sewage treatment.

Photos courtesy of Environmental Protection Agency.

Left photo: Sewage treatment at a plant in Des Moines, Iowa. This trickling filter contains two rotating (four-armed) distributors. Waste water then goes through a finer trickling filter, three clarifiers, grease removal tanks, and sludge digesters before being returned to the river. *Right photo:* An experimental advanced waste treatment plant. The purpose is to develop an economically feasible method to restore the effluent to a quality equal to that in the original domestic water supply. Here, powdered activated carbon is being added to mixing and settling tanks for the removal of organic constituents. Chemical coagulation and sedimentation are then needed to remove the powdered carbon from the water.

A few of the treatment plants heat-dry the activated sludge, usually after some form of mechanical water removal, called *dewatering,* and sell the product as fertilizer. The process is costly and there has been only a limited demand for the product. It is less expensive to incinerate the sludge in furnaces or to use it as landfill. Incineration sterilizes the sludge and reduces its volume. However, incineration has some disadvantages. It creates an air pollution problem and leaves an ash that must be disposed of. Still, disposing of a small amount of ash is easier than getting rid of a large amount of sludge.

Final disposal

The methods used for disposal of sludges are usually those that are the least costly. Whether disposal on land or sea is selected depends on the proximity of the treatment plant to suitable disposal locations. Dewatered sludges are commonly used as landfill. Trucks, trains, barges, and pipelines are used for transporting the sludge. Liquid sludges are disposed of either on land or in bodies of water. Liquid sludge is used to fertilize or condition agricultural land. However, problems of odor, water pollution, and stimulation of insect and algal growth, as well as other aspects of public health and aesthetic values, must be considered.

Tertiary treatment

Any treatment to further purify waste water beyond the methods commonly employed for secondary treatment may be called "advanced treatment." The term *tertiary treatment* is often used to refer to certain advanced treatment procedures for treating the effluent coming from secondary treatment. Although many of the procedures are in the research or development stage, some of them are in use on a fairly large pilot-plant scale and a few are used on an operational level. The purpose is the removal of the contaminants of waste water that remain after secondary treatment. These contaminants consist mainly of suspended solids, dissolved organic compounds, and the inorganic plant nutrients nitrogen and phosphorus. A thorough job of cleaning up sewage waste water is difficult and costly. However, there is a steadily growing need for advanced procedures that will provide a product capable of being re-used for various purposes, thereby relieving the dual problems of waste disposal and the ever-diminishing supply of fresh water.

A waste water treatment plan that has been operating since 1962 at Whittier Narrows, near Los Angeles, produces an effluent that is used to replenish the ground water supply. A total of 15 million gallons of chlorinated effluent is run onto spreading beds where it percolates through the soil and into the ground water. Percolation through the soil is a sort of tertiary treatment. However, the possibility that dissolved salts, organic chemicals, and heavy metals may find their way into the water supply is cause for concern. Another advanced treatment plant is in operation at Lake Tahoe where 7.5 million gallons of waste water from an activated sludge plant is treated each day. An advanced treatment plant on a larger scale is at the Salt Creek Water Reclamation Plant of the Metropolitan Sanitary District of Greater Chicago, intended to handle 30 million gallons per day.

REUSE OF WASTE WATER

Many of the molecules of water that travel down the rivers in densely populated areas pass through the human physiological system several times—possibly 10 to 40 different bodies[13]—before reaching the ocean. Even if the drinking water is treated to kill most of the bacteria, we are repelled by the thought of drinking in the afternoon the molecules that were flushed into the sewer by our upstream neighbors in the morning. Yet that is literally what many people who live along the waterways are doing. The rivers are used as repositories for vast quantities of domestic and industrial waste products having an unknown fate and largely unknown biological effects.

Some use is made of waste water from sewage plants for irrigation, mainly for nonfood crops and golf courses, but in some countries, untreated or poorly treated waste is used for vegetable farming. Along part of the southern California coast, pumping of well water for the burgeoning population has lowered the natural water table to the extent that underground intrusion of sea water is a serious threat to the water supply. When the underground water is removed by pumping, the pressure is reduced and seawater fills the void by seeping into the area through underground strata of sand. Treated sewage waste water was pumped underground experimentally to replenish the natural water supply and create an artificial barrier against seawater intrustion.

At Santee, California, treated waste water is fed into several lakes that are used for recreational purposes. Treatment involves running the secondary waste water into an oxidation pond and then into percolation beds from which the lakes are fed by underground flow. There have been problems of eutrophication. Excessive aquatic growth results in oxygen deficiencies and subsequent fish kills. Even so, the project is considered generally successful. Similar plans for using treated waste for recreational

---

[13]G. Borgstrom, *The Hungry Planet.*

USDA photo.

Winter application of sewage effluent for irrigation and fertilization creates ice
statuary.

purposes are in progress. Though no public health problems have been
detected, the need remains for careful study of chemical contaminants
and their potential hazard to vacationers, sportsmen, and wildlife.

It cannot be assumed that tertiary treatment would render the water
safe for drinking even if the effluent were free of pathogenic bacteria.
Waste water may contain viruses as well as high concentrations of nitrates, phosphates, and other mineral salts. Organic degradation products
of detergents and other domestic and industrial synthetic chemicals of
largely unknown composition and biological activity may be present. The
only place where waste water from sewage treatment is recycled directly
into the drinking water is at Windhoek, South Africa, where treated sewage effluent normally makes up 14 percent of the water supply, increasing
to 40 percent during the winter. South Africa is desperately short of water
sources; most cities are unwilling to stand the expense of extensive tertiary treatment or to run the risks of the undetermined hazards that remain. Methods of advanced treatment that will permit large-scale reuse of
waste water for domestic use have not yet been developed.

Viruses   We can effectively cope with nearly all the water-borne bacterial diseases by conventional methods of treating supplies of domestic water.
Sedimentation, filtration, chlorination, or combinations of treatments are
generally reliable in providing drinking water that is potable and bac-

teriologically safe. Pathogenic viruses, however, present a more difficult problem.

The only one of the enterovirus diseases that has been proved to be water-borne is infectious hepatitis, a dangerous liver disease that is incapacitating and sometimes fatal.[14] The disease is prevalent in many parts of the world, although there have been no large-scale epidemics in the United States. The spread of infectious hepatitis, as well as other viral diseases, by the reuse of treated waste water is an eventuality that must be thoroughly explored before widespread reuse of processed sewage waste can be contemplated with confidence. Evidence indicates that the chlorination of secondary effluent in accordance with current practices does not remove or inactivate all of the viruses. It is possible to filter out some viruses by adsorption on activated carbon, but the viruses are readily released and remain infectious. Removal by adsorption on precipitates is also under study. Irradiation by gamma rays from radioactive isotopes can kill viruses as well as bacteria, but the effectiveness of disinfectants such as high-level treatment with chlorine or other disinfectants requires further research. Many of the reaction products of such chemicals acting on the organic materials also remain to be determined (see Chemical Quality).

Research

The renovation of waste water to a quality that would permit its reuse for a variety of purposes is a major objective of current research projects. Some industrial users can tolerate more impurities than agricultural, recreational, or domestic users. Needed improvements include removal or inactivation of viruses, removal of salts or their dilution to lower the salt content to acceptable levels, continuous monitoring for a wide variety of chemical pollutants, and probably the development of ways to remove toxic substances.

Supply and demand

Water usage in the United States is about 25 percent of the fresh water supply that is now economically available from all sources. It is estimated that we could provide 600 billion gallons per day out of a total run-off of 1,200 billion gallons. But there would still be serious shortages in some areas, especially in the drier sections of the west. Among more than 200 million people in the United States, there are at least 90 million who are in short supply of water. As the demand for water increases and the price goes up, increasingly costly methods for the recovery of waste water become more feasible.

The problem is twofold: water in good supply and water of good quality. Future additional drinking water supplies may be obtained by several processes:

- Desalting of brackish and ocean waters, involving, however, high costs and transport problems.
- Complete recycling–requiring, however, careful and extensive study of fail-safe systems for pollution control and removal of contaminants.
- Dual supplies, involving one source for drinking, cooking, and bathing and another source for other domestic and horticultural uses. The latter would probably consume at least 80 percent of the domestic water supply.

---

[14]Enterovirus is a virus infecting the gastrointestinal tract.

- Greater economy in the use of water. For example, backyards and gardens are getting smaller in waterpoor areas of high population density. And with appropriate water-closet designs, it would not be necessary to use 3 gallons of water to flush a pint of urine into the sewer system.

The total United States demand for water in the year 2000 is estimated at 805 billion gallons per day, whereas only 600 billion gallons per day will be economically available from "usable surface water." However, as much as 1,200 billion gallons a day could be made available if the cost were in line with the demand. Sooner or later, much of the sewage water will have to be recycled in order to meet the demand.

PROBLEM AREAS

Civilizations thrive where water is abundant, for clean water is one of the most basic requirements for human existence. But water has typically been used for almost any convenient purpose besides drinking and irrigation. Waterways, in particular, are favorite receptacles for industrial and human wastes. It has long been evident in some areas of high population density that the effluents of human activity were having such seriously disruptive effects on water quality that it would eventually reach intolerable proportions.

An example of the magnitude of the problem is to be seen in the Great Lakes region where deterioration in water quality was recognized by fishermen and biologists as early as 1925 when the cisco fishery of Lake Erie collapsed. This decline of an important commerical fish was only the first of a series of crises in the long history of deterioration in the water quality of the Great Lakes. It had seemed to the people along the shores of the Great Lakes that they were to enjoy an almost limitless supply of fresh water for their coastal cities and industries, so they used the lakes freely as a convenient disposal system for sewage and industrial wastes. By 1938, the city of Green Bay had to close its bathing beaches for health reasons. They have never been reopened.

Neighbors often suffer more than the polluters. Because Chicago takes its drinking water from Lake Michigan, it wisely (from the Chicagoan's viewpoint) does not dump its sewage back into the lake but, instead, sluices the wastes into the Mississippi River through an artificial canal that is flushed with water from Lake Michigan. Many cities downsteam take their drinking water from the Mississippi River, and in turn put it back into the river along with domestic and industrial wastes. It is sometimes hard to determine who is polluting whom. The state of Illinois sued the city of Milwaukee for polluting Lake Michigan, but ironically both Milwaukee and Chicago take their drinking water out of the same lake.

Besides the cost in time and money of fighting one another in the courts, the cost of getting clean water is rising rapidly in the face of increasing populations, congestion, industrial activity, greatly increased recreational usage, and a dramatic overall increase in the demands for water. In 1975 it was estimated that industry would have to invest 8 billion dollars to meet 1977 requirements for water pollution control at existing plants alone, and that controlling thermal pollution discharges would require an additional 9.5 billion dollars. All told, about 50 billion dollars' worth of new sewers and treatment plants are needed. Of course, the costs to industries and cities will be passed on to the people in higher prices and taxes, and they will pay the price or suffer the consequence— water polluted beyond the point of its usefulness.

A recent solution to the over-pollution problem in some local areas is to stop construction of new residential projects or declare a moratorium on hooking up to sewer lines. But artificial constraints on where a person chooses to live have never fit comfortably with the philosophy of democratic societies and are sometimes regarded suspiciously as attempts to keep the "ins" in and the "outs" out. Therefore, such measures, however expedient, are of temporary value only, to buy time for construction of treatment and disposal facilities.

The finding that chlorination of drinking water to remove bacterial contamination is probably, at the same time, contributing to the formation of toxic organic chemicals, some of which may be carcinogenic or worse, poses problems that require vigilance far beyond any practiced in the past. When the discovery was announced, EPA administrator Russell E. Train said that while people should be aware of the problem, they should not react with any sense of panic because chlorination remains the most effective means of preventing several serious diseases including typhoid, cholera, and dysentery. However, alternatives to chlorination are needed. Those that should be pursued vigorously include such disinfectants as ozone, carbon dioxide, and ultraviolet radiation, and filtration with activated carbon in place of the sand filtration systems now in almost universal use.

READINGS

Water Pollutants

Anon., "Drinking Water; Is It Drinkable?" *Env. Sci. & Tech.* 4(10): 811–813 (1970).

BALDRY, P. E., *The Battle Against Bacteria*. Cambridge, Mass., Cambridge Univ. Press, 1965.

Clean Water for the 1970's. A Status Report, 1970. Federal Water Quality Administration, US Dept. Interior, Washington, D.C.

"The Continuing Tale of the Torrey Canyon," *Env. Sci. & Tech* 1(5): 391–393 (1967).

FERGUSAN, F. A., "A Nonmyopic Approach to the Problem of Excess Algal Growths," *Env. Sci. & Tech* 2(3): 188–193 (1968).

McKEE, J. E. and H. W. WOLF, *Water Quality Criteria*. Publication 3-A, 1963. State Water Quality Control Board, Sacramento, Calif.

Oil Pollution. A Report to the President. Secr. of the Interior and Secr. of Transportation, February, 1968, pp. 1–131. USGPO, Washington, D.C.

SIGERIST, HENRY E., *Civilization and Disease*. Chicago, Univ. of Chicago Press, 1943.

SPERRY, K, "The Battle of Lake Erie: Eutrophication and Political Fragmentation," *Science* 158: 351–355 (1967). (Reprinted in K. E. Maxwell, *Chemicals and Life*. Belmont, Calif., Dickenson Publishing Co., 1970).

Water and Health

A Primer on Waste Water Treatment, 1969, pp. 1–24. Federal Water Pollution Control Administration. U.S. Dept. Interior. Government Printing Office, Washington, D.C.

Cleaning Our Environment. The Chemical Basis for Action, 1969, pp. 1–249. American Chemical Society, Washington, D.C.

Clean Water for the 1970's. A Status Report, 1970, pp. 1–80. Federal Water Pollution Control Administration, U.S. Dept. Interior, Washington, D.C.

Problems of Combined Sewer Facilities and Overflows, 1967. U.S. Dept. Interior, Federal Water Pollution Control Administration WP–20–11.

Public Health Service Drinking Water Standards 1962, Publ. Health Serv. Publ. 956, pp. 1–61. U.S. Dept. Health, Education and Welfare. Government Printing Office, Washington, D.C.

Waste Management and Control, Publ. 1400, 1966, 257 pp. National Academy of Sciences-National Research Council, Washington, D.C.

Water Quality Criteria. Report of the National Technical Advisory Committee to the Secretary of the Interior, April, 1968. Federal Water Pollution Control Administration. U.S. Dept. Interior, Washington, D.C.

EARTH, SOIL, AND MINERALS

SOLID WASTES

Few features of the environment are more abrasive to the senses than the discarded objects of civilized life. Litter along the roadsides, trash on the beaches, heaps of refuse in the streets, the pungent piles of rotting garbage, and similar unwelcome debris are unpleasant reminders of untidy human habits. Thousands of alleys, backyards, and vacant lots contain the remains of discarded refrigerators, defunct washing machines, and parts of shattered automobiles—rusting remnants of the chrome-plated glory of a technological society. Open dumps scar the landscape; and highway travelers view a countryside spotted with the incongruity of beauty and blight.

Much of the discarded waste is swept under the rug, so to speak, in accordance with the anaesthetizing but naive philosophy, "out of sight, out of mind." But less visible parts of the iceberg-like problem remain. Solid wastes are dumped in the ocean; mining wastes are produced at the rate of millions of tons a day; slag heaps and mill tailings accumulate near processing operations; and industrial refuse contaminates streams and lakes. The growing mass of solid waste produced annually in the United States includes 30 million tons of paper and paper products, 4 million tons of plastics, 30 billion bottles, 60 billion cans, 100 million tires, 8 million junked and abandoned automobiles, and millions of major appliances of innumerable makes, sizes, and kinds. In addition, each year there are millions of tons of miscellaneous debris from the demolition of such items as grass, tree shrubbery and trimmings, food wastes, and sewage sludge.

No one knows the total amount of the world's waste. It is fairly closely estimated that in the United States solid waste is produced at the colossal rate of more than 20 billion pounds a day, or 4.34 billion tons annually. This includes 250 million tons of municipal and institutional waste, 110 million tons of industrial waste, 2,280 million tons of agricultural waste, and 1,700 million tons of mineral waste (see Figure 8.1).

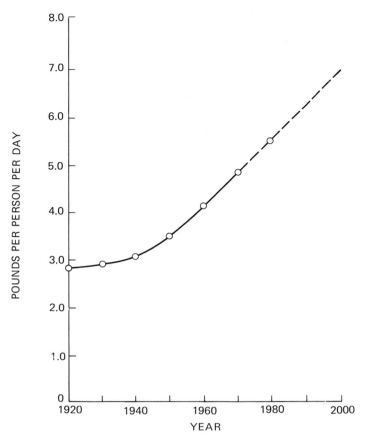

From W. C. Gough, *Why Fusion?*, U.S. Atomic Energy Commission, 1970.

FIGURE 8.1 Refuse production per person in the United States.

GARBAGE AND URBAN WASTE

The most commonly used methods for final disposal of solid wastes are sanitary landfill and open dumping. They are cheaper than other disposal methods and have the advantage of avoiding the acute pollution problems associated with discharging wastes into waterways or polluting the air from incineration.

Landfills, if properly planned, can be used later for construction sites or recreational facilities such as parks and golf courses. But there are major disadvantages to open dumps, which are a public health problem, breeding flies, rats, and other vermin. And burning combustible wastes produces mixtures of particulate and gaseous matter that are noxious. The blight of open dumps is aesthetically intolerable and depresses land values.

Land disposal of any kind requires large tracts of land. The increasing shortage of land and the ever-growing mountain of waste make the cost prohibitive to many communities. Leaching by surface water or ground water may contaminate streams and other water supplies. Air pollution

Photo courtesy of HEW, Bureau Solid Waste Management.

Sanitary landfill.

may result from the decomposition of wastes, especially in open dumps but also in landfills if the wastes are improperly covered.

The million-odd tons of urban waste produced every day in the United States must be disposed of at an estimated average cost of 1 cent per pound or about 7 billion dollars a year. Since fewer than one-half of the cities have adequate refuse disposal systems, the cost of satisfactory disposal would at least double. More importantly, many of the disposal sites are aesthetically undesirable. They are in locations where they create nuisances with respect to other human activities. Space and appropriate locations for disposal of the waste produced on the extravagant scale enjoyed by the technological society are becoming critical.

The adverse economics of the disposal of solid wastes as commonly practiced are far-reaching. Mineral resources such as copper, zinc, lead, and tin are limited and the ores are nonrenewable. These metals are not destroyed by disposal but are often scattered beyond recovery. Even iron ore is in finite supply in economically desirable deposits. The squandering of these resources that occurs when wastes are dumped indiscriminately in spoil areas will inevitably produce a disruptive economic impact at some time in the future unless improvements are adopted. Development of salvage techniques will, therefore, have direct economic effects. Salvage and recycling will also alleviate other problems associated with the rapidly increasing density in populations, which results not only in the generation of greater quantities of wastes but also in wastes of greater complexity. (See Tables 8.1 and 8.2.)

Not all of the components of solid waste are valuable. Some are useless and others are toxic. For example, selenium, a poisonous element that occurs naturally in the soil, is taken up by plants that are used for making paper and is present in newspaper in the amount of about 8.6 ppm. This element is found in incinerator stack gas, raw solid wastes, incinerator residue, and in water that has been used for removing fly ash. Many other toxic chemicals of various kinds and amounts are present in municipal and industrial waste.

TABLE 8.1   Percent of Solid Waste Materials Recycled in the United States

Materials used	Percent recycled
Paper	19
Iron and steel	31
Aluminum	18
Copper and lead	50
Glass and textiles	4
Rubber	26

From *Solid Waste Management. Recycling and the Consumer*, pp. 1—12. 1974. U.S. Environmental Protection Agency.

ALTERNATE DISPOSAL METHODS

Ocean dumping   Ocean dumping of wastes, from garbage to radioactive wastes and discarded chemical warfare agents, has been, in the past, a favorite disposal method under the prevalent notion that the ocean in an inexhaustible sink. We are now learning that the ocean does not have an infinite capacity for covering, holding, absorbing, or decomposing materials, nor the capacity for anywhere near the load of pollution that would result if the rate of ocean disposal were continued as in the past.

A study made in 1968 identified 126 ocean disposal sites used by twenty coastal cities. There were seven major categories of waste: harbor dredgings, industrial wastes, municipal sewage sludge, refuse and garbage, construction and demolition debris, military explosives and chemicals, and miscellaneous wastes—a total of 38 million tons per year in coastal and estuarine areas. Other sources of waste include agricultural land run-off, pesticides, fertilizers, sewage and garbage from vessels, and oil spills. Often overlooked is the fact that a large part of the air pollutants eventually settle out and end up in the ocean. A major concern is that ocean dumping has increased fourfold during a period of twenty years and is expected to greatly increase relative to other means of disposal in the future. Based on a 1970 study, the Council on Environmental Quality recommended a series of drastic actions to control ocean dumping.

Little is known about either the immediate results or the cumulative effects of ocean disposal. By the time adverse effects become evident, it may be too late to discontinue the disposal practices in time to prevent long-term pollution problems. We have a preview of what could happen in the case of DDT, plasticizers such as the polychlorinated biphenyls (PCBs), and more long-lasting and potentially dangerous substances such as lead, mercury, cadmium, and selenium. The amount of wastes transported and dumped into the ocean is now relatively small compared to the total volume of pollutants that reach the ocean. Industrial and municipal liquid wastes are the largest source of pollutants.

The oceans cover 70 percent of the earth's surface. The 140 million square miles of water surface are crucial to the equilibrium of the world's atmospheric environment—the oxygen and carbon dioxide balance and the global climate. The oceans are a major part of the earth's hydrologic system. They are an important part of the biosphere, supporting a rich biota of marine plant and animal life, and are economically valuable to man for the production of food, minerals, and other necessities.

Incineration

Incineration reduces the volume of waste by 60 to 80 percent and reduces the public health problems associated with the accumulation of refuse. This method is adaptable to a wide range of capacities from small domestic incinerators to large centralized municipal plants. It can handle a mixture of garbage and rubbish, and the physical nature of the residue—called *clinker*—aids in the disposal of the waste. Much of the heat that is generated can be recovered with modern equipment. However, other forms of pollution are created. Air pollution from the products of fuel combustion must be taken into account, but the principal air pollutant is fly ash. Several methods are used to reduce the emission of fly ash. One such method entails the use of water with devices called *scrubbers*; but the large amount of water needed introduces a water pollution problem. The increasing quantities of plastic products and plastic packaging materials aggravate the air pollution problem due to the formation of gaseous decomposition products that will require research on methods for special treatment of the stack gas. The ultimate requirements are difficult to predict.

Chemical processing

Chemical processing of solid wastes is appealing because of the prospect that energy and usable materials can be recovered (see Table 8.2). However, chemical processing is usually costly, even where technically feasible. In addition to the economics of processing and marketing, there are often problems of uniformity, impurities, and the quantity and com-

Left photo courtesy of Environmental Protection Agency. Right photo courtesy of Monsanto Commercial Products Company.

Left photo: Burning dump is a source of air pollution. *Right photo:* A kiln for a solid-waste disposal and resource recovery system in Baltimore, designed to process 1,000 tons of garbage per day, while producing energy and recovering by-products.

TABLE 8.2   Average Composition of Municipal
Refuse (% by Weight)

Rubbish (64%)	
paper, all kinds	42.0
wood and bark	2.4
grass	4.0
brush	1.5
cuttings, green	1.5
leaves, dry	5.0
leather goods	0.3
rubber	0.6
plastics	0.7
oils, paint	0.8
linoleum	0.1
rags	0.6
street refuse	3.0
dirt, household	1.0
unclassified	0.5
Food Wastes (12%)	
garbage	10.0
fats	2.0
Noncombustibles (24%)	
metals	8.0
glass and ceramics	6.0
ashes	10.0
	100.0

From W. C. Gough, *Why Fusion?* 1970, pp. 296–331.
Washington, D.C., U.S. Atomic Energy Commission.

position of the residue. However, the salvage of some wastes by chemical methods is now economically feasible. The reclamation of nonferrous (non-iron) scrap metals has long been an established industry amounting to many millions of dollars annually, involving thousands of businesses engaged in the collection and processing of scrap metal and metal-containing products (see Table 8.3).

Composting   Composting municipal refuse to convert it into a fertilizer and soil conditioner is appealing not only because it appears to be a good way to recycle the resources in solid wastes, but also because of the beneficial nature of the compost. Several composting operations have been attempted in the United States, but they have been generally unsuccessful. The majority of those started during the 1950s and 1960s were forced to shut down or greatly curtail their operations primarily because of an inability

TABLE 8.3   Foreign Metals in a U.S. Automobile

	Total Pounds	From Foreign Countries
Iron	3,705	36%
Copper	52	38%
Lead	24	58%
Aluminum	48	89%
Zinc	123	59%
Other materials	415	—
	4,367	

From W. C. Gough, *Why Fusion?* 1970, pp. 296–331. Washington, D.C., U.S. Atomic Energy Commission.

to sell the compost at a price that would sustain the operation. One reason for this is that the fertilizer is low-grade with respect to its content of basic plant nutrients as compared to other sources of plant foods. Composting demonstrates the difficulties in taking a low-grade waste material that nobody wants and changing it into a resource so costly that few people will buy it. (See Figure 8.2.)

New Methods   New methods of waste disposal are under trial in various parts of the world. Some of the more promising methods are:

- The transformation of the organic content into sugars or proteins.
- The heating of organic refuse under anaerobic conditions (out of contact with air) to convert it into useful gases such as methane, which is usable as fuel, or into liquid products.
- The compression of refuse into building blocks that can be sheathed with more durable materials. One such plant is in operation in Japan.
- The compression of refuse into briquets that can be used as fill. A mixutre of shredded refuse, fly ash, dried sewage sludge, incinerator residue, and river and lake dredgings is under study.
- The use of degradable plastic bags instead of trash cans in the collection and transport of municipal refuse. Collection is speeded up; the bags resist water and keep out flies.
- The transport of refuse as a liquid slurry in pipelines, a method now under study.
- The transport of dry refuse in pneumatic tubes for short distances, a method now being developed in Sweden.
- The use of trains to haul solid urban refuse to land reclamation areas, a method now under study in the United States.

One of the most promising methods of waste disposal is to convert it into energy. Several methods are under intensive study. Conversion to methane has already been mentioned. Another method is to burn the garbage as a supplement to oil and coal for the production of electricity. As a start, burning garbage during 1975 was estimated to generate electricity for 45,000 homes and to conserve about 100,000 tons of coal. In St. Louis, 25,000 homes are served with electricity generated by burning 300 tons of low-sulfur household garbage per day—one third of the city's daily output. Even more promising is the near-commercialization of ob-

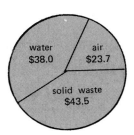

total = $105.2 billion

Redrawn from the Second Annual Report of the Council on Environmental Quality, 1971.

FIGURE 8.2   Costs of air, water, and solid waste pollution control, 1970–75. Disposal of solid waste is the most costly.

Photo by Wells, Ambassador College; courtesy Environmental Protection Agency.

Eight million automobiles are junked or abandoned each year in the United States.

taining oil and other by-products from garbage. In 1975, a pilot plant was started near San Diego to process 200 tons of refuse per day into low-sulfur fuel oil, metals, and glass. Cost of the plant, about 9 million dollars, was shared by the county, the federal government, and the industry developer of the process.

If the trend continues, we will be creating a "recycle society," in which waste and scrap will become our major resources.

MINERALS

Inorganic substances are so prevalent in the natural environment that one might think all living organisms would find them invariably harmless if not universally beneficial. But we know from experience that this is not necessarily so. The concentrations of the mineral elements differ greatly from place to place, and even amounts that occur naturally are sometimes poisonous. The margin of safety is often a narrow one, and in such cases it is difficult to determine what is safe and what is unsafe, especially when the element is one that is essential in small amounts.

Accidental poisoning from exposure to trace elements has long been recognized. Arsenic, lead, cadmium, antimony, and beryllium have been the cause of accidental deaths in industry. Illnesses have been caused by many of the other trace elements (Table 8.4). A reason for concern is that the trace elements may go undetected or their physiological effects may be attributed to other causes. In addition to mercury, now known to be a widely distributed environmental contaminant having injurious effects of uncertain magnitude, 30 elements are listed as potentially dangerous: chromium, cadmium, zinc, arsenic, nickel, lead, antimony, copper, thorium, cobalt, ruthenium, selenium, tellurium, boron, tin, strontium, cesium, barium, manganese, silver, beryllium, magnesium, thallium, yttrium, rubidium, cerium, molybdenum, osmium, vanadium, and bismuth.

The burning of fossil fuels, including gasoline in motor vehicles, appears to be the primary source of environmental pollution by the trace elements. Cadium, lead, nickel, mercury, and selenium are examples of toxic elements that are emitted to the atmosphere in quantities amounting to thousands of tons annually. Polluted water is another important source of environmental contamination (see Figure 8.3). The Texas Water Quality Board found that the daily discharge into the Houston Ship Channel included 1,600 pounds of lead, 7,900 pounds of zinc, 5,000 pounds of cadmium, and 300 pounds of chromium. The resulting concentrations were 63,000 times the natural level of lead, 15 times that for cadmium, and 108,000 times that for chromium.

The toxic elements are not degradable. Thus, in some soils and waters, and therefore in some food supplies, there is an accumulation of one or more of the toxic elements in injurious amounts. Among those found most frequently in natural sources, as well as in man-made contamination, are

TABLE 8.4   Trace Metals May Pose Health Hazards in the Environment

Element	Sources	Health effects
Nickel	diesel oil, residual oil, coal, tobacco smoke, chemicals and catalysts, steel and nonferrous alloys	lung cancer (as carbonyl)
Beryllium	coal, industry (new uses proposed in nuclear power industry, as rocket fuel)	acute and chronic system poison, cancer
Boron	coal, cleaning agents, medicinals, glass making, other industrial	nontoxic except as boran
Germanium	coal	little innate toxicity
Arsenic	coal, petroleum, detergents, pesticides, mine tailings	hazard disputed, may cause cancer
Selenium	coal, sulfur	may cause dental caries, carcinogenic in rats, essential to mammals in low doses
Yttrium	coal, petroleum	carcinogenic in mice over long-term exposure
Mercury	coal, electrical batteries, other industrial	nerve damage and death
Vanadium	petroleum (Venezuela, Iran), chemicals and catalysts, steel and nonferrous alloys	probably no hazard at current levels
Cadmium	coal, zinc mining, water mains and pipes, tobacco smoke	cardiovascular disease and hypertension in humans suspected, interferes with zinc and copper metabolism
Antimony	industry	shorted life span in rats
Lead	auto exhaust (from gasoline), paints (prior to about 1948)	brain damage, convulsions, behavioral disorders, death

Note: Bismuth, tin, and zirconium are also present as pollutants from industry, coal, and petroleum, respectively. C & EN has no data on their possible health effects. Titanium, aluminum, barium, strontium, and iron are air pollutants that occur naturally in dust from soils. Other metals known to be in the environment—chromium, manganese, cobalt, copper, zinc, and molybdenum—are essential to human health and probably pose no danger at current levels.
Sources: Battelle Memorial Institute, Dartmouth Medical School
Reprinted by permission from "Trace Metals: Unknown, Unseen Pollution Threat," C & EN, July 19, 1971, p. 30.

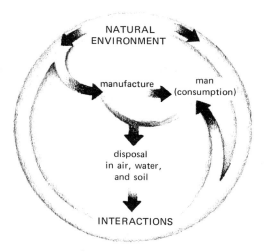

Adapted from *Toxic Substances,* Council on Environmental Quality, 1971.

FIGURE 8.3 Pathways for toxic substances.

arsenic, fluorine, selenium, and the heavy metals. It is beyond the scope of this book to discuss the physiological effects of all the elements that are toxic in nature or found as contaminants. The more common ones that are presented here will illustrate the extent and magnitude of the problem.

NONMETALLIC POISONS

Arsenic   The poisonous effects of arsenic are notorious. Severe poisoning of humans can be caused by as little as 100 milligrams (about $1/300$ ounce) and 130 milligrams has been known to be fatal. The element accumulates in the body with the result that decreasingly small doses can be lethal. As long as ten days may be required for the body to completely eliminate a single dose of arsenic. Repeated or prolonged intake, therefore, has a cumulative toxic effect.

Arsenic is an element that is intermediate between the metals and the nonmetals, having properties of both. It does not have any recognized role in human nutrition, although it is widely distributed in nature and is present in both water and food. It is found in many vegetables and fruits. Some marine organisms, especially shellfish, tend to concentrate arsenic within their bodies, which may contain more than 100 ppm. Examples are: 174 ppm in prawns, 42 ppm in shrimp, and 40 ppm in bass. Before the advent of modern insecticides, arsenic compounds were widely used to treat food crops, and the U.S. Food and Drug Administration set a limit of 3.5 ppm of arsenic from such sources. In 1959, the Food Standards Committee for England and Wales placed a limit for arsenic of 1 ppm in food and 0.1 ppm in beverages.

Arsenic is present in most soils, though highly varied in concentration, ranging up to 500 ppm. Arsenic often stimulates plant growth in very low concentrations, but it is injurious in excessive quantities. Destruction of the chlorophyll (the green pigment of the foliage) appears to be the main effect. As little as 1 ppm of arsenic trioxides in the water has caused injury to plants. Beans and cucumbers are especially susceptible to arsenic injury. Turnips and grasses, including cereal crops, are more resistant.

Arsenic is present in toxic concentrations in many water supplies. In New Zealand, cattle have died from drinking water containing a natural content of arsenic. There are also several areas that have a high incidence of skin cancer among people drinking water from wells with high arsenic content. In 1942, a safe limit of .05 ppm was set by the U.S. Public Health Service, and in 1962 the same agency recommended a maximum of .01 ppm in drinking water, more severe than an earlier British limitation on arsenic in beverages. However, in 1971, the Geological Survey issued a report that the raw water supply of several cities contained arsenic that exceeded the Public Health Service standards. A survey of 720 waterways found that 2 percent of the samples were above the standard of .05 ppm for arsenic.

Fluorine   Few aspects of public health have engendered more controversy than the fluoridation of water supplies. Most people accept the importance of fluorine to the growth of healthy teeth in children. On the other hand, the fact that fluorine compounds have been used as pesticides and are known to be toxic to plants and animals causes some people to be concerned about the practice of adding fluorides to public water supplies. The fervor with which both sides of the dispute are defended tends to be more enlivening than enlightening.

Fluorine belongs to a group of elements which includes chlorine, bromine, and iodine. Collectively they are called the *halogens,* from a Greek word meaning "salt producing." (They all form chemical combinations consisting of salts, such as sodium chloride, the common table salt.) Fluorine in its elemental state is, like chlorine, a gas. However, in nature it is always found in various combinations. For the most part it is in the form of the mineral fluorspar (calcium fluoride, CaF_2) and in large deposits of the mineral cryolite (sodium aluminum fluoride, Na_3AlF_6). Cryolite is also made synthetically.

Fluorine makes up about 0.1 percent of the earth's crust. Plants take it from the soil and water; and animals derive it from food, water, and minerals. Thus, fluorine is found in the bones of most vertebrate animals. The fluorine in rock phosphates used as fertilizers and for animal feed supplements adds to the naturally occurring levels in plants and animals, and the addition of fluorine to water supplies is also a source of human intake.

Fluorine is commonly found in natural water supplies, but is present in highly varied concentrations. The effect of tooth decay from drinking the water of certain districts was noted by Vetruvius Pollio, the great Roman architect and builder of the first centruy B.C. The water at Susa, the ancient capital of Persia, was for this reason said to be unift for human consumption, although the relationship to low fluorine content was not then known. Tooth impairment of a different nature called *denti neri,* or black teeth, was observed in later times among people in the vicinity of Vesuvius and other places in Italy where the soils of volcanic origin contain large amounts of fluorine. It was not known until relatively recently that symptoms of dark or mottled teeth are caused by an excess of fluorine.

When fluorine is ingested in excessive amounts for a prolonged period of time, the result is faulty formation of the teeth involving dark discolorations and changes in the structure of the bones, a condition called *fluorosis*. These undesirable effects on teeth were noted early in the century by Dr. F. S. McKay, a dentist who practiced in Colorado Springs, where the water supply contained about 2 ppm fluoride. The condition was called "Colorado brown stain." Since then, other areas with water containing fluorine in concentrations of 2 to 3 ppm have been characterized by the prevalence of mottled teeth and abnormal bone structure. Fluorine in concentrations as high as 10 to 14 ppm in water have been noted, although such quantities are unusual (soils average about 300 ppm). Livestock are also affected. Cattle are the most susceptible of the farm animals, but susceptibility also occurs in sheep, swine, horses, turkeys, and chickens in that order of decreasing effect. Cattle and other livestock have been poisoned by eating forage containing fluorine that had been emitted as an atmospheric pollutant in the manufacture of phosphate fertilizers and aluminum.

In areas where fluorosis occurs in England, the Punjab of India, South Africa, and Soviet Asia, there appears to be a correlation with a high incidence of endemic goiter. The harmful effects are intensified by a high intake of calcium but are alleviated by an increase in iodine.

The margin between the optimum amount of fluorine for dental health and the amount that will cause some tooth discoloration is not great. The approximate effect of concentrations of 1 ppm and above are as follows:

1ppm	May cause mild mottling in 10% of the children.
1.7 ppm	Mottling in 40 to 50% of the children.
2.5 ppm	Mottling in 80% of the children, ¼ of which is severe.
4–6 ppm	100% mottling with a marked increase in severity.[1]

Because the intake of water by people is highly variable depending upon climate, temperatures, salt intake, quantity and quality of food, age, and individual temperament, an attempt is usually made to add only enough fluoride to deficient water supplies to bring the concentration up to a minimum level. In the southern parts of the United States the optimum is generally considered to be between 0.5 and 0.7 ppm. In some cities the amount is lowered in the summer and raised in the winter.

Selenium

Selenium is an excellent example of the fine balance that sometimes occurs in nature between the beneficial and injurious effects of a naturally occurring substance. The balance tips one way or the other depending on the concentration in the environment. Selenium is an element that chemically resembles sulfur and can replace sulfur in amino acids, but unlike sulfur, which is distinctly a nonmetal, selenium tends to have metallic characteristics. Sulfur and selenium also differ greatly in their biological properties. Although selenium appears to be an essential micronutrient for at least some plants, it is one of the most toxic substances to occur natu-

---

[1]L. S. Goodman and A. Gilman, *The Pharmacological Basis of Therapeutics,* third edition (New York: The Macmillan Company, 1965).

rally in the environment. Concentrations only slightly above those needed for growth of plants may become poisonous to animals (see Figure 8.4).

This element has a number of industrial uses, and can be recovered in large quantities during the production of sulfuric acid inasmuch as it occurs in mixture with sulfur. However, it is often an unwanted contaminant and a hazard to industrial workers who may not be aware of the danger. The disposal of selenium waste and its compounds is a problem. Dumping it into waterways causes a serious public health hazard unless great dilution can be accomplished. Burning the material cannot be done in or near residential areas because of the bad odor, described as "rotten horseradish," from the selenium dioxide that is formed. The only practicable method to prevent water and air pollution is to reclaim the material.

A selenium compound was used at one time to kill mite pests in fruit orchards, but the practice was discontinued because of the danger of leaving a poisonous residue in the fruit. It is the only inorganic pesticide known to have appreciable *systemic* action, that is, the selenium is taken up by the root system of the plants and becomes toxic to susceptible organisms that feed on the foliage or fruit.

A chronic form of poisoning in livestock results from feeding on species of plants that accumulate selenium. Relatively low levels of selenium—about 25 ppm—produce the poisonous effect when the forage is eaten for several weeks or months. The ailment is called *alkali disease,* characterized by dullnes, lack of vitality, and emaciation. Cattle lose the long hair from the tail and horses lose the hair from both tail and mane.

Most grains and forages low or very low in selenium

Variable levels of selenium—low and adequate

Most grains and forages adequate to high in selenium

• Local spots where some plants have excess selenium

From W. H. Allway, "Selenium–Vital But Toxic Needle in the Haystack," *Yearbook of Agriculture,* 1968.

FIGURE 8.4 Selenium uptake by crop plants in relation to animals needs.

There is often abnormal hoof growth and sometimes severe pain. Often the animal becomes too lame to obtain food and water and soon dies.

The alkali disease has long been recognized in livestock that graze on some of the soils of South Dakota and Nebraska. Large areas of land with selenium in concentrations that produce toxic vegetation have been found in the area surrounding the Black Hills, as well as in western Colorado, part of western Kansas, and portions of Montana, Texas, and Utah. However, not all of the vegetation is poisonous and the degree of toxicity varies widely.

Selenium is abundant primarily in rocks that were formed during the cretaceous era when the formation of the ancestral Rocky Mountains took place more than 100 million years ago. Poisoning can occur where these rocks have been exposed and eroded. Soils that are deposited at some distance from the rocks may contain high concentrations of the element. Since selenium compounds are leached by water, high concentrations are not often found in areas of high rainfall but more usually in arid and semi-arid areas. Certain species of range plants are known to be efficient accumulators of selenium and are considered to be indicators of seleniferous soils.

Studies of people living in seleniferous areas of South Dakota and Nebraska were made as early as 1936. There were many complaints of gastrointestinal symptoms among the inhabitants, and many of the individuals were found to excrete quantities of selenium in their urine. Selenium was found in meat, eggs, milk, and vegetables. And in an area near Irapuato, Mexico, an ailment of hitherto unknown cause had afflicted the inhabitants for 200 years. Selenium poisoning was finally demonstrated to be the cause. Vegetables from the public market varied in selenium content, and some were found to contain as much as 70 ppm of selenium. Selenium is also present in cigarette, pipe, and cigar tobaccos.

Because selenium is taken up by plants that are used to produce paper, it is found in various paper products including cigarette paper. It appears in newspaper at a concentration of about 8.6 ppm and is given off in incinerator stack gas when the paper is burned. The American Conference of Governmental Industrial Hygienists recommended a limit for selenium of 0.2 milligrams per cubic meter in air and 0.01 ppm in water.

Asbestos One of the most controversial problems related to hazardous minerals concerns the health effects of asbestos. At least forty years of study, intensive during the last ten or twelve years, shows that inhaled asbestos fibers cause severe lung disease and that certain types of cancer are apt to follow heavy intake such as that from occupational exposure. There may be a latent period of as much as thirty to forty years before the cancer shows up. Case histories of one type of cancer showed that there was a high frequency among those who had been under heavy exposure as children.

The possibility of environmental hazards from asbestos was of little interest to the general public until the finding that western Lake Superior contains significant amounts of asbestos-like fibers and that a large part of the contaminant has its origin from mine tailings—67,000 tons each day—dumped into the lake at Silver Bay by a mining company that processes taconite, a low-grade iron ore. In a law suit culminating in a

139-day trial, the judge ruled that the 200,000 people who live around the western part of Lake Superior were in danger from asbestos, not only by water contamination but also by emissions from the plant's stacks. The plant was ordered to close down until the situation could be corrected. But the order, issued in April 1974, was in effect for only two days; a three-judge panel of the U.S. Court of Appeals stayed the order. Later, the U.S. Supreme Court refused to order the plant closed, although the possibility remained that a two-year limitation could be placed on the operations after which the emissions would have to be corrected. The case is a classic example of a fight over profits and jobs versus environmental quality.

While the effects of heavy inhalation exposure are well known, what is not clear is the effect on the general population from relatively small amounts of asbestos-like fibers in the environment. There are over a thousand uses of asbestos; therefore, there are many possible sources of exposure. Animal studies show that while inhaled asbestos can be found in various parts of the body, asbestos taken by mouth is very poorly absorbed, if at all, through the intestinal tract under normal conditions of ingestion. But there is some evidence that ingestion of asbestos fibers are potentially harmful.

The term *asbestos* is an imprecise designation for a whole group of silicate minerals that exist in fibrous form. To add to the confusion, each of six recognized main types exists in both fibrous and nonfibrous forms, the latter often called *asbestiform* and sometimes considered an asbestos mineral and sometimes not. Different types of asbestos do not seem to be equally hazardous. It may be found that the most serious effects from low-level exposure are not from asbestos itself but from the capacity of the fibers to combine with other substances and carry them to susceptible tissues in the body. For example, it is known that the risk of cancer is very high in cigarette smokers who are also exposed occupationally to asbestos.

Asbestos-like materials in small amounts are probably widely distributed in water supplies and may be taken into the body regularly by other routes as well. It is urgent to know what effect routine exposure to these particles has on human health, and whether an increase in the amount from man-made pollution is an important health hazard.

HEAVY METALS

Seven metals were known to ancient metalworkers: gold, silver, copper, mercury, lead, tin, and iron. Two of them, mercury and lead, have long been recognized as poisonous in the forms and products commonly used in both ancient and modern technology. Mercury and lead are major environmental contaminants. (They are discussed separately later.) The other ancient metals are hazardous only under special circumstances. Some of them are used in modern technology in large amounts, though they are toxic in small quantities and potentially injurious to the environment. However, the environmental effects of the more obscure metallic poisons have not been thoroughly studied. Cadmium and copper will serve to illustrate the nature of the problem.

Cadmium

In Japan, along the Jintsu River and its irrigated basin, there is a debilitating human disease that affects the liver, kidneys, and bones. It is often fatal. The Japanese call the disease *itai itai,* which translated means "ouch ouch." The cause of the ailment was a mystery for a long time. Beginning during World War II, the waste water from a cadmium, zinc, and lead mine was discharged into the Jintsu River. Rice farmers who used the water from the river to irrigate their crops blamed the mine for the poor harvests. However, they were not aware of any connection between the water and their ailments. It was not until Dr. Jun Kobayashi of Okayama University's Institute for Agricultural and Biological Sciences analyzed the bones and other tissues of people who suffered from the disease that the cause was recognized.

It became apparent that *itai itai* was brought on by the people's consumption of heavy metals, primarily cadmium, either by drinking the water or by eating rice which had accumulated the metal from the irrigation water. The onset of symptoms comes after several years of ingesting the metal. The affliction is characterized by kidney malfunction, a drop in the phosphate level of the blood serum, loss of minerals from the bones, and a condition called *osteomalacia,* which is a rickets-like condition characterized by pathologic bone fracture and intense pain.

Experiments with laboratory animals produced typical *itai itai* symptoms and also showed that prolonged ingestion of zinc, lead, and copper caused a loss of minerals from the bones. The investigators found high levels of soil pollution from cadmium, zinc, and lead in the vicinity of smelters. The waste-water pollution was controlled by the construction of a lagoon to permit settling of the ore and metal particles before discharging the water to the river. When this was done, the incidence of *itai itai* decreased.

Cadmium belongs to the same family of elements as zinc and mercury and is often found as an impurity in zinc and lead ores. It is used in metallurgy for the manufacture of alloys of coppers, lead, silver, aluminum, and nickel. It is also used in electroplating and in making pigments, ceramics, photographic equipment, and nuclear reactors; and it is found in the wastes of factories engaged in these processes, as well as those engaged in textile printing, lead mines, and various chemical industries. Cadmium is present in cigarette smoke. Analyses of "market" milk showed that some of the samples contained cadmium in excess of the standard for drinking water. The source of the contamination was unknown or not reported.

The results of a U.S. Geological Survey of 720 waterways showed that 4 percent had concentrations of cadmium that were above Public Health Service standards. The U.S. Public Health Service, in its 1962 Drinking Water Standards, set a mandatory limit of 0.01 ppm, or 10 ppb (parts per billion), for cadmium, and the World Health Organization (WHO) prescribes a limit of 0.05 ppm for the metal in drinking water.

There is no evidence that cadmium has any useful biological function. It is toxic in relatively small amounts and has reportedly caused a number of human deaths from oral ingestion of the metal present as a contaminant in food or water. In animals and humans, cadmium tends to accumulate in some of the important organs of the body, including the liver, kidneys, pancreas, thyroid, and bones. It is not easily eliminated, and once in the body, it tends to remain. Many plant and animal tissues contain small amounts. Seafood and grains appear to be the main source of cadmium in foods.

One of the dangers of cadmium is that it acts synergistically with other substances to increase their toxicity. Such synergism has been noted with zinc, cyanide, and possibly selenium. Full knowledge of the toxic effects of the metal would require further scientific investigation.

Copper The biological properties of copper are responsible for its having a useful purpose specifically for its toxic effects. Various chemical combinations of the metal are efficient fungicides and algacides. The copper used for these purposes alone amounts to about 15 million pounds annually in the United States and accounts for about 40 percent of all chemical uses of the metal.

Copper is essential for the growth, development, and health of most forms of life, and the element must be supplied as a micronutrient for plants growing in soils that are deficient. Relatively large concentrations can be tolerated by most forms of animals, including vertebrates, but the effect of copper on aquatic organisms varies greatly among species and is also affected by the physical and chemical characteristics of the water. Many forms of algae, fungi, and bacteria are highly susceptible. Most of the higher organisms are less susceptible, and copper poisoning of people is rare.

Copper salts occur in natural waters only in small concentrations, from a trace to about 50 ppb. Therefore, its presence in injurious amounts is almost always due to pollution. The following limits are indicated for the metal in water that is intended for specific uses:

Domestic supply	1.0 ppm
Irrigation	0.1 ppm
Fish and aquatic life	0.02 ppm
Sea water	0.05 ppm

Lead and mercury The most important metallic pollutants with respect to environmental contamination are lead and mercury. Lead is a serious public health problem, especially among children who are exposed to the metal, and mercury is often present in certain seafoods at concentrations at or near the danger level. These two metallic poisons are discussed in the following pages.

LEAD

Lead serves no known useful purpose in the nutrition of animals and plants and is highly toxic to many living organisms, including man. The element is prevalent in the natural environment. The earth's crust contains an average of about 10 to 15 ppm lead, though the content in rocks, soil, and water is extremely variable.

Lead is a powerful though insidiously deceptive poison. An adult person can tolerate a relatively large single dose without fatal or even serious effects. Though the acute effects may or may not be alarming, chronic poisoning from prolonged or repeated exposure can cause permanent damage to the nervous system and other organs. Continual ingestion of relatively small amounts can be fatal.

Child victims

Lead generally affects children more severely than adults. Over 500 confirmed cases of lead poisoning occurred in children during one year (1964) in New York City. In Chicago during a three-year period (1959–1961), lead poisoning accounted for 4.7 percent of 9,853 cases of accidental poisoning in children and was responsible for 79 percent of the total deaths due to accidental poisoning. The mortality rate in children is high. In Cleveland, Ohio, during the period 1952 to 1958, 30 percent of the children with confirmed lead poisoning died. Many of the survivors were permanently injured.

Most of the lead poisoning in small children is caused by eating flecks of lead-containing paint from the walls and woodwork of old houses that were painted before lead pigments for interior paints were largely discarded. This problem is not restricted to the poor, although it is especially prevalent in slum areas of large cities.

The outlook for a happy life is grim for many of the children who survive lead poisoning. A follow-up study of 425 children in Chicago for periods of 6 months to 10 years showed that 82 percent of the 59 with encephalopathic (brain disorder) symptoms were left with serious nervous system impairment. Of these, 54 percent had recurrent seizures, 38 percent were mentally retarded, 13 percent had cerebral palsy, and 6 percent had optic atrophy. Mental retardation was the most serious consequence of lead poisoning.

General hazard

Lead poisoning is common in adults and is a major industrial hazard. Though lead paints are a greater hazard to children, who are prone to ingest and chew on painted articles, painters also may run a risk from continual use and exposure. A chronic ailment of painters, formerly known as "painter's colic," may have been caused partly by paint thinner fumes. Severe poisoning has also been caused by fumes resulting from the removal of lead paint with a blow torch and by the dust from scrapings. Lead-containing solders are also a hazard to plumbers and industrial workers. Battery factories use large quantities of lead, the handling of which is a hazard to workers. In printing plants and casting rooms, the workers may be affected by the suboxide of lead that forms on the surface and makes a dust that is soluble in weak acids. Some European manufacturers of plastic pipe use lead as a stabilizer. The amount of lead that is extracted from the pipe by water exceeds the maximum limit for lead in drinking water set by the U.S. Public Health Service. Lead in glazing putty may be another source of poisoning, especially for children.

Lead monoxide, also called *litharge,* is an orange-yellow pigment that is used in glazing pottery. Sometimes glycerine is used as a cement, since the mixture sets to a solid lead glyceride. Improper fixing of lead-glazed pottery, resulting in dull glazes, can be dangerous. Acid foods and beverages stored in such pottery can leach out the lead, and if the pots are used repeatedly for such purposes, they may cause chronic poisoning.

Other domestic sources of lead exposure include solder and ammunition. Cigarette smoke contains lead, although the hazard is less than when lead arsenate was extensively used as an insecticide.

Left photo: Lead pigments in paint were a common cause of poisoning of children. *Right photo:* Pewter is a soft metal alloy made of tin and other metals and often containing lead, copper, antimony, or bismuth. It was used extensively in Europe during the 1600s and 1700s for household utensils. Pewter is hazardous as a food container because it is a potential source of heavy-metal poisoning.

MANY FACES OF LEAD

The symptoms   Lead poisoning is a multifaceted affliction. It causes liver and kidney damage, reduction in hemoglobin formation, mental retardation, and abnormalities of fertility and pregnancy. A bewildering variety of symptoms may occur and the identity of the cause is often not recognized until damage to health from chronic poisoning is beyond repair. The poisoning may mimic other ailments and be elusively difficult to identify.

Central nervous system effects, or "CNS syndrome," are the most serious manifestations of chronic lead poisoning. Symptoms include a panoply of nervous system disorders which in severe poisoning may culminate in delirium, convulsions, coma (profound unconsciousness), and death. Among survivors there are often permanent effects on behavior and intelligence as well as physiological damage to the kidneys and the vascular system, including blood abnormalities.

Fetal effects   It was known for a long time that there was a high rate of miscarriages among women who worked in pottery and white-lead factories where exposure to lead was liable to be high. In the early days of female employment in the pottery industries in England exposure to lead was often excessive. Studies showed that women so employed were more apt to be sterile than those in the general population; and if they became pregnant, the pregnancy was more apt to result in a miscarriage or stillbirth; and if the child was born living, there was a higher infant mortality. These observations were confirmed in other countries. There is also some evidence that plumbism (lead poisoning) in the father affects the survival, vigor, and fertility of the offspring. One investigator concluded that among families from the industries of Milan and the galena mines of Sardinia in which the fathers had plumbism,[2] 44 to 82 percent of the pregnancies resulted in stillbirths and miscarriages.

---

[2]Galena is lead sulfide, the principal lead ore.

LEAD AND THE FALL OF ROME

Lead colic was well known to Greek and Roman physicians, and later to Arabian doctors. The ancient Romans suffered severely from ailments that were apparently caused by heavy metal poisoning. There was a flourishing lead mining and processing industry and extensive occupational exposure to the metal. The Romans loved color. The favorite paint for their richly colored structures was Pompeian red, which contained red lead, a mixture of lead oxides known as *minium*. Some of their paints may have contained mercury pigments, even more poisonous. White lead has been used in paints since 400 B.C. White lead[3] is soluble in gastric juices and now known to be a common cause of many cases of poisoning in children. The Romans also developed the use of lead water pipes. The use of lead pipe in more recent times is believed to have been the cause of lead poisoning in epidemic proportions.

Lead poisoning may have been extensive, though unrecognized, among the wealthy Romans who used lead cooking and eating vessels. It was common to add lead to wine as a sweetener and preservative, a practice that may have been derived from the Greeks. Grape syrup was made by boiling down unfermented grape juice in a lead vessel. The syrup, called *sapa* or *defrutum,* depending on the degree of concentration, was used in foods and beverages for flavoring and sweetening. Lead was also used in ointments and various medicinal concoctions.

A theory has been proposed that the diminishing birth rate and the large number of mentally incompetent children born to the patricians during the decline of Rome was the result of the extensive use of lead by the wealthy people. By contrast, according to this theory, the poorer classes and slaves, who could not afford the expensive metal cookery, used pottery for their domestic needs and probably drank their wine, when they could get it, unsalted with lead.[4]

The poisonous effects of lead on the citizens of Rome is only one of many theories concerning the downfall of the Roman Empire. That lead might have played a role is an interesting speculation, but at this distance in time, we will probably never know how serious its effects were or if it played a significant part in the decline of the Empire.

LEAD POISONING IN RECENT TIMES

Lead poisoning continued to afflict Europe during more recent times. In parts of France the vintners made a practice of adding lead to the wine to prevent spoiling and to promote fermentation. The unhappy effects of drinking the wine came to be known as "colic of Poitou," after the district where the method was used. A treatise was written by des Planches Tanquerel describing 1,217 cases of plumbism in the Hospital La Charite in Paris during 1831 to 1839. In Britain there were cases of poisoning from cider contaminated by lead vessels, and in Spain poisonings occurred from eating food prepared in lead-lined cooking utensils.

---

[3]Lead basic carbonate $[Pb(OH_2)PbCO_3]$.
[4]S. C. Gilfillan, "Lead Poisoning and the Fall of Rome," *Journ. Occup. Med.,* 7(2):53–60 (1965).

The most dramatic incident of suspected mass lead poisoning in modern times occurred during the year 1900 in England when nearly 6,000 people were suddenly afflicted with a mysterious ailment that baffled medical practitioners. At least 70 persons died of the malady in Manchester alone. Officials at first suspected a disease, but they finally concluded that the epidemic, in reality, was mass poisoning from beer. Contamination with arsenic or lead was suspected. Some people thought that the beer contained lead picked up from the lead pipes that carried the beer from the barrels under the bars to the taps.

Beer has long been a ritualistic beverage in Britain, and the shock of poisoned beer alerted officials to the dangers of poisonous residues in food and beverages and had a great deal to do with the British taking the forefront in establishing tolerances (maximum limits) on insecticides. Beginning in 1892 when lead arsenate was invented to combat the gypsy moth and until the advent of DDT, lead arsenate was one of the most widely used pesticides on fruit trees. Apples and pears often had heavy coatings of lead and arsenic (as lead arsenate) and created difficult problems in marketing.

DANGER TO WILDLIFE

Lead pollution is a hazard to some types of wildlife. Hunters scatter 12 million pounds of lead shot each year over the favorite habitats of water fowl. The shot settles in the feeding areas where some of it is ingested by the birds. The loss is estimated at one million birds per year from lead poisoning. Ammunition manufacturers and conservation organizations are attempting to develop shot that would be relatively nontoxic, and recently shells containing iron shot have become commercially available.

The lead content of seawater is about 0.03 ppb, but the concentration in some waters including drinking water is often much higher. It is found in some marine plants at about 8 ppm. Lead in some marine animals has been found at a level of about 0.5 ppm, concentrated most heavily in calcareous tissue. Lobsters died within 20 days in lead-lined tanks, but others lived for 60 days or longer in steel-lined and other types of tanks. Lead was also shown to be toxic to the Eastern oyster. Concentrations of 0.1 to 0.2 ppm produced changes in the mantle and gonads.

In laboratory experiments on mice and rats, a concentration of 5 ppm from lead acetate in the diet for a duration of 29 months significantly reduced longevity when the diet was also deficient in chromium. The animals had also lost hair and weight by the end of the experiment, and rats that had been administered lead were more susceptible to bacterial endotoxins.

LEAD IN THE ATMOSPHERE

One of the main sources of lead in the environment is believed to be from the leaded gasoline used as fuel for the internal combustion engine (see Figure 8.5). According to the U.S. Bureau of Mines, about 570 million pounds of lead were used to make gasoline antiknock compounds during

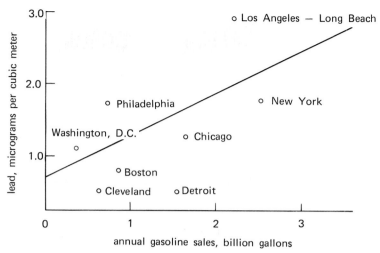

From Sixth Report of the Secretary of HEW, 1967.

FIGURE 8.5 Lead in the atmosphere in relation to the use of gasoline.

1970, a peak year. Worldwide release of lead to the environment, mainly from automobile exhaust, is estimated at about 2 billion pounds per year.

Leaded gasoline

The form of lead usually added to gasoline is the organic compound *tetraethyl lead* (TEL), a form that is extremely poisonous. Therefore, leaded gasolines are always colored as a warning. Because TEL is more efficient in gasoline that is already high in octane rating, the quantity of TEL needed to produce the desired effect is variable. The average amount is probably less than 0.05 percent by volume, or about 2 grams of lead per gallon, at which rate a tank holding 15 gallons of gasoline would contain about 1 ounce of lead. However, the average amount of lead in gasoline has been progressively reduced as engines have been modified to use gasoline having a lower octane rating. The goal is to eliminate lead completely from gasoline, although the primary incentive is to control other exhaust emissions, and only secondarily to eliminate lead pollution of the environment (see Chapter 6).

Gasoline containing TEL as the sole additive has the undersirable property of resulting in the formation of lead oxide as one of the products of combustion. The oxide is quickly reduced to metallic lead which is deposited in the combustion chamber and causes pitting. In order to prevent the corrosion, an organic bromide and usually an organic chloride are added. With the bromide and chloride present, the combustion products are lead bromide, lead chloride, and combinations of lead bromochloride, which are resistant to reduction. The lead halides (bromides and chlorides) are volatile and are emitted with the exhaust. A typical formula for TEL gasoline additive is:

> 65%   tetraethyl lead (TEL)
> 25%   ethylene dibromide (EDB)
> 10%   ethylene dichloride

Ethylene dibromide and some related compounds are also pesticides, used as soil fumigants for their action against insects and other pests.

The atmosphere in the northern hemisphere is estimated to contain 1,000 times more lead than if the activities of people were not contributing, and the amount is increasing. More lead is found in the air of cities than in nonurban areas. During a short period of heavy traffic, the following concentrations were found:

	micrograms per cubic meter
Cincinnati	14
Los Angeles freeways	25
In a tunnel	44

About 75 percent of the lead burned in gasoline is exhausted to the atmosphere in the form of aerosols (finely divided particles).

In studies of urban versus nonurban atmospheric pollution, there are correlations between the aerosol lead and gasoline consumption. There is also a correlation between the amount of lead in precipitation and the amount of gasoline consumed in the area. In a study of the San Diego region in California, lead aerosols accounted for 3 to 4 percent of the total suspended particulate material. More than 75 percent of the particles are less than 1 micron in size and more than 90 percent are under 2 microns. It is estimated that particles of less than 1 micron are easily retained in the lungs from which absorption may take place into the blood. Particles larger than 2 microns would probably be expelled from the lungs by the action of the mucous lining and either excreted or swallowed.

The earth's soils are being contaminated by lead from gasoline. Much of the particulate matter is carried in the air currents and distributed around the globe, eventually settling as dust or in rainfall (San Diego dust contained 0.84 percent lead). Lead deposits in the snow of high mountains and in the surface waters of the ocean were identified as the isotopes of lead in leaded gasolines. The results showed that there is widespread lead contamination of the environment.

MERCURY

In Roman mythology, Mercurius was the god of merchandise. The merchants were called *mercuriales* or *mercatores*. Mercurius was later identified with the Greek god Hermes. Perhaps mercury, the metal, was named after the god because he was considered tricky and deceitful or because in another role, he was also the swift messenger of the gods. The ancient name *"quicksilver"* came from the bright metallic luster of mercury and its peculiarly elusive nature. A liquid at ordinary temperatues, mercury does not wet objects but rolls freely about on the surface in slippery fashion, and at the slightest disturbance scatters in droplets that are difficult to collect.

Alchemy　　Mercury was known and used at least as early as the fourth century B.C. Concentrations of the ore in the form of red mineral *cinnabar,* mercuric sulfide (HgS), have been worked for centuries in Italy, Spain, and Austria (in the latter at Idria, now part of Yugoslavia). The alchemists were fascinated with mercury in their attempts to transmute the base metals into gold. Early physicians attributed unusual powers to it.

A versatile poison

Mercury is an impressive example of a naturally occurring substance that is toxic at such low levels that it leaves but a small margin of safety to living organisms.[5] The element is poisonous in the metallic state, as inorganic salts or mercury, or in the form of organic mercury compounds. It does not have to be ingested to be poisonous. Metallic mercury gives off vapors at room temperature; some of the metal even vaporizes at the freezing point of water.

There are many records of fatal poisoning from mercury vapors. A dramatic incident occurred in 1810 when the British cargo ship Triumph had some flasks of mercury broken in its hold. Mercury poisoning affected 200 sailors and three of them died. All the birds and cattle on board died. In another incident there were two deaths from poisoning by diethyl mercury, an organic compound used to treat grain for the fungus disease known as *smut*. The victims were not workmen but stenographers who worked near the storage area. The air in the room was analyzed and on the basis of the results it was concluded that slightly more than 1 milligram of mercury per cubic meter of air, if breathed daily for three months, could be fatal.

Vapor is the most common form of occupational exposure. Smoking is dangerous for those who handle mercury or mercury compounds. The miners of Spain, who were in the habit of rolling their own cigarettes, were said to be poisoned from the mercury fumes from the ore that rubbed off on the paper. Cinnabar is reduced to free mercury at temperatures slightly over 300°C, so experienced miners recommend that new men give up smoking and take up chewing tobacco instead.

Some compounds of mercury are more toxic than others. But unlike many of the synthetic chemicals of commerce that decompose at various rates, mercury is almost infinitely persistent. It may remain in the environment as the free metal or as one of the mercury compounds. In the habitat of living organisms, mercury can be modified by metabolic action, stored in the tissues of an organism, and passed along to other organisms in the food chain. Small amounts can be hazardous to man and other vertebrates if they regularly consume contaminated food.

USES OF MERCURY

Ancient industry

The ancient Romans regarded mercury poisoning as a disease of slaves because only slave labor was used in their rich mines at Almaden in Spain. The emperor Justinian (527–565 A.D.) said that an assignment to the mercury mines was equivalent to a death sentence. After slavery was abolished, criminals were used in the mines. When free labor finally came into use, there was so much illness and disability that the men were permitted to work no more than 8 days of 4½ hours each during an entire month. The ores at Almaden were especially dangerous because they

---

[5]Mercury is present in the earth's crust at an average concentration of about 0.00001%, or 0.1 ppm.

contained native quicksilver in addition to cinnabar, arsenic, lead, and sulfur.

*Mercury in
medicine*

When syphilis went on a rampage in Europe, apparently following its introduction from either the Orient or the Americas, quacks took to treating the disease with mercury concoctions. Physicians promptly picked up the practice, with some measure of success, though it is difficult to tell which was more harmful, the disease or the remedy. Paracelsus, a Swiss alchemist and physician of the early sixteenth century, was a strong advocate of heavy metals as medicine. He named himself after Celsus, a famous Roman physician—Paracelsus meaning "equal to Celsus"—but his real name was Theophrastus Bombastus von Hohenheim. Mercury was one of his favorite remedies, which he continued to advocate even after his own use showed that it was highly poisonous. He was influential in promoting the widespread use of mercurials. In America, pirate surgeons mixed sarsaparilla with mercurials for syphilis. The attractions of strong medicines containing mercury continued until modern times.

Napoleon's physicians treated him liberally with heavy metal medicinals. The symptoms of his ailments during his later years suggest that he was steadily poisoned most of his life by taking arsenic, antimony, and mercury. In one instance, it is reported, he became suspicious upon becoming ill after drinking a glass of lemonade containing tartar emetic (antimony tartrate). The next time, he handed the lemonade to his attendant, who was also made violently ill. Napoleon reportedly thereafter dispensed with the services of his physician.

The use of mercury in medicine has diminished steadily in recent years because of the discovery of nonmercurial diuretics, the advent of antibiotics, and the development of synthetic antibacterial ointments. One of the most popular antiseptics for many years was mercurochrome (merbromin), an organic substance containing mercury and bromine. It is poisonous when taken internally, and mercury poisoning has been reported from its use on large burn areas.

The mad hatters

About 1685, French hat makers invented "le secret," a method for softening the animal hairs that were used in the manufacture of hats. The method involved a solution of mercury nitrate and was kept as a valuable trade secret by the hatters' guild. The French called the process *secretage,* and in England it was called *carrotting* because the solution turned white fur to a reddish-brown color.

The hatter's trade was notoriously unhealthful. The expressions "mad hatter," "mad as a hatter," and "hatters' shakes" acknowledged well known neurological symptoms of chronic mercury poisoning. In some parts of Europe the making of felt hats was a home industry. Entire families were sometime afflicted with mercury poisoning. The practice of using "le secret" continued for centuries. The French eventually solved the problem by substituting nontoxic chemicals for mercury nitrate, but the transition was slow.

MINAMATA DISEASE

Minamata Bay is an indentation on the rough coast of Kyushu, the southernmost of the main islands of Japan. It was the setting for one of those human tragedies brought about inadvertently by one of man's activities that science and society were unprepared to cope with. A mysterious epidemic of disabling illnesses, consisting of a series of incidents from 1953 to 1961, struck young people especially hard. It was called, for want of a better name, "Minamata disease." Studies eventually led to the inescapable conclusion that the people were being poisoned by mercury. But the source as well as the symptoms involved a previously unknown dimension of mercury poisoning (see Table 8.5).

On the shores of Minamata Bay there are eleven fishing villages and a chemical plant. The disease appeared when several people, both young and old, were struck with incapacitating disorders. Some of them could not use their chopsticks or button their clothes. As the ailment progressed, some could no longer walk and some became demented. The behavior of some became infantile, while others became hopeless invalids. Not all of the functions of the nervous system were impaired; many of the victims suffered unremitting and agonizing pain. Even cats in the villages were severely affected. More than a hundred young people and adults either died or suffered serious neurological damage. Among 111 persons that were stricken with the ailment, more than 20 were congenitally defective infants.

Fish and shellfish comprised a large part of the villagers' diet. Investigations showed that the fish contained excessive amounts of mercury and that the mercury came from the chemical plant near the bay. The chemical company produced vinyl chloride, a material used in the manufacture of plastics for plastic sheets, phonograph records, and other plastic items. Mercuric chloride is used as a catalyst (promoter of the reaction) in the production of vinyl chloride and the product is then washed to remove the mercury. The waste was sluiced into the bay. As a result, shellfish from the bay contained 27 to 102 ppm of mercury with an average of 50 ppm, based on dry weight.

A similar disaster struck at Niigata, Japan, where in 1965, 120 persons were poisoned. The families of affected persons ate fish with an average

TABLE 8.5   Japanese Epidemics

Minamata Bay		*1953–61*
114 cases		
44 deaths		
22 brain damage/400 live births		
Niigata		*1965*
30 cases		
6 deaths		
Symptoms		
malaise		
numbness		
visual disturbance		
dysphasia		
ataxia		
mental deterioration		
convulsions		
death		

From E. Kahn, "Public Health Aspects of Environmental Contamination With Mercury," in J. E. Swift, ed., *Agricultural Chemicals–Harmony or Discord* (Berkeley, Calif.: Univ. of Calif. Div. Agricultural Sciences, 1971), pp. 85–90. Reprinted by permission.

TABLE 8.6   Toxicity of Methyl Mercury

Several hundred times more toxic than inorganic or
  aryl compounds
98 percent absorbed from gastro-intestinal tract
Strongly bound to certain tissue proteins
Irreversible damage (brain)
Biological ½ life in man: 70 to 100 days
Affinity for fetal tissues
Genetic damage
May be a nonthreshold substance
Narrow margin of safety

From E. Kahn, "Public Health Aspects of Environmental
Contamination With Mercury," in J. E. Swift, ed.,
Agricultural Chemicals—Harmony Or Discord (Berkeley,
Calif.: Univ. of Calif. Div. Agricultural Sciences, 1971),
pp. 85–90. Reprinted by permission.

frequency of 0.5 to 3 times a day, and the fish were found to contain as much as 9 to 24 ppm of mercury.

Methyl mercury   Nearly all the mercury found in the seafood at Minamata Bay and Niigata was in the form of the organic compound *methyl mercury,* a form that is several hundred times as toxic as inorganic compounds of mercury. It was present as methyl mercury chloride in the sediments of the polluted bay. Inorganic or organic mercury compounds in industrial wastes find their way to the bottom muds of lakes and other bodies of water where anaerobic microorganisms metabolically convert the mercury to alkyl mercury compounds, of which methyl mercury is the most common troublemaker.[6] (See Table 8.6.)

$$CH_3Hg^+ \quad \text{or} \quad (CH_3)_2Hg$$

methyl mercury ion   methyl mercury

Methyl mercury is volatile and soluble in lipids (fatty materials). The methyl mercury ion, CH_3Hg^+, is formed from methyl mercury under acid conditions, and is soluble in water. Methyl mercury is readily taken up by aquatic organisms, and in common with many other lipid soluble substances, is readily stored in the fatty tissues. The complete role of methyl mercury in the food chain has not been determined, but apparently fishes can accumulate the methyl mercury ion directly, the concentration in pike being as much as 3,000 times that in the water (see Figure 8.6).

Delay in recognizing the seriousness of mercury in the environment was due largely to the fact that the symptoms of methyl mercury poisoning are not those of classical mercury poisoning familiar to the medical profession. The relatively stable alkyl mercury compounds penetrate quickly to the central nervous system and have an effect that is different from that of either the nonalkyl organic mercurials or inorganic mercury. The urine and especially the feces are the most important means of mercury elimination, but it is secreted in the milk if the mammary glands are functioning.

[6]The alkyl compounds are those that have *alkyl* groups, i.e., methyl (CH_3), ethyl (C_2H_5), etc., attached to the mercury atom. Examples of nonalkyl organic mercurials are the phenyl mercury and methoxyethyl mercury compounds.

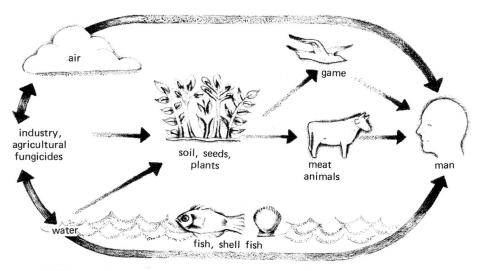

FIGURE 8.6 How mercury invades the environment and reaches living organisms.

HOW MERCURY POISONS

Mercury combines with certain enzymes in the body and makes them ineffective. Thus, mercury is described as an *enzyme inhibitor*. Some, but not all, of the important enzymes in the body contain sulfur in their active centers. Because sulfur is present with hydrogen in the form of a thiol, or sulfhydryl, group (−SH), such enzymes are called *sulfhydryl* enzymes.[7] Mercury has the property of readily forming covalent bonds with sulfur and in so doing replacing the hydrogen atom in the sulfhydryl group, thus destroying the action of the enzyme. There may be other effects from mercury of unknown importance in mercury poisoning.

Fetal effects   Methyl mercury has an affinity for the fetus and is teratogenic in its effect.[8] It readily penetrates through the placenta to the fetus where the concentration of mercury in the blood and in the brain of the fetus is about 20 percent higher than in the mother. Infants whose mothers are exposed to large amounts of methyl mercury are liable to be afflicted with mental retardation, cerebral palsy, and convulsions.

In the Minamata area of Japan, there were about 400 births during the period 1955–1959 of which 22 infants showed evidence of brain damage. There was a high level of mercury in the hair of some of the mothers and children, and most of the mothers experienced numbness during pregnancy. Although all of the mothers were heavy seafood eaters, few of them had symptoms that were characteristic of Minamata disease. Methyl mercury is much more potent as a cytogenic agent than inorganic mercury. Because of potential damage to the fetus, even in mothers who show no symptoms of mercury poisoning, women of childbearing age should avoid the risk of occupational exposure to alkyl mercury compounds.

---

[7]Enzymes are natural substances that promote biochemical reactions. They are sometimes called *biological catalysts*.

[8]*Teratogenic* means producing deformations in the offspring, in this case notably affecting the nervous system (see Chapter 13).

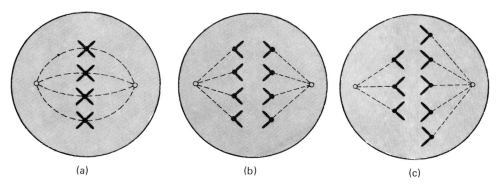

FIGURE 8.7 Methyl mercury is a potent inhibitor of mitosis, and is presumed to cause genetic damage because of chromosome aberrations during germ cell division (a) Normal cell. (b) Normal division. (c) Abnormal behavior of chromosomes, liable to cause birth defects.

Genetic effects   The teratogenic effects among the children born in the Minamata area occurred at an earlier stage of fetal development than would be expected simply from methyl mercury damage to the central nervous system. Therefore, the early effects may have resulted from induced chromosome aberrations. This view is supported by experiments on plant and animal cells.

Although there is no direct evidence based on human experience that mercury has genetic effects, laboratory studies on plant and animal cells show that mercury compounds cause chromosome breakage and interfere with cell division, resulting in *polyploidy*[9] or abnormal distribution of chromosomes (see Figure 8.7). When the common fruit fly *Drosophila melanogaster* is given food containing 0.25 ppm methyl mercury, the cells of the offspring frequently have one extra chromosome. The cytogenic nuclear toxicity of methyl mercury is more potent than colchicine, a substance that is widely used as a standard for producing chromosome aberrations.[10] In view of the safe level (tolerance) for mercury in fish set by the U.S. Public Health Service at 0.5 ppm, studies of the relation between mercury intake and chromosome aberrations in humans would be of great importance.

THE FOOD CHAIN

Mercury levels in sea animals appear to be 500 or more times greater than in comparable volumes of seawater, based on an assumed 90-percent water content of living organisms. The accumulation of mercury in the food chain is potentially disastrous. The great blue heron and the tern, fish-eating birds, have shown decreased fertility and poor egg quality from mercury in the diet.

It is difficult to determine how much mercury in food products comes from natural sources and how much of it is attributable to man's activities.

---

[9]*polys* (Gr) = many; *aploos* = onefold; hence, reduplication of the chromosome number giving rise to three, four, etc., times the normal haploid number in the gametes.

[10]*Cytogenic* means producing a permanently injurious effect on cells.

Photo courtesy of Standard Oil Company of California.

Mercury in fish is a major problem that threatens the economy of segments of the fishing industry.

Fish caught in the open sea usually contain less than 0.1 ppm of mercury. During the 1930s and 1940s, the mercury content ranged from 0.15 to 0.25 ppm. Fish taken more recently from Scandinavian waters contained generally 0.2 to 1 ppm, the highest being 9.8 ppm from southern Sweden. The mercury content of several meat products from Sweden had a higher mercury content than similar products from Denmark. In 1964, eggs from Sweden averaged 0.029 ppm as against an average of 0.007 ppm for six European countries. The prevalence of mercury contamination in Sweden has been attributed to the widespread use of mercurial fungicides for treating seeds.

A tolerance for mercury of 0.5 ppm in fish was set by the U.S. Food and Drug Administration. Because the limit of 0.5 ppm was often exceeded, fishing in several areas of the Great Lakes region had to be banned until the problem could be solved. (Analyses of fish in the Great Lakes region of the United States and Canada have shown concentrations up to 7 ppm in fish, a level that would be dangerously high for toxicants much less poisonous than mercury.) Commercial and sport fishing were adversely affected in recreational areas where mercury pollution was detected. Analyses of fish captured and preserved many years before the industrial use of mercury reached present proportions showed the mercury content to be higher in some cases than the official tolerance. Questions regarding the relative amounts of mercury derived from natural sources and from industrial pollution remain unresolved. In the meantime, much of the tuna, bonita, and swordfish must remain off the market because the mercury content often exceeds 0.5 ppm. Because the amount of mercury in fish is known to be exceptionally high in the catch taken from certain fisheries and low in fish taken from other areas, it is usually possible to keep the amount of mercury in human food, such as canned tuna, within safe levels by blending at the processing stage. But the sale of

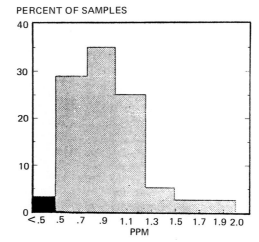

Redrawn from *Pesticides Monitoring Journal*, March 1974.

FIGURE 8.8 Distribution of mercury in swordfish.

swordfish had to be abandoned in the United States because of generally high mercury levels (Figure 8.8).

MERCURY IN THE ENVIRONMENT

More than 150 million pounds of mercury have been used in the United States alone during this century. More than 6 million pounds were used in 1969 (see Table 8.7). There is little information about its final disposition or where it is in the environment. Mercury is found in soil, air, and water. The amount of mercury found in the air depends on conditions. In San Francisco on a normal breezy day the atmospheric concentration of mercury has measured 2 nanograms per cubic meter.[11] High concentrations of mercury have been found in sediments near sewage outfalls. The use of mercury has declined so much since clean-up operations began that some of the mercury mines in the United States had to close down for lack of demand.

Sources of mercury

Large amounts of mercury are introduced to the ecosystem. Much of it comes from the erosion of soil and rocks. It is estimated that normal release from weathering accounts for one-half of the mercury entering bodies of water. Industrial use of mercury varies between about 5 and 8 million pounds per year. The Secretary of the Interior reported in 1970 that 50 manufacturing plants in the United States were dumping 287 pounds of mercury per day. Prompt action reduced the amounts drastically from those plants that were discharging mercury in their wastes (see Table 8.8). Sources include a wide variety of everyday products and processes:

- The amount of mercury released to the environment by burning coal is estimated at 3,000 tons per year, approximately the same amount emitted as waste from industrial processes. Coal contains up to 0.5 ppm mercury.

---

[11] A nanogram is one billionth of a gram; a gram is about 1/28 of an ounce.

TABLE 8.7   1969 Mercury Consumption in U.S.
Topped 6 Million Pounds

(Thousands of pounds)	
Electrolytic chlorine	1572
Electrical apparatus	1382
Paint	739
Instruments	391
Catalysts	221
Dental preparations	209
Agriculture	204
General laboratory use	126
Pharmaceuticals	52
Pulp and paper making	42
Amalgamation	15
Other	1082
Total	6035

Note: Consumption based on 76-lb. flasks.
Source: *Chemical & Engineering News*, June 22, 1970.
From "Mercury in the Environment," *Environmental Science
and Technology* 4(11): 892 (November 1970). Reprinted by
permission of the American Chemical Society.

- The chlor-alkali industry (manufacture of chlorine and caustic) uses large amounts of mercury in the electrodes. Some plants were found to be discharging into the environment 0.25 to 0.5 pound of mercury per ton of caustic soda produced. Recovery procedures were in use but the methods were inefficient.

- Mercury is used as a catalyst in the production of the plastic raw materials urethane and vinyl chloride (as at Minamata Bay) and in the manufacture of the chemical acetaldehyde.

- More than one million pounds of mercury per year are used by the electrical manufacturing industry in switches, batteries, fluorescent light tubes, high-intensity street lamps, and silent switches.

- Organic mercury preservatives, mildew inhibitors, antibacterial agents, and fungicides are used in water-base paints and cosmetics and by commercial laundries.

TABLE 8.8   Sources of Environmental Mercury

Mercury Mines and Ore Deposits	
Gold Mining	
Industry	
chlorine-alkali	chemical
paper-pulp and lumber	paint manufacturing
Other	
laundries	burning of coal
swimming pools	burning of paper
cooling towers	
Agricultural Uses	

From E. Kahn, "Public Health Aspects of Environmental
Contamination With Mercury," in J. E. Swift, ed.,
Agricultural Chemicals—Harmony Or Discord (Berkeley,
Calif.: Univ. of Calif. Div. Agricultural Sciences, 1971),
pp. 85–90. Reprinted by permission.

- Some vaginal deodorant suppositories and some spermicidal creams contain PMA (phenyl mercuric acetate). The products have been in use for fifty years, though no information is publicly available on the absorption of mercury from such uses.
- Metallic mercury is used primarily for making thermometers and electrical switches. One of the dangers in the manufacture of devices and instruments that contain mercury is the great penetrating power of mercury when spilled on benches and floors, from which dust and vapors of mercury can result in chronic exposure.
- The amalgams of tin, silver, and gold are used by dentists for filling teeth. Amalgams in the mouth release large amounts of mercury initially but only small amounts continuously thereafter and probably account for only a small proportion of the daily intake, although further studies are needed. Poisoning of dental workers has been reported from mercury-silver amalgam used for dental fillings. The amalgam contains as much as 40 percent mercury.

PROBLEM AREAS

Almost all the mineral pollutants occur naturally in the environment. This fact has often been interpreted to mean that plants and animals, including people, have a degree of tolerance for many of these substances that they do not, in fact, possess. Whether a substance normally present in small amounts is injurious under normal circumstances, such as mercury in fish, arsenic in shellfish, or selenium in plants, can be determined by laboratory tests in which animals are given food or water contaminated with different amounts of the poisons. But such experiments are not simple matters to perform because it is not easy to find and prepare feedstuffs that are completely free of any given element for use as controls.

Recent findings that mercury can be concentrated in the food chain, and that amounts only slightly above those ordinarily found in sea foods are dangerous, increase the suspicion that other naturally occurring minerals may have as yet undetected injurious effects on living organisms. Though we now have enough evidence to implicate mercury, cadmium, and a few other elements as potent environmental contaminants, the complete returns are not yet in on lead and some of the other metals, even though we know that lead, for example, has been increasing in the environment for many years as the result of human activities.

Unfortunately, people are misled into believing that there is no reason for concern when the concentrations of mineral poisons in food, water, and air are below those that can cause symptoms of acute illness. The subtle, manifold nature of the chronic effects from long-term exposure to metallic substances obscures the fact that some of the mineral poisons have profoundly injurious effects at very low levels of intake. We do not know whether these substances have a *threshold* level of concentration, that is, a level of intake below which they are harmless. Evidence points to the conclusion that x-rays and other forms of ionizing radiation do not have a threshold, but do some injury even in the smallest amounts. The heavy metal poisons may have similar effects, but this has not yet been determined.

Historically, one of the problems with heavy metal poisoning is that the symptoms can be easily mistaken for some common ailment; thus it goes undetected or receives treatment that is ineffective or even injurious. The effects of some of the mineral toxicants are so subtle, varied, and elusive that the cause may go undetected even when it is perceived that something may be wrong. Better understanding of the symptoms can help to locate and correct sources of contamination.

READINGS

Solid Wastes

BREIDENBACH, A. W. and E. P. FLOYD. *Needs for Chemical Research in Solid Waste Management,* 1970, pp. 1–29. U.S. Dept. Health, Education and Welfare, Washington, D.C.

Ocean Dumping. A National Policy. A report to the President prepared by the Council on Environmental Quality, 1970, pp 1–45. U.S. Government Printing Office, Washington, D.C.

Solid Waste Processing. A state of the art report on unit operations and processes. Public Health Service Publ. No. 1856, 1969. U.S. Dept. Health, Education and Welfare, Environmental Control Administration, Washington, D.C.

SORG, T. J. and H. L. HICKMAN, JR., *Sanitary Landfill Facts.* SW-4ts, 1970. Bureau of Solid Waste Management, Public Health Service, U.S. Dept. Health, Education and Welfare, Washington, D.C.

VAUGHN, R. D., *Solid Waste Management: The Federal Role* (Reprint 1970). Public Health Service, U.S. Dept. Health, Education and Welfare, Washington, D.C.

Nonmetals in the Environment

BYERS, H. G., *Selenium Occurrence in Certain Soils in the United States with a Discussion of Related Topics.* Second Report. Tech. Bul. No 530, 1936. U.S. Dept. Agriculture, Washington, D.C.

KINGSBURY, J. M., *Poisonous Plants of the United States and Canada.* Englewood Cliffs, N. J., Prentice-Hall, 1964.

McCLURE, F. J., *Water Fluoridation. The Search and the Victory,* 1970, pp. 1–302. U.S. Department of Health, Education and Welfare, Bethesda, Maryland, U.S. Government Printing Office, Washington, D.C.

RADELEFF, R. D., *Veterinary Toxicology.* Philadelphia, Lea & Febiger, 1964.

Lead

"Air Quality and Lead," *Environmental Science and Technology* 4(3):217–253(1970). A Symposium containing a series of articles.

CHOW, T. J. and J. L. EARL, "Lead Aerosols in the atmosphere. Increasing Concentrations," *Science* 169:577–580(1970).

GILFILLAN, S. C., "Lead Poisoning and the Fall of Rome," *Jour. Occupational Med.* 7(2):53–60(1965).

LIN-FU, J. S., "Lead Poisoning in Children," Children's Bur. Publ. 452, pp. 1–25, 1967. U.S. Dept. Health, Education and Welfare. Superintendent of Documents, Washington, D.C.

PATTERSON, C. C., and J. D. SALVIA, "Lead in the Modern Environment. How Much is Natural?" *Scientist and Citizen,* April, 1968, pp. 66–79.

PATTERSON, C. C., "Contaminated and Natural Lead Environments of Man," *Arch. Environ. Health* 11:344–361 (1965).

Mercury

Anon., *Maximum Allowable Concentrations of Mercury Compounds.* Archives Environmental Health, Vol. 19, pp. 891–905, 1969.

Anon., *Maximum Allowable Concentrations of Mercury Compounds.* Archives Environmental Health, Vol. 19, pp. 891–905, 1969.

CELESTE, A. C. and G. G. SHANE, *Mercury in fish.* FDA Papers, November, 1970, pp. 27–29.

KLEIN, D. H. and E. D. GOLDBERG, "Mercury in the Marine Environment," *Environmental Science and Technology* 4(9)765–768(1970).

MARTIN, H., "The Mad Hatter Visits Alice's Restaurant," *Today's Health,* October, 1970, pp. 39–43, 79, 80, 83, 84.

MARX, W., *The Frail Ocean.* New York, Ballantine Books, 1967.

NOVICK, S., "A New Pollution Problem," *Environment* 11(4):2–44(1969).

PESTICIDES

In medieval times, diseases that had reached epidemic proportions were called *pestilences,* and places where diseased people lived or were kept were called *pestholes* and *pesthouses.* They were frightening words that reflected the broad meaning of the word *pest.* The modern custom is to categorize the crawling, walking, or flying creatures separately from the pestiferous microorganisms that cause diseases. We call the latter, collectively, microbes or *germs,* and the chemicals that are used to control them are medicinals and related toxicants in the categories of disinfectants, antiseptics, germicides, and internal medicines such as the antibiotics, for killing bacteria and other internal parasitic organisms. Most of the larger enemies of man—the so-called pests—are usually combatted in similar ways.

PESTS IN THE ECOSPHERE

Parasites and predators, undesirable though they may be to the victims, are a normal part of the ecosphere. Few organisms escape being victimized by others higher in the food chain. Humans are among the world's worst pests from the viewpoint of the victims. In turn, we have been pestered by other living organisms since the dawn of history, and before that, judging from what takes place in the animal and plant world around us, our ancestral relatives were almost certainly plagued with numerous kinds of parasites, predators, competitors, and destroyers of their food, health, and comfort. Jonathan Swift, one of the greatest English satirists, commented on this state of war in nature:

> So, naturalists, observe, a flea
> Hath smaller fleas that on him prey;
> And these have smaller still to bite 'em;
> And so proceed *ad infinitum.*[1]

---

[1]From "On Poetry, a Rhapsody."

Pests played a prominent role in ancient literature and culture. The Israelites viewed with contempt the Chaldean god Baal; they called him Beelzebub, meaning "Lord of the Flies," probably in reference to the flies that must have swarmed around the animal sacrifices. Eight of the ten plagues of Egypt, as related in the Book of Exodus, were pest and related health problems:

> . . . the water turned blood red [algae?] and the fish died, polluting the water . . . frogs came out of the water and died in heaps, where they stank . . . lice infested man and beast . . . flies [mosquitos?] swarmed all over . . . all the cattle of the Egyptians died . . . an epidemic of boils afflicted men and animals . . . an invasion of grasshoppers ate all the plants and fruit trees . . . all the firstborn children and all the firstborn cattle of the Egyptians died.

Aristophanes (about 445 B.C.) wrote about the buds of the grapevines being nipped off by locusts and the figs being eaten by ants and pestiferous gall flies. He spoke of "death to creepy-crawly creatures":

> To safeguard fruit and flower
> From tribes of savage power
> Whose ravening jaws devour . . .
> Feeding on the crops of earth.[2]

Dysentery was a common affliction in ancient Greece and Rome and it was perceived by early writers that it was spread partly by flies falling into food. Later, in 1498, Bishop Kund of Denmark associated high populations of flies with an approaching plague. Thomas Sydenham, a seventeenth-century English physician, noted that an abundance of flies in the summer was an unhealthy sign.

PESTS AND HUMAN WELFARE

The variety of injurious effects caused by pests—from the human viewpoint—is almost beyond description. Among the major pests are those that consume crops and stored food products; others destroy fibers, forests, and structural materials; and, among the worst offenders are those that cause or carry diseases.

Carriers of disease

More than one hundred human diseases are transmitted to man by arthropod pests (arthropods are invertebrate animals such as insects, having "jointed" legs and other appendages). Malaria is now a minor problem in most parts of the United States, but it was not always so. Thirty years ago, there were between fifty thousand and one hundred thousand cases per year. Worldwide, as late as the 1950s, at least 2½ to 3 million people died from the disease, and millions more were afflicted with its debilitating effects. Even today, millions of people suffer from malaria, and the disease is on the increase again due to ineffective control measures against the malaria mosquito carriers (several species of the genus

---

[2]From "The Birds," *The Complete Plays of Aristophanes*, trans. by R. H. Webb (New York: Bantam Books), 1962. Reprinted by permission of the University Press of Virginia, Charlottesville.

Anopheles). Malaria mosquitos become infected with the malaria blood-parasite, *Plasmodium,* when they suck blood from a malarious person; they then transmit the disease to a healthy person when they bite a new victim. Historically, malaria is believed to have caused more widespread disease and death than any other infectious disease, and despite popular notions, more people now die each year from mosquito-borne diseases than from any other single cause.

The tsetse fly has been for centuries one of the most dangerous animals in Africa. It ranges across 4½ million square miles of tropical Africa, an area that is nearly 1½ times that of the United States. Together with the organism it carries, the trypanosome blood parasite that causes sleeping sickness in man and nagana in animals, the tsetse fly dominates much of this vast domain by destroying people and their animals, and choking the economy. An epidemic of sleeping sickness that swept the shores of Lake Victoria killed over 200,000 people, two-thirds of the population, while the nagana disease of domestic cattle is the direct cause of much of the malnutrition in Africa today.

Bubonic plague is always present in the Orient and other parts of the world, and is under close watch by health authorities guarding against its spread. This was the disease of the great epidemics of the Black Plague that occurred during the Middle Ages when nothing was known about its cause. We now know that the black rat[3] teams up with an accomplice, the oriental rat flea,[4] which carries the plague germ[5] from rats to humans. Rats are called the *reservoirs* of the disease, and fleas the *vectors*.

Pests and food

Pests consume much of the food intended for humans. In some parts of the world, pests harvest as much as 25 to 50 percent of the crops before the crops ever reach maturity, and afterward, insects and rodents destroy much of the remaining food while in storage. In a world where one-fourth of the people suffer from hunger, much depends on who gets to the dinner table first. Hundreds of pests—insects, mites, fungi, worms, weeds, and rodents and other mammals—cut down the harvest of crops by reducing the yield, cause losses in storage, and degrade the quality of cereals, fruits, and vegetables. All this increases the cost of food in the markets. (Food problems are discussed in detail in Chapter 16.)

The food consumed by some species of pests is prodigious. A single tobacco worm larva (a cousin of the tomato hornworm) consumes in 28 days of growth, food weighing about 50,000 times its birth weight, during which time the larva increases 12,000 times in size. A single cabbage aphid weighs about 1 milligram (1/28,000 of an ounce). If it could find enough cabbage, this tiny aphid could produce, in one season, enough aphids to form a mass weighing 822 million tons, five times the weight of the total human world population. This does not happen. But, because of the vagaries of the natural checks and balances, the cabbage aphids and other pests often get out of hand. The nourishment need not be human food. A single rainbarrel has been observed to produce more than 100,000 mosquitos in one season.

---

[3]*Rattus rattus.*
[4]*Xenopsylla cheopis.*
[5]*Pasteurella pestis.*

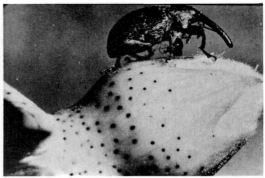

Left photo: Grasshoppers have periodically devastated crops since biblical times. Right photo: The cotton boll weevil invaded the United States from Mexico or Central America and caused disastrous losses. Many farmers were forced to grow other crops.

A disease of potato plants called "late blight," caused by the fungus *Phytophthora infestans*, was responsible for one of the greatest social upheavals in history when it wiped out the potato crops in Ireland and caused the frightful four-year famine during 1845–1848, now known as the Irish Potato Famine. Many thousands of those who survived moved to the United States in one of the largest migrations in history. The fungus is still present in potato fields throughout the northern hemisphere, and waits for suitable weather conditions when it turns the plants black almost overnight.

Agriculture was developed about ten thousand years ago. With it came concentrations of single crops, irrigation, disturbance of the soil, and other practices that intensify old pest problems and create new ones. Highly intensified agriculture brings about drastic changes in the environment by which minor pests may become major pests or become so destructive that the effort must be abandoned unless practical means can be found to combat them.

Until recent times pest control was simple: either mechanical means were employed, or the pests were allowed to share in the harvest. Since the pests had first choice, there was hunger and periodic famine. Even today, in many areas where primitive methods of agriculture are still followed, there is chronic malnutrition and starvation, partly due to pests.

Thomas Malthus, a British economist, published a treatise in 1798 in which he said that people would outbreed their capacity to feed themselves. His prediction has long since come true for many parts of the world. Except for the relatively small part of the earth tilled by highly industrialized agricultural methods, the world is overpopulated and underfed. The failure of his prognostication to come about among the affluent people can be attributed partly to the use of machinery and chemicals in the growing of food. The past, present, and future role of pesticides in demolishing Malthus's prediction of doom is today a much discussed topic.

Insects   The insects occupy a special place in the category of pests because of their abundance and their impact on human health and nutrition. Three-fourths of all species of animals are insects. All the land animals on the face of the earth—including man—do not weigh as much as the earth's

Photo courtesy of Enterprise Chamber of Commerce.

Said to be the only monument glorifying a pest, a cast-iron statue of a woman holding a cotton boll weevil stands in Enterprise, Alabama. Farmers in the surrounding area were forced to switch from cotton to peanuts, which became a major money-making crop.

insects. Some of these insects are carriers of disease. Others destroy food, fiber, stored products, and man-made structures. Insects steal or destroy one-third of everything that people grow, build, or store. Except for man, they are more destructive to the natural animal and plant environment than any other group of animals.

Left photo courtesy of Michigan Department of Natural Resources. Right photo courtesy USDA.

Left photo: The parasitic sea lamprey destroyed most of the sport fishing in the Great Lakes. After a way was found to treat the parasite's breeding grounds while it is still in the larval stage, trout and salmon were successfully reintroduced. *Right photo*: The United States has about as many rats as people. These filthy and sly rodents destroy about a quarter of a billion bushels of grain in the United States every year.

There are about 10,000 species of insects that are considered to be pests in the United States. Most of them are minor pests much of the time, but they go on sporadic rampages when conditions are favorable for a surge in numbers. About 100 species are major pests year in and year out, and these types cause 80 to 90 percent of the insect-pest damage. About 80 to 90 percent of the total insecticide use is for their control.

THE BALANCE OF NATURE

The *balance of nature* is a concept that applies to the natural condition in which various checks and balances maintain dynamic equilibrium in animal and plant populations. Taken too narrowly, the interpretation of *equilibrium* can be misleading, for the equilibrium often exists only within wide latitudes of population fluctuations. Like a swinging pendulum, the population of a species under favorable conditions may suddenly increase with explosive acceleration, then as suddenly lose out in its battle for numerical expansion and descend to precariously low levels of survival against the onslaught of unfavorable forces (see Figure 9.1).

Charles Darwin, in explaining the "struggle for existence and natural selection," pointed out the inherent capacity of a species of plant or animal for geometrical increase. He calculated that even the elephant, the slowest breeder of all animals, would populate the earth with 19 million elephants in 750 years, starting with only one pair. Darwin took pains, however, to point out the limitations on such increases that bring about sweeping changes in populations. He marked a piece of ground 3 feet long and 2 feet wide and tagged the seedlings of native weeds as they came up. Out of 357 seedlings, 295 were destroyed, mainly by slugs and insects. Darwin claimed that if the shooting of game on the large estates of England were stopped and if at the same time the predators that keep them naturally in check were not destroyed, in 20 years there would be less game than if shooting were continued, though hundreds of thousands of game animals were then being taken annually.

The principle of the balance of nature is critically important for agriculture and the structure of civilization, which for ten thousand years has depended on intensive culture of food plants for most of its nutritional

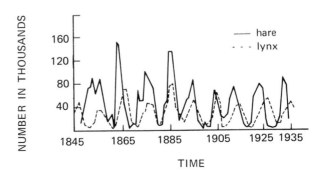

Redrawn from D. A. Maclulich, *Univ. Toronto Studies Biol. Ser.* No. 43, 1937.

FIGURE 9.1 A classic example of cyclic oscillation in population density is shown by the changes in abundance of the lynx and the snowshoe hare. Abundance is shown as the number of pelts received by the Hudson's Bay Company over a period of 90 years.

Left photo: Weeds can reduce crop production by more than 50 percent. Here a strip through a field of safflower has been treated to test a herbicide. *Right photo:* Water hyacinth, introduced from tropical America as an ornamental plant, grows in ponds and streams. It became a serious pest, choking waterways with its heavy growth of floating leaves.

requirements. The objective of agriculture is not to maintain the balance of nature but to nullify Darwin's "checks to increase" which would take possibly 295 of 357, or 83 percent, of the plants.

Unfortunately, the cultivation of large numbers of a single species of plant crowded into the most highly congested conditions possible does nothing to alleviate Darwin's 83-percent loss. This kind of plant manipulation, practiced since agriculture began, is called *monoculture*. It not only intensifies the existing checks, but creates new problems that would totally destroy the farmer's crop and sometimes does despite his most strenuous efforts to enlist the forces of nature on his behalf.

Pests and man   The role of pests without the intervention of man can be seen in those remote areas of the world that are uninhabited by humans. Populations of plants, animals, parasites, and predators are usually well balanced before man's arrival. But when people appear, even in moderate numbers, they change the course of rivers and streams, dig new channels, make artificial barriers, impound water, and irrigate fields where only rain sufficed before. They cut down the trees, burn the shrubbery, rip open the soil that supports a great diversity of wild plants and animals, and put in their place expansive fields of one kind of crop. The disruption of the natural condition upsets the balance and encourages the buildup of what then comes to be known as pests. Instead of people living in harmony with nature, there ensues a battle, and forever after there is unceasing strife between man and pest. (See Figure 9.2.)

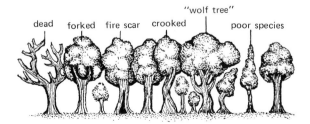

dead   forked   fire scar   crooked   "wolf tree"   poor species

Redrawn from M. M. Bryan, USDA.

FIGURE 9.2 Undesirable trees—weed pests by definition. Removal improves the forest.

Left photo courtesy of Ciba-Geigy.

Left photo: The codling moth larva—the "worm" in "wormy" apples—would leave little of the fruit if not controlled. *Right photo:* A tomato hornworm, or its close relative the tobacco hornworm, can eat in 28 days food weighing 50,000 times its original weight. During that time it increases in size 12,000 times.

An example was the opening of the prairies to settlement and agricultural development. The "corn belt" was originally a vast grassland with what is called an Andropogon Climax vegetation. The grass was the natural food of a small sucking insect called the chinch bug,[6] which lived off the juices of the plants and hibernated during the winter in the clumps of dead grass. People saw the potentialities of this rich grassland and quickly found that here they could grow luxuriant corn and wheat. They plowed out the Andropogon, planted crops of one or two species, cultivated and fertilized them, and created a succulent growth that was highly favorable to the chinch bug. Wheat in the spring and corn in the summer were made to order for the habits of the insect, which has two generations a year, and enabled it to build up to enormous populations. Despite the settlers' annoyance and their vigorous battle against the bugs, it was inevitable that the insects would destroy the crops.

COPING WITH PESTS

As soon as people could think of better ways to deal with pests than by scratching or clubbing them to death, they took recourse to whatever means were available. Flailing with bundles of sticks, a method of attack still in use against grasshopper invasions, was supplemented by the use of scoops, ditches, and traps of various kinds.

People of primitive cultures placed great store in ceremonial practices to control pests. When Europeans came to the east coast of America, they found natives with an advanced agriculture who were engaged in a continuous fight against insects, birds, and large animals. The Indian remedy for cutworms was for the squaw, who was responsible for planting and tending the garden, to step around the garden naked under a full moon, dragging her garment in a magic circle that would keep out the pests. A

---

[6]*Blissus leucopterus.*

Right photo courtesy of the Metropolitan Museum of Art.

Left photo: The chinch bug's natural habitat is the prairie. When settlers plowed up the prairie grass, the two-generation habits of the chinch bug *(Blissus leucopterus)* were made to order for wheat in the spring and corn in the summer. The chinch bug became a serious pest. *Right photo:* "Cleansing the Scalp" of lice, from *Hortus Sanitatus,* Mainz, Meydenbach, 1491.

similar custom was practiced in ancient Rome, but there, the quarter of the moon made no difference. However, there was an advantage if the woman was menstruating. The ritual was used, according to Pliny, to keep out canker worms, caterpillars, beetles, and "all such worms and hurtful vermin." The invention, said Pliny, came from the Cappadocians who, however, were more mannerly inasmuch as they had their women go only barefoot with their hair hanging loose. But they had to be careful not to do it at sunrise or the crop would wither away to nothing.

The early American settlers were inclined to experiment with more sophisticated remedies than those used by their neighbors to protect their crops and fruit trees. Vinegar, cow dung, and urine were applied in various ways. John Josselyn wrote of his observations during a visit to the New World in 1638:

> Their fruit trees are subject to two Diseases, the Meazles, which is when they are burned and scorched by the sun, and the Louziness, when the Woodpeckers eat holes in their bark; the way to cure them when they are louzy is to bore a hole in the main root with an augur, and pour in a quantity of Brandie or Rhum, and then stop it up with a pin made of the same tree.

There is no evidence that the extravagant measure recommended by Josselyn caught on with the thrifty, and thirsty, Yankees.

CHEMICAL CONTROL

The use of chemicals for pest control was an early innovation. An Egyptian scroll dating about 1500 B.C. gives formulas for preparing pesticides. Homer spoke in the Iliad, about 1000 B.C., of "pest averting sulfur," perhaps in reference to the early practice of burning sulfur to rid the premises of pests and disease germs. Sulfur dioxide, which is evolved from the burning of sulfur, is highly effective and is still used for fumigation of stored fruit. The magically curative effect of sulfur when pounded into a fine powder was inevitably discovered in the early days of the need for a palliative against crawling creatures. Elemental sulfur in finely divided form, as well as sulfur in inorganic chemical combinations, is still used in large quantities as a pesticide.

Democritus, a Greek philosopher of about 400 B.C., described a remedy for "blight" of plants. Cato (184 B.C.) told of rubbing crude oil, or the waxy portion of it, to cure scabies or sarcoptic mange, a skin ailment caused by a small mite. He described a pesticidal formulation consisting of a mixture of chemical substances for burning to produce an aerosol for killing pests of the vineyard. Dioscorides, a Greek physician of the first century A.D., said that scorpions could be paralyzed by putting larkspur on them. Pliny, a prominent Roman of the first century and the most popular science writer of ancient times, told how to control "mildew" of wheat seeds with a pesticide and described several concoctions for protecting plants and killing lice, nits, rats, and mice. Among Pliny's favorite remedies were sulfur and bitumen, as were materials of botanical origin such as olive extracts, powdered larkspur seeds and flowers, wild mint, and hellebore. The latter continued in use until modern times.

Many of the control measures for pest infestations in use during the

USDA photo.

Aircraft are used for many purposes in agriculture and forestry. This specially equipped airplane is one of many that are used to spray areas with high populations of grasshoppers.

past three hundred years have included ancient herbal remedies as well as more recently discovered botanicals. The more modern botanicals include the ground-up roots of certain tropical plants containing *rotenone*, the ground flowers or extracts of *pyrethrum* (a daisy originating in the Near East), and the powdered leaves or water infusions of *tobacco*. Urine had been a popular cure-all throughout the ages but was too rustic for most modern pest fighters. The antibiotic property of urea was rediscovered in 1942 by two American plant pathologists, George Zentmeyer and James Horsfall.

Upon discovery of the pesticidal effect of the more poisonous minerals, farmers and gardeners often turned to substances other than those of plant and animal origin. The Chinese were using arsenic to kill garden pests more than one thousand years ago. Arsenic, mercuric chloride, copper sulfate, soap, and turpentine came into use during the seventeenth and eighteenth centuries. By the time lead arsenate was invented in 1892 for use against a devastating invasion of the Gypsy moth that was defoliating forest and ornamental trees in Massachusetts, the beleaguered populace had little compunction against using all manner of known poisonous materials to protect vegetables and fruit trees against insects and fungi.

About fifty years after the introduction of lead arsenate, a fresh attack was made on the Gypsy moth with DDT, a material much safer to humans. The widespread revulsion against its indiscriminate use from airplanes in residential areas contributed to a popular distrust of pesticides and eventually the elimination of the use of DDT and other persistent pesticides from most parts of the United States and Western Europe.

MODERN PESTICIDES

There are about 600 different basic chemical compounds used for pest control, and as of June 1974, there were 34,029 formulations of these registered for use in the United States by the Environmental Protection Agency. They are mainly insecticides, herbicides, fungicides, algicides, and rodenticides. But they also include chemicals that are used against other pests, for example, nematocides (microscopic worms), molluscicides (snails and slugs), acaricides (mites and spiders), avicides (birds), piscacides (fish), lampreycides (parasites of fish), and echinocides (sea urchins and sea stars). Materials related to pesticides include repellents, attractants, and synergists. The latter are substances that enhance the effectiveness of a toxicant.

Pesticides include a variety of chemical types. The most commonly used are among the following groups.

Botanicals

Botanicals—materials derived from plants—although among the oldest of pesticides, still find useful applications because of their unique biological effects. The active ingredients called *pyrethrins*, extracted from the flowers of a species of chrysanthemum,[7] are used widely in household insect

---

[7]*Chrysanthemum cinerariaefolium* Vis. (Compositae).

sprays because they are practically nontoxic to humans, though extremely small amounts will kill fish in aquaria. Nicotine, an alkaloid extracted from the leaves and stems of tobacco,[8] was once a favorite insecticide for use on both agricultural and ornamental plants because it kills a wide variety of pests quickly. Though nicotine has the disadvantage of being extremely poisonous to humans and other higher animals, it decomposes or volatilizes within a few days to a few weeks and is no threat to the environment. Rotenone, which in pure form is a crystalline material, is extracted from the roots of several species of plants found growing in parts of Southeast Asia, Indonesia, and Central and South America.[9] It was discovered by the aborigines who used the material to catch fish by beating the sides of their canoes with shredded roots, releasing the poison into the water. The paralyzed fish could be easily picked up by hand. Other botanicals that find limited use are ryanodine, contained in the roots and stems of ryania,[10] a plant growing in Trinidad, and the alkaloids in the seeds of sabadilla,[11] a plant that grows in the highlands of northern Mexico.

Organochlorine Compounds

Organochlorine compounds include DDT, chlordane, dieldrin, aldrin, endrin, lindane, 2,4-D, 2,4,5,-T, and other widely used, though less well-known, pesticides. Many of the chlorinated hydrocarbons, a group of organochlorine compounds which includes DDT and its relatives, are highly stable chemical compounds. Many of them are *broad spectrum* pesticides, so-called because they are effective against a large number of species. This property gives a pesticide an economic advantage in that it can be broadly useful for controlling a wide variety of pest problems; but it has the disadvantage that the lack of a high degree of specificity may make it seriously injurious to beneficial organisms such as parasites and predators.

Characteristically, the chlorinated hydrocarbons are persistent in the environment, have an affinity for lipids (fatty tissues), are stored in the body fat, are metabolized (decomposed) within the body slowly, and therefore eliminated slowly, and are prone to be taken up by organisms and passed along in the food chain. Some of them, such as DDT, can be found in living organisms at concentrations many times that in the environment.

Organophosphorus Compounds

Organophosphorus compounds include parathion, malathion, diazinon, DDVP (dichlorvos), trichlorofon, and a number of others in this class of compounds. Nearly all the organophosphorus, or OP, compounds are used exclusively for their effectiveness against insects and mites. They

---

[8]*Nicotiana tabacum* L. (Solanaceae).

[9]Mainly from *Derris elliptica* (Wall.) Benth. (Leguminosae).

[10]*Ryania speciosa* Vahl. (Flacourtiaceae).

[11]*Schoenocalon officinale* (Schlecht. and Cham.) A. Gray (Liliaceae).

differ greatly in their insecticidal and miticidal action. Most of them are broad spectrum with only limited specificity. Parathion, for example, tends to be disruptive of the ecosystem by reason of its effectiveness against beneficial species. However, the chemical decomposes relatively quickly and does not constitute a significant long-term hazard to the environment.

The OP compounds also differ in their safety to users. Parathion (See Figure 9.3) is extremely poisonous to humans and other vertebrates as well as to invertebrate pests. Users must be instructed in appropriate precautions and safety measures. Malathion, on the other hand, is about as toxic as aspirin and is therefore an effective insecticide that can be used safely around the garden and on household pets. Most of the other OP insecticides are intermediate in their mammalian toxicity.

The OP insecticides indirectly affect the nervous system and for that reason have been called "nerve gases" or "nerve poisons." The primary action of the OP toxicants is to inactivate the enzyme acetylcholinesterase, the enzyme that is responsible for breaking down acetylcholine (ACh), one of the nerve transmitter substances, after it has performed its function of bringing about transmission of a nerve impulse across the junction between two nerve cells or between a nerve cell and an affector organ such as a muscle fiber. Thus it is accurate to speak of the OP

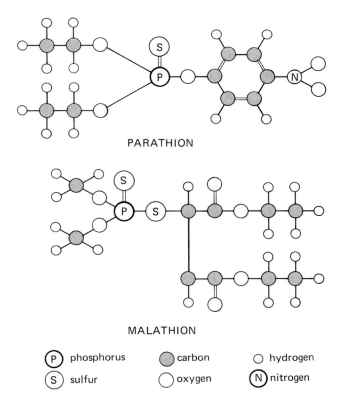

PARATHION

MALATHION

(P) phosphorus	⬤ carbon	◯ hydrogen
(S) sulfur	◯ oxygen	(N) nitrogen

FIGURE 9.3 Two insecticides belonging to the same chemical group (organophosphorus compounds) but differing greatly in their biological effects. Parathion is highly toxic and dangerous to handle; malathion is safe enough for backyard use.

pesticides as "enzyme poisons," although they are customarily called *enzyme inhibitors* or, more specifically, *cholinesterase inhibitors*.

Other classes of pesticides include a miscellany of chemical types, having a variety of effects on living organisms. It would be possible to classify them according to their biological effects, but the classification would be complex and overlapping.

BIOLOGY OF PESTICIDES

Selectivity   The primary consideration in the use of a pesticide, as with any biocide, is to find a substance that will have maximum effect against the offending organisms and minimum effect against the organism to be protected: plants, animals, other beneficial organisms, or humans. This property of being effective but safe is called *selectivity*. It is an ideal that is rarely achieved to perfection. In medicine the undesirable toxic effects are called "side effects." In pest control the injurious effects are referred to as phytotoxicity (effect on plants), mammalian toxicity (effect on mammals), human toxicity, or whatever specific undesirable effect may be of importance. The goal of future research on pesticides is to find materials that will be more highly specific in their action against the particular pest under attack, while being noninjurious to beneficial species, safe to humans, and nonpersistent in the environment.

Resistance   Many pests have the ability to develop resistance to the effects of certain chemicals (see Figure 9.4). Populations of organisms commonly become resistant to chemicals following repeated treatments that do not completely wipe out the population but leave some survivors to reproduce. This biological trait is a direct threat to health when disease-causing

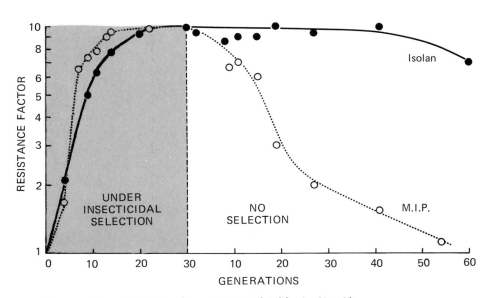

Figure courtesy of G. P. Georghiou, University of California, Riverside.

FIGURE 9.4 Development of resistance in colonies of house flies treated with insecticides Isolan or M.I.P. (m-isopropylphenyl methylcarbamate). Treatment was discontinued after 30 generations.

bacteria become resistant to medicinal drugs (see Antibiotics, Chapter 10). As with medicines and bacteria, pest populations often become resistant to chemical treatment, and when this happens, it creates difficult problems that may make it necessary to discontinue using the material and to find an effective substitute.

More than 230 species of insects have become resistant in different degrees to various pesticides. It is a major problem in malaria control in areas where mosquitoes have developed resistance to commonly used materials. However, the ability to adapt to such artificially imposed conditions as the intrusion of chemicals in the environment is not restricted to insects. Rats in one part of England became resistant to the anti-coagulant poison Warfarin. Resistance to pesticides has been developed by insects, bacteria, fungi, mice and other rodents, and by six species of weeds. Malaria took 2,500,000 lives annually in the mid-1950s. It seemed that control of the number one killer was in sight, but today thirty-eight species of anopheline mosquitoes, carriers of the protozoan parasite that causes the disease, are resistant to insecticides.

Resistance is also characteristic of bacteria and fungi. During the summer of 1970, a new race of the fungus[12] that causes southern corn leaf blight reached epidemic proportions, and before the season was out, destroyed an estimated 710 million bushels of corn worth one billion dollars. Reliance had been placed on a strain of corn (with T-type cytoplasm) widely used in hybrid seed production. The incident was a dramatic demonstration of a shift in the genetics of the host-parasite balance.

A familiar example is the danger of penicillin and other antibiotics resulting in the appearance of resistant strains of bacteria for which no control is available. Resistant strains of staphylococcus bacteria can develop within a few hours in the laboratory by artificial selection. Only 5 to 15 percent of staphylococcus strains found generally outside hospitals are resistant to penicillin, but astonishingly, 90 to 95 percent of those found in hospitals are resistant. It was found that 45 percent of a large number of children studied under the age of five years had penicillin-resistant staphylococci acquired from hospital nurseries. Penicillin resistance in *Staphylococcus aureus* is common enough to cause the abandonment of the drug in all serious systemic "staph" infections (see also *Resistance*, Chapter 10).

Many strains of microorganisms formerly sensitive to streptomycin also have become resistant in some degree to the drug. Infections from resistant strains are more dangerous because higher doses of less desirable drugs may have to be substituted. With pesticides, second choice materials may require very heavy applications, involving greater dangers to users and the environment.

Mechanism of resistance The development of resistance on the part of a species or strain is *genetic* in nature. In some cases a single pair of genes is responsible. A population of houseflies, for example, consists of individuals that differ in their susceptibility by reason of inherited characteristics. Even though most of the members may be highly susceptible to a particular insecticide, there will be a few individuals that have the ability to survive. The survivors will pass that ability on to some of their offspring, while the least

---

[12]*Helminthosporium maydis.*

resistant individuals of the original population will not have the opportunity to do so because they do not survive. The process is one of *Darwinian selection*, though artificially so by the introduction of a human-made substance into the flies' environment.

This selection process taking place over several generations of insects has been known to develop resistance to a pesticide many fold over the normal resistance of the original wild population. For example, DDT was initially highly effective against houseflies. Now, strains of houseflies that are completely unaffected by 100 percent DDT are used in research to find new insecticides that can be used against resistant insects. The result of selective pressure often involves simultaneous resistance to chemical toxicants other than the one that brought about the selection process. This is called *cross resistance*. But following removal of the toxicant from the insects' environment, the original susceptibility returns to the population slowly.

The problem of resistance is not easily solved, with medicines or pesticides. Resistance has developed to all pesticides that have systemic action. Even those materials that exert their effect by physical action, such as petroleum oil, may in time manifest genetic pressure to bring about strains that are more difficult to kill. However, in general, physically acting pesticides, cultural methods, and other ecological approaches to controlling pests are apt to have more long-range dependability than chemicals that exert highly specific modes of physiological action involving enzymes or other biochemical reactants.

Biomagnification

The chlorinated hydrocarbons are notoriously persistent under most conditions in the environment, and even living organisms generally find them resistant to breakdown and hard to get rid of. Any chemical that fails to degrade to harmless end products is apt to give trouble. A persistent chemical may be taken up by plants or animals and thus enter the food chain of both man and wildlife. When this happens, there is often a progressive increase in the concentration of the chemical as it passes from one organism to another in the food chain, resulting in what has been called *biomagnification*. Sometimes extremely high concentrations are found in the tissues of animals high in the food chain. For example, analyses for chlorinated hydrocarbons in several parts of the ecosystem of Lake Michigan showed that these substances were several hundred times as concentrated in herring gulls as in the fishes on which they fed, and several thousand times as concentrated as in small invertebrates that served as food for the fishes. Even the small invertebrates contained within their bodies almost 50 times as much as that in the bottom sediments of the lake. The biological importance of biomagnification is discussed further under DDT.

"DEFOLIANTS"

Few pesticides have caused more divisive controversy than the so-called defoliants (actually weedkillers) used as chemical warfare agents during the Vietnam war. During that episode of jungle warfare, the United States flew 6,539 missions between 1965 and 1971, including 4,561 defoliation missions and 858 crop destruction missions, for a total of 17.6 million gallons of weedkiller spray. Several formulas were used in which the most important active ingredient was a weedkiller compound based on 2,4,5-T.[13]

There were many complaints from people in and outside the United States. At first, objections to the use of defoliant sprays were based on the fundamental opposition to the use of chemical warfare agents in any form or for any purpose. But the opposition became even stronger and more vocal when it was discovered that 2,4,5-T could cause deformed offspring of animals in laboratory experiments. Subsequent studies showed that nearly all of the fetal effects were due to an impurity called dioxin, or TCDD,[14] that was formed during the manufacture of the weedkiller. TCDD is one of the most toxic substances known, and the fact that a potent chemical capable of causing fetal abnormalities was being sprayed over the Vietnamese countryside in wholesale quantities was cause for concern.

Several surveys were made in an effort to determine whether reported birth defects in children and animals born in the sprayed jungle areas were related to the use of the defoliants. The most comprehensive study was conducted by an international committee of seventeen scientists under the auspices of the National Academy of Sciences, which was instructed to proceed by Congress in 1970. The panel found "no conclusive evidence" linking defoliants to birth defects, but did not rule out a relationship between defoliants and symptoms of human illness, reports of which, in the words of the panel, "are so striking it is difficult to dismiss them as simply the effects of propaganda, high normal death rates, or faulty understanding of cause and effect."

The study group found that the ecological damage to the war-ravaged country of Vietnam was easier to assess, and that it was extensive. They concluded that the mangrove forests would not spontaneously recover for nearly a century, if at all, but that a massive reforestation program might cut the recovery time to two or three decades.

THE DDT LESSON

DDT is well known to most people. Few man-made substances have had greater impact on our worldwide attack on disease, the production of food and fiber, the public's interest in environmental contamination, our sociological outlook on science, and even on our scientific approach to technological problems.

DDT (Dichloro Diphenyl Trichloroethane[15]) was first made in 1874 by a German chemist who did not, however, become aware of its biological potency (Figure 9.5). The compound was synthesized again in 1939 by Paul Müller, a Swiss chemist, during a systematic search for new insecticides. It proved to be amazingly potent against a wide variety of insects.

---

[13]2,4,5-trichlorophenoxyacetic acid and derivatives, a family of herbicides that came into commercial use after World War II.

[14]2,3,6,7-tetrachlorodibenzodioxine.

[15]1,1-trichloro-2,2-bis(p-chlorophenyl) ethane.

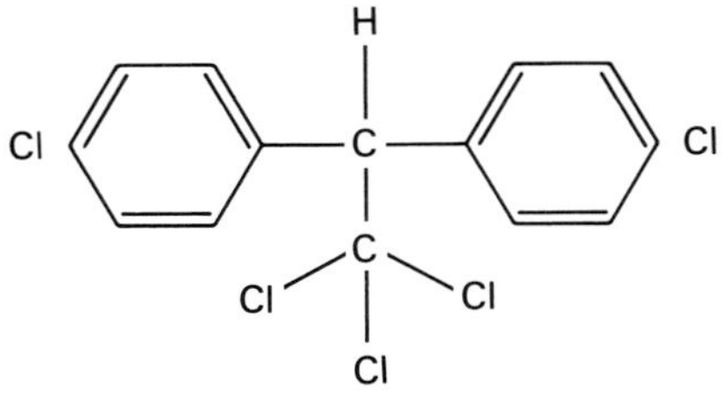

FIGURE 9.5 DDT (Dichloro Diphenyl Trichloroethane). The precise chemical designation is 1,1,1-trichloro-2,2-bis (p-chlorophenyl) ethane.

The material also appeared to be safer to humans than any known synthetic insect killer, a greatly desired property in view of the poisonous nature of lead, arsenic, and fluorine compounds, components of the most extensively used pesticides at the time.

Prevention of pestilence

The first large-scale use of DDT was by the United States in a campaign to prevent an impending epidemic of typhus fever in Italy during World War II. Typhus is a disease that is carried by the body louse, historically causing great loss of human life, especially during periods of social stress such as wars and famines. In Italy, people were lined up and the insecticide was applied directly to their heads and bodies in liberal amounts. The feared epidemic was prevented with no observable adverse effect from DDT on the treated population. Subsequently, DDT was used with dramatic success to kill the *Anopheles* mosquito carrier of malaria, the all-time number one human disease, causing 2½ to 3 million deaths a year and untold misery in sickness, lowered vitality, and economic deficiency. A representative of the World Health Organization of the United Nations credited DDT with saving the lives of possibly 50 million people from malaria alone. The new insecticide also came to be used extensively to protect food crops from the attacks of a wide variety of insect pests throughout the world. Some of the older, more poisonous materials were replaced. The Swiss discoverer of DDT, Paul Müller, was awarded the Nobel Prize in 1948 for his contribution to public health and nutrition.

The wicked angel

V. B. Wigglesworth, a well-known British insect-physiologist, warned in 1945 that DDT was a "two-edge sword." He was referring to the fact that DDT is a broad-spectrum insecticide and, therefore, could kill beneficial parasites and predators as well as the target pests. Unknown at the time, the more serious problem of accumulation of DDT in the environment and magnification in the food chain eventually caused widespread concern about the long-term effects on the environment and the ability of susceptible forms of wildlife to survive. Chemicals that upset the balance of nature had been known before. DDT was merely the latest and one of the most violent. Twenty-five years after Wigglesworth's warning, during which more than a billion pounds of DDT were introduced into the environment and during which DDT became the subject of more scientific investigation as well as public controversy than any other synthetic organic chemical ever known, DDT and related persistent insecticides were finally laid to rest and replaced generally by less persistent materials.

DDT IN THE FAT

The first signal of trouble was the finding in 1946 that DDT is stored in the fatty tissues and is eliminated very slowly from the body. Because of DDT's extremely low solubility in water (about 2 ppb [parts per billion]), the body, which is essentially a water system, cannot handle the fat-loving substance as it does most other foreign materials, but instead "sweeps it under the rug" by depositing it in the fat. For the same reason, DDT is deposited in the butterfat of milk—a potential danger to infants and others whose diet may be mainly milk. DDT is eventually broken down by the body and eliminated, but the process is slow and may require a period of several months. If the intake is continuous, a level may be reached at which the concentration in the fat remains relatively stable. Children may take 5 to 10 years to reach storage equilibrium.

In humans, the concentration varies. The average DDT in the fat of adults, samples of which were taken during the period 1954 to 1956, was about 5 parts DDT per million parts of fat. In a study of DDT in persons accidentally or violently killed in Dade County, Florida, during the period 1965 to 1967, the average concentration ranged from 5 ppm to 22 ppm in the fat. There was more in adults than in children, more in non-whites than in whites, and more in non-white males than in non-white females. Samples from four different locations in the United States from 1961 to 1962 gave an average of 6.7 ppm. Quantities as high as 109 ppm have been found in people sampled from the general population. A reasonable estimate of the average DDT load in industrialized or highly developed agricultural areas where DDT usage may have been high would be an average of 5 to 10 ppm DDT in the fat of adults.[16] However, it has not been demonstrated that these amounts are harmful to human health.

DDT IN THE DIET

The U. S. Food and Drug Administration initially established a tolerance[17] of 7 ppm in most foods and a zero tolerance in dairy products. The tolerance was later changed to 0.05 ppm in milk because of the impossibility of achieving zero residues with the refined methods of analysis finally developed. The belief by scientific investigators that DDT is relatively harmless to humans was supported by studies on volunteer prisoners who consumed large quantities of DDT with their food (200 times that in the normal diet) for periods of several months. No adverse effects could be detected.

Studies on animals, however, showed that prolonged feeding of low levels of DDT resulted in cellular changes in liver tissue as well as other pathological and histological effects. Therefore, the suspicion that DDT

---

[16]Because the human body consists of about 20 percent fat, give or take a considerable amount, the total body concentrations can be roughly estimated at about one-fifth the figures given. In general, the concentration of DDT in the blood is about $1/1,000$ of that in the fat.

[17]A tolerance is the maximum allowed (see Chapter 10).

A tractor-pulled mechanized sprayer applies DDT in an apple orchard. DDT came to be the most widely used insecticide on fruit trees and vegetables as well as to control insect carriers of human diseases.

might eventually cause damage to human physiology was kept alive and prompted investigators to further pursue the problems of potential DDT injury.

Residues of chlorinated organic pesticide chemicals were found in all diet samples and in all food classes except beverages in a comprehensive study made during the period from 1964 to 1966.[18] Based on the findings, the daily intake would be 0.0014 milligrams per kilogram of body weight, equal to less than 0.1 milligram per day for a 150-pound person or 36 milligrams per year. The expected intake for an entire year was calculated to be only a small fraction of that required for a lethal dose. DDT accounted for one-third of the total. The average lethal amount for DDT in a single dose is usually taken to be about 8,000 to 14,000 milligrams per 150-pound person, based on animal experiments. Because the use of chlorinated hydrocarbon pesticides has been curtailed sharply since the study, it is expected that smaller quantities are being taken in with food and that residues in human fat will decline. (For pesticides in people, see Food Contaminants, Chapter 10.) This may avert a controversy because more recent studies showed DDT to have carcinogenic properties.

DDT IN THE ENVIRONMENT

The second warning flag came many years later with the slowly revealed knowledge that DDT is exceptionally long-lasting under natural condi-

---

[18]R. E. Duggan and J. R. Weatherwax, "Dietary Intake of Pesticide Chemicals," *Science* 157:1,006–1,010 (1967).

tions and cannot easily be decomposed by microorganisms and physical forces in the environment. It was already known that accumulation of arsenic in the soils of some orchards sprayed with the older insecticide lead arsenate had reached 1,400 pounds[19] per acre (an acre is about 208 square feet). The arsenic was more injurious to growing plants than the accumulated lead which reached similarly high levels. Accumulations of DDT in orchard soils have been found up to 100 pounds per acre. In soils treated with DDT for control of the grubs of Japanese beetles, the loss of DDT over a ten-year period was only 80 percent for heavy applications and 90 percent for light applications.

Biomagnification of DDT

The extent of DDT accumulation in the environment was not fully realized until analyses of fish, birds, and other wildlife showed high concentrations in the fat and other tissues of some species. Large amounts were found especially in those species that are several steps along the food chain, such as fish-eating birds. For example, DDT is deposited in the coastal areas of the ocean, where it is taken up by algae and other microorganisms; higher marine organisms eat the algae, and a succession of prey can pass the DDT along with "magnification" of DDT concentrations at each step in the food chain. Alarmingly high concentrations— thousands of parts per million—have been found in the fat of some predaceous birds and sea animals.

DDT is widely distributed in man and animals. It is found in the fat of native Alaskans in isolated areas, in reindeer in the arctic, in penguins and seals in the antarctic, and in fish, birds, and other animals in widely separated parts of the globe. Some shellfish can magnify DDT 70,000 times the concentration in the environment. As much as 3,000 ppm have been found in the fat of wild pheasants. Deaths of the western grebe, a fish-eating bird, resulted when Clear Lake, California, was treated with DDD,[20] a close relative of DDT, to control gnats. The chemical was applied three times over a ten-year period to give a concentration of 0.02 ppm each application. The treatment resulted in 5 ppm DDD in plankton; as much as 800 ppm in the visceral fat of blackfish, a phytophagous (plant-eating) fish; 2,275 ppm in the fat of the large-mouth bass, a carnivorous fish; and 1,600 ppm in the western grebe, a fish-eating bird. Another classic example of biomagnification of DDT in the food chain was brought about by the spraying of elm trees, which resulted in 99 ppm of DDT in the soil, 140 ppm in earthworms, and more than 400 ppm in robins that fed on the worms.

REPRODUCTION OF BIRDS

The third danger signal was the discovery that high DDT concentrations in the tissues of certain birds were associated with thin eggshells and loss of reproductive capability. Reproduction of brown pelicans[21] at their an-

---

[19]Expressed in arsenic trioxide.

[20]Also called TDE, 1,1-dichloro-2,2-bis (*p*-chlorophenyl) ethane.

[21]*Pelecanus occidentalis*.

 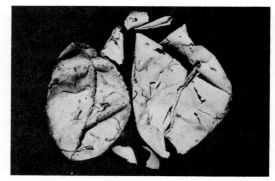

Left photo by Daniel W. Anderson, U.S. Fish and Wildlife Service.

Left photo: California brown pelicans *(Pelecanus occidentalis californicus)* at a rookery on Isla Pelícano (Pelican Island) in the Gulf of California. *Right photo*: Eggs of the brown pelican that never hatched. Environmental contaminants that get into the food chain, such as DDT and PCBs, cause thin egg shells and poor reproduction of some birds of prey.

cestral rookery on Anacapa Island off the coast of southern California is drastically reduced. The finger of guilt pointed toward DDT by reason of its presence in high concentrations in eggs, chicks, and adults and by the discovery that many toxic substances, including the chlorinated hydrocarbons to which group DDT belongs, bring about an increase in the production of detoxifying enzymes by the microsomes of the liver cells. For further discussion on the decline of the pelican, see *Endangered Wildlife*, Chapter 19.

The way DDT acts in susceptible species of birds is not known for certain. It is known that DDT and its principal metabolite, DDE,[22] are powerful inducers of the production of detoxifying enzymes by hepatic (liver) cells. The detoxifying enzymes are not specific in regard to the compounds that they attack. They also readily decompose some of the naturally occurring steroid hormones that are involved in reproduction and, in birds, in the withdrawal of calcium from secondary bony structures called medullary bones at times when large quantities of calcium are needed for deposition in the eggshells.

DDT is not the only substance that causes these effects. Many chemicals that are foreign to the system induce the production of enzymes, the apparent function of which is to accelerate the breakdown and elimination of the extraneous substance. Numerous drugs such as phenobarbital, the carcinogenic agent benzpyrene, and a number of chlorinated hydrocarbon pesticides cause the production of nonspecific detoxifying enzymes in laboratory animals. A group of products known as PCBs (polychlorinated biphenyls) are extensively used in plastics to modify and control their physical characteristics. Such ingredients are called *plasticizers*. PCBs are present in the environment and are found, often in company with

---

[22]Metabolites are products of biochemical degradation resulting from the attack of naturally produced enzymes, often part of the body's mechanism to more easily rid itself of a foreign substance. DDE is 1,1-dichloro-2,2-bis (p-chlorophenyl) ethylene.

DDT, in the fatty tissues of wildlife. Laboratory experiments indicate that PCBs may be even more potent than DDT as inducers of nonspecific detoxifying enzymes.

The contribution of each of the potential troublemakers is not yet known. However, the long life of DDT in the environment and its known biological effects were compelling reasons for putting the brakes on the use of the compound except where there was no satisfactory substitute for the protection of nutrition, human health, or natural resources. Unfortunately, more dangerous materials such as dieldrin and endrin were often substituted because there was relatively little opposition to their use. Eventually, even these fell by the wayside when, on August 2, 1974, the Environmental Protection Agency ordered a halt in the production of aldrin and dieldrin and prohibited their use except for specific minor purposes. Evidence had been developed, ironically in the manufacturer's own laboratory, that the pesticides caused an increase in cancerous tumors of mice and rats. On July 30, 1975, a ban was announced on two more chlorinated hydrocarbon insecticides, chlordane and heptachlor, because they were showing up in unborn babies, mothers' milk, and the body tissues of 97 percent of the people. They were considered a potential cancer threat.

THE DECLINE

The virtual disappearance of DDT from the domestic scene was a process that took place over a period of several years. Peak production of DDT in the United States occurred in 1959 when 156,741,000 pounds were manufactured of which 78,682,000 pounds were used within the country, the difference being the amount exported or carried over. During the 1960s the domestic use of DDT decreased steadily, mainly because of replacement by more effective or otherwise more desirable insecticides. By 1969 the quantity of DDT used in the United States had dropped to 30,256,000 pounds, less than half that used a decade earlier, and by 1971 its use had further declined to 18,000,000 pounds. However, DDT continued to be used in large quantities in other parts of the world. The total volume probably remained higher than for any other single insecticide.

Meanwhile, efforts were made to curtail the use of DDT by legislative and court action. Arizona placed a moratorium on the agricultural use of DDT beginning in January of 1969, and other states followed. West Germany placed a ban on DDT, except for forest use, effective May 1971. Sweden, Norway, and Denmark had already banned DDT. Finally, in the United States, the Environmental Protection Agency announced an almost total ban on the use of DDT effective December 31, 1972.

In the end, it was not any specific injurious effect from DDT that led to the rulings which drastically restricted the use of DDT. There were several reasons for the concerted attack on DDT from several directions.

One of the lessons to be learned from the DDT experiences is that similar environmental pollution could occur with other potentially dangerous materials about which we have inadequate information. Lead, mercury, polychlorinated biphenyls, and certain radioactive substances are now

Left photo by Karl Kenyon, courtesy of the Bureau of Sport Fisheries and Wildlife, U.S. Dept. of Interior. Right photo courtesy of Latter Day Saints, Salt Lake City.

Left photo: The American bald eagle, an endangered species of large birds. DDT was suspected of contributing to the decline of several predatory birds. *Right photo*: A devastating plague of Mormon crickets in the Salt Lake Valley of Utah in 1848 was eliminated by flocks of seagulls, now the state bird of Utah. This monument stands in the grounds of the Mormon temple at Salt Lake City.

familiar examples of environmental contaminants that are highly toxic, persistent, and known to be either concentrated or passed along in the food chain.

BIOLOGICAL CONTROL

According to Dr. E. F. Knipling, science advisor at the U.S. Department of Agriculture, emphasis has been placed on finding biological control methods as alternatives to pesticides since 1955, and for several years three-fourths or more of the federal research activities have been in that direction. Progress has been slow, partly due to the fact that alternate methods are necessarily more selective, often more costly, and not as easily adaptable to individual operations as, for example, spraying an apple orchard with a spray-oil. Some of the methods of counterattacking insects are discussed in the following chapters.

The prospect of controlling pests by pitting one organism against another has fascinated people for centuries. Aristophanes (about 445 B.C.), in *The Birds*, wrote of hopes for controlling pests of the fig:

> The locusts and mites will no longer delight
> In the buds of the fruit they devour.
> A squadron of robins will swoop from the sky
> And clean out the pests in an hour.

One of the best-known instances of natural control of a devastating pest in modern times was that of the Mormon cricket,[23] an insect belonging to the group of Katydids and green grasshoppers. In 1848, a plague of the insects threatened to ruin the crops of the newly arrived Mormon settlers in the Salt Lake Valley of Utah. As starvation seemed imminent, flocks of seagulls suddenly appeared and ate enormous numbers of the insects. A monument now stands in the grounds of the temple at Salt Lake City in honor of the pioneer Mormons' faith, and the seagull is the state bird of Utah.

ECOLOGY OF VICTORS AND VICTIMS

Weeds, insects, rodents, and many of the other groups of organisms that include pestiferous species multiply with incredible rapidity. Insects may increase in number a millionfold or more in the course of a few months. Their numbers would overwhelm the earth if it were not for the natural forces that hold them in check. Changes in the seasons, sudden fluctuations in heat and cold, drought and flood, disappearance of the food supply, and pressure from competitors each contribute their part to natural population control. Moreover, the crawling creatures, as well as their larger counterparts, are engaged in unceasing strife with natural enemies. Thus the population of a species in a locality may undergo large fluctuations in which periods of high population density alternate with periods of greatly reduced numbers.

As a rule, however, the abundance of a species does not change very much, unless there is intervention by an outside agency such as man. Though the female of an insect may lay several hundred eggs, it is not often that all of them come through to maturity. Perhaps only two or three of the young ones that hatch will survive the rigors of the environment long enough to reach maturity. Thus, survival may be only a fraction of a percent of the total reproductive potential. Parasites and predators play a major part in this colossal natural mortality. Organisms that feed on their prey externally are predaceous; those that lay their eggs within the bodies of their prey, or in other ways manage to work from the inside and devour their victims slowly, are *parasites*. Microorganisms such as bacteria and fungi that destroy their victims by causing diseases are called *pathogens*. The biological pressure of parasites, predators, and pathogens is often a major factor in bringing about a decline, or maintaining a low level, in the population of a species and is sometimes crucially important in pest control.

The ladybird and the cottony cushion

One of the most spectacular cases of managed biological control was that of the cottony cushion scale,[24] an insect that feeds by settling down on a plant, inserting its slender beak into the tissues of the host, and sucking the juices. The common name of the insect comes from the mass of fluffy white wax that it secretes from pores in its upper surface.

The cottony cushion scale attacks citrus trees (oranges, lemons, grapefruit) as well as other kinds of trees and shrubs. The scale was first noticed in the United States in 1872 and was thought to have been accidentally introduced from Australia three years earlier. By 1883 the insect

---

[23]*Anabrus simplex*.

[24]*Icerya purchasi*.

had invaded the orange-growing areas of California with such ferocity that it threatened to wipe out the industry.

Most of the bizarre insecticides of the time were tried: lye, soap, kerosene, bitter aloes, sal soda, turpentine, lime water, pyroligneous acid, and Paris green (a highly poisonous arsenical). Pyrethrum was tried, as well as fumigation with various poisonous gases, the most effective of which was hydrogen cyanide. Charles Valentine Riley, chief of the Federal Bureau of Entomology, decided to put a team to work on finding natural enemies. Albert Koebele was sent to explore Australia, which was believed to be the pest's native habitat, while his teammate, Daniel Coquillet, was to receive any parasites and predators that Koebele might send back and find ways to culture them in captivity in preparation for releasing large enough numbers for them to become established in their new home.

Koebele was almost instantly successful. He collected and sent back about 12,000 living specimens of a well-known parasite. But more importantly, Koebele also discovered a then unknown predator, a small ladybird beetle called *Vedalia*. He was able to find a few specimens and send them back. They were carefully cultivated by Coquillet, and the progeny were distributed throughout the orange-growing areas of California. They thrived in California and their effectiveness in cleaning up the pest was phenomenal. Within a year the cottony cushion scale was no longer a threat. The Vedalia beetle,[25] in conjunction with the original parasite and another predator, were subsequently used in other parts of the world with similar success. The Vedalia, which arrived in the United States in 1888, marked the beginning of an era in insect pest control during which attention would be increasingly given to the utilization of natural enemies.

The prickly pear

The spread of the prickly pear in Australia is a superb example of the uninhibited spread of a pest in a new land where it found favorable conditions and few, if any, enemies. Cactus is not only a spectacular plant that is an example of biological control but is also of great interest because of its ecological importance. The family Cactaceae is native to North, Central, and South America. A number of species have been transported to various parts of the world as ornamental plants and curiosities. Several of them were introduced in Australia where they became too well established. The first introduction of cactus was in 1788. The species that became established as the main pest in the country was introduced in 1839 when a plant was brought in a flower pot to Scone, New South Wales.[26] Plants or cuttings were transported to other areas and it came to be grown as hedges around homesteads. It eventually got beyond control and in 1895 was listed as a noxious weed in Queensland. By 1920 the prickly pear covered 60 million acres and it was spreading at the rate of about a million acres per year, rendering much of the land useless for grazing or agriculture. Following many years of destruction of grazing land, the

---

[25]*Rodolia cardinalis.*

[26]*Opuntia Bentonii* is one of more than 350 species of the genus *Opuntia,* many of which have become pests.

weed was brought under reasonable control by a combination of bulldozing and the use of a tunneling caterpillar,[27] a cochineal scale insect, and other species of introduced cactus enemies. Cactus diseases also assisted.

The rabbit virus   The rabbit was intentionally introduced into Australia from England in 1859. It thrived, multiplied enormously, and quickly became a serious pest. It was estimated that by 1950 the animals were eating as much grass as 40 million sheep. Attempts were made to subdue the rabbits by a virus disease called *Myxomytosis*. This is a disease that normally produces only a mild tumor in wild rabbits of Brazil. But when a colony of European rabbits was imported to Brazil and came in contact with the disease in 1896, the virus caused a severe form of tumor that was fatal to these domestic European types.

It was seen that the virus might be useful for controlling the rabbits that had become pests in Australia. Diseased rabbits were taken to Australia and liberated in 1926 without success, but the method was tried again in 1950 in areas where the insect vectors (transmitters) of the virus were prevalent. The mosquito carriers[28] spread the disease initially along waterways, later into the drier areas. The disease spread rapidly over a half million square miles. There was a tremendous reduction in the rabbit population, resulting in an increase in the abundance of grazing plants.[29]

There are numerous examples of biological control that are either completely successful or sufficiently so to be significant factors in keeping the infestations down to a level that can be dealt with easily and economically by supplementary means.

LIMITATIONS AND PROBLEMS IN BIOLOGICAL CONTROL

Insecticides and beneficial insects   Insecticides, if improperly used, can be counter-productive. Following the dramatic success in biological control with the Vedalia ladybird, trouble was not long in coming. Those who had placed their faith in biological control had failed to reckon with the potentially counter-productive effects of insecticides. The cottony cushion scale, defying the parasites, was found attacking pear trees in destructively large numbers. Large quantities of lead arsenate were being sprayed on the trees to kill codling moth larvae, the insect that causes "wormy" apples and pears. It was found that lead arsenate stuck to the surfaces of the cottony cushion scale insects in amounts that were poisonous to the ladybird predators.

The pear growers were faced with a dilemma. They needed the lead arsenate, without which the crop would be almost a total loss as they knew from previous experience, but they also needed to get rid of the cottony cushion. They eventually cleaned up the scale insects by knocking them off the trees with jets of water under 300 to 400 pounds pressure and by spraying with oil during the winter dormant period of the trees.

---

[27]*Cactoblastis cactorum.*

[28]*Culex quinquefasciatus* and other species.

[29]H. L. Sweetman, *The Principles of Biological Control* (Dubuque, Iowa: William C. Brown Co., 1958).

TABLE 9.1   Control of Insect Pests by Biological Control

Degree of Control	Number of Locations	Number of Pest Species*
complete	25	21
substantial	66	53
partial	68	51

*There is some overlapping in the number of species controlled, i.e., some species that are substantially controlled in one location are only partially controlled in another location.
Compiled from data in Paul DeBach, ed., *Biological Control of Insect Pests and Weeds* (New York: Reinhold Publishing Corp., 1964).

As with other methods of dealing with pests, biological control is not always successful. The fact that pests, from the human viewpoint, are abundant in nature is impressive evidence that natural control is seldom complete. If natural control were perfect, the victims would die out and become extinct, and so would their natural enemies from lack of food. Parasites, predators, or pathogens may result in various degrees of effectiveness (see Table 9.1). Sometimes two or more natural enemies are required, and often supplementary methods such as mechanical or chemical treatment are needed. John Curtis wrote of this in 1860:

> . . . . for it is a wise dispensation of providence to keep every animal in check by some other that is either more powerful or more sagacious than itself. . . . so that in a greater or less space of time the destructive power may be rendered no longer formidable. . . . This natural process, though never failing, is often too slow in its operation to secure immediate relief. . . . The farmer must, therefore, devise means if possible for the more speedy destruction of the enemy.[30]

Natural biological control takes place to some degree with almost all pests. When left alone, parasites and predators usually keep their victims in check to various degrees depending on the ecological balance between the biotic potential and the environmental resistance of both eaters and eaten. But stable equilibrium does not always hold true. Sometimes an animal or plant finds its way to a new geographic region and manages to leave its enemies behind. For example, larvae of the Japanese beetle,[31] a beautiful metallic green insect, were accidentally brought to the eastern United States from Japan on the roots of nursery stock. From the time of the beetle's arrival in New Jersey about 1916, it spread rapidly, and promptly became a destructive pest.

In other cases, plants and animals have been deliberately introduced for a supposed useful purpose, but later became pests. The starling,[32] a common bird in Europe and parts of Asia and north Africa, came to the United States in 1890 when 60 of the birds were released in Central Park in New York City. About 40 more were set free in 1891. Millions of starlings now live in the United States from coast to coast. Because they eat

---

[30]J. Curtis, *Farm Insects* (Glasgow, Edinburgh and London: Blackie & Son, 1860).

[31]*Popillia japonica* (Scarabaeidae).

[32]*Sturnus vulgaris* (Sturnidae)

fruits as well as insects, they become seasonal pests to farmers and gardeners, and they roost in trees and buildings in such numbers that they are a nuisance even to city-dwellers. The virtual elimination of the American chestnut is another example of the destruction that can be brought about by an introduced parasite, in this case a fungus[33] that wiped out the large chestnut forests of the eastern United States. At least half of the most destructive insects in the United States originated in foreign lands.

It is not unusual for these invaders from foreign lands to run amok with uninhibited frenzy, often causing great damage to food crops, livestock, or human health. Thus the most spectacular cases of biological control, and in general the only cases where managed biological control is practicable, are those cases in which the pests are not native to the area and have, therefore, escaped from their worst natural enemies until such time as human intervention can restore the balance.

Successful use of biological control requires an intelligent understanding of the pest and its potential natural enemies as well as careful screening to determine whether the introductions are capable of doing more harm than good. The introduction of a new species can cause great harm if it has varied food habits or multiple hosts.

The mongoose[34] was introduced in Puerto Rico in 1877 to control rats. For a time the rats were reduced, but they soon learned to escape the mongoose by taking to the trees. The mongoose turned to other food and attacked poultry and wild birds. Worst of all, it destroyed the subterranean lizard, *Ameiva exsul*, which fed on white grubs, the larvae of June beetles. The white grubs became pests. Partial balance was reestablished by the introduction of the giant toad, *Bufo marinus*, to feed on the insects. Elsewhere, the effects of the mongoose have varied, with good to bad results against snakes, rats, mice, birds, lizards, toads, and frogs.

Rabbits infected with the Myxoma virus were intentionally released in France in 1952 and in England in 1953, with serious consequences. There was a distressing reduction in the rabbit population in parts of Europe where rabbits are an important game and food animal. The disease spread rapidly in both France and Britain, and each season destroyed about 90 percent of the afflicted population. (In Europe the means of transmission from rabbit to rabbit was not a mosquito vector but the rabbit flea.) The losses continued year after year for ten years, after which it declined, either because the rabbits developed resistance to the disease or because the virus became less virulent. But the problem continued to be serious and numerous attempts were made to check the virus.

In Europe, it was fortunate that the virulence of the pathogen had a tendency to subside, thereby saving the rabbit population from annihilation. But in Australia, where the rabbits were pests and the goal was to eradicate them, there was also a decline in the effectiveness of the virus. The disease was at first 98 percent effective, but the effectiveness dropped, either due to the appearance of an attenuated strain of the virus or, as some people believed, because immunity among the rabbits developed by natural selection.

Insects can be no less tricky. When Koebele went to Mexico in 1902 to look for natural enemies of lantana in its native habitat in the warmer parts

---

[33]*Endothia parasitica,* the causative agent of chestnut blight.

[34]*Mungos birmanicus.*

of America, he found 225 species of insects feeding on lantana and sent 23 of them to Hawaii. It is not surprising that some of the pests attacked other kinds of plants.

In another case, the black scale,[35] a major pest of many kinds of plants in tropical and subtropical areas throughout the world, was brought under partial control in parts of California by a parasite found in South Africa and Australia. Another parasite[36] from Australia was introduced later; but instead of attacking the black scale it turned out to be an enemy of the first parasite. Thus, the work of the primary parasite was disastrously nullified by the secondary parasite.

The introduction of the gypsy moth[37] in New England set off a chain of events finally culminating in upheavals in scientific, political, sociological, and economic trends that are still taking place. The gypsy moth is generally thought to be the most destructive pest of forest and shade trees in the northeastern United States, sometimes doing millions of dollars worth of damage each year. It is one of many examples of an organism that causes only moderate or infrequent damage in its native region but, when introduced into another part of the world, becomes vicious and costly.

The gypsy moth was accidentally liberated by a scientist who hoped to develop a hybrid race of silk-producing insects that would be resistant to the silkworm disease which was then causing great losses among silk producers. In 1869 he brought egg clusters of the European insect to Medford, Massachusetts, where he intended to try crossing the adults with silkworm moths. Some of the gypsy moth larvae escaped and soon became well established in surrounding woodlands. By 1889 forest and shade trees were being defoliated over 360 square miles. Massachusetts officials tried to exterminate the pest but relaxed their efforts when it appeared to be under control. Unfortunately, the infestations subsided only temporarily, and by the time control measures were resumed the insect had spread widely (see Figure 9.6). A new insecticide, arsenate of lead, was developed to combat the pest. Numerous attempts were made to "reeradicate" the pest during the next half century with indifferent success, although it was believed by federal agencies that the insect had been retarded. Finally DDT was used on such a massive scale by aerial spraying that the residents of sprayed areas became alarmed and angry. The controversy stimulated a wave of scientific investigations unparalleled by studies of any other synthetic substance and eventually led to the banning of DDT, discussed earlier.

ALTERNATE PEST CONTROL METHODS

Some of the more successful methods of controlling harmful organisms do not involve the use of either chemicals or parasites and predators. Such control measures have been in practice for many years. Some of them are

---

[35]*Sassetia oleae.*

[36]*Quaylea whitteri.*

[37]*Porthetria dispar.*

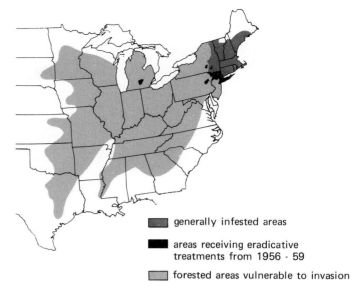

Redrawn from J. O. Nichols, *Pennsylvania Misc. Bul. 4404*, 1961.

FIGURE 9.6 Gypsy moth eradication efforts.

only partially effective; some are effective only under favorable conditions; and some of them can be relied upon for complete protection under suitable circumstances. Time-tested methods include:

- Cultural practices such as the rotation of crops to provide a period during which the pest has no susceptible host plant to attack.
- The selection of varieties of plants that are resistant to attack or that are more tolerant of attack.
- The use of resistant root-stock on which the desired variety is grafted.
- Timing the planting date to come before or after the period of high susceptibility to pests; for example, avoiding those periods when the soil is so damp that fungus growth is encouraged, or when insect pests are normally emerging.
- Plowing or otherwise disturbing the overwintering environment needed for completion of the life cycle of the organism.
- Flooding the area temporarily until pests such as soil insects and fungi are killed.
- The drainage of water from the breeding areas of aquatic organisms, such as mosquitos, which require water for the development of the immature stages.
- The removal of diseased plants from a field to prevent or impede further spread of the causative organism.
- The construction of barriers which the invading organisms can cross only with difficulty, such as ditches and fences of various sizes and kinds to impede insect migrations, and protective devices around the trunks of trees to keep out rodents.
- Weeding to eliminate alternate host plants on which the pests may be breeding either in the field or nearby.
- Traps, such as attractive light-traps, bait-traps, and electrocuting traps.
- Heat, such as that used for the direct sterilization of clothing and bedding.

Photo courtesy of Paul DeBach and Van Nostrand Reinhold Publishing Company.

Studies on biological control have been conducted in these facilities constructed in the early 1930s for rearing insect parasites at the University of California, Riverside.

- The removal of breeding material such as manure, lawn clippings, or other organic material in which pests, such as flies, multiply.
- Covering breeding material, such as that in garbage cans, with lids or other devices to prevent access to organisms.

NEW APPROACHES TO PEST CONTROL

Ingenious and imaginative approaches to the problems of safer and more effective pest control have been under intensive study during the past two decades. Some of the new methods have become established as practicable procedures.

Integrated or managed control

Many of the most destructive pests are those which attack food crops. Agriculture, by its nature, involves high-density domestic cultivation of plants or husbandry of animals. Pests are liable to find congenial breeding grounds in such artificially produced food supplies. Pesticides are widely used to improve yields and to prevent deterioration in the quality of the product caused by insect feeding or infection by fungi and other micro-organisms. But there are numerous examples of adverse effects of pesticides on beneficial organisms. Therefore, efforts are made to use pesticides when needed with maximum effectiveness for their intended purpose while causing the least amount of damage to friendly parasites and predators. This is called *integrated control* or managed control. Integrated control involves an evaluation of the total ecology of the plants or animals to be protected.

Sexual sterilization

An imaginative new approach was the control of a pest population by the release of sterilized males. It is sometimes referred to as a method of *genetic* control. The first successful eradication program by means of liberating sexually sterilized organisms was on the West Indies island of

Curaçao off the coast of Venezuela, in 1954, in a campaign against the screw-worm fly, [38] a damaging pest of livestock. Males were sterilized by exposing the pupae[39] to cobalt-60 radiation. The sterile insects were released over the entire island at an average rate of 435 males per square mile per week. After 8 weeks, 100 percent of the eggs laid by female flies were infertile, and after 13 weeks no egg masses at all could be found. Before the test was started it had already been determined in the laboratory four years earlier that when screw-worm pupae were irradiated with x-rays within 2 days of the time the adults were to emerge from the pupal stage, the males could be sterilized with a dose of 2,500 roentgens while it took twice that dosage to sterilize the females.

The stage was set for the Curaçao experiment many years earlier by experiments that had been conducted on the effects of x-rays on insects in the laboratory. It was noted as early as 1916 that if adult cigarette beetles were irradiated with x-rays, they laid eggs that were infertile. About ten years later it was found that mutations followed x-ray treatments of the laboratory fruit fly *Drosophila melanogaster*, and that if the dosage was high enough, the males and females became sterile. It was theorized in the late 1930s that if the males of the screw-worm flies could be sterilized without impairing their mating ability, it might be possible to eradicate an isolated population because it was known that the females mate only once during their lifetime. A large number of sterile males mixed in with the normal male population would inevitably have a disastrous effect on reproduction. Several attempts were made in the field to prove the hypothesis before the Curaçao experiment, but with negative results. Thus, the success of the sterilization method was not a spontaneous discovery but was the consummation of tedious trial and error over a period of twenty years directed toward the specific goal of eradication of a pest by taking advantage of its habits and behavior combined with the application of sophisticated scientific technology.

Following the Curaçao success, a more ambitious program was conducted from 1958 to 1962 in an attempt to eradicate the screw-worm from its range in the southern part of the United States. This campaign was also successful, although it was found to be necessary to release sterile males periodically because of the influx of reinfestations from south of the Mexican border, where the flies are abundant. Also, after several years, the method no longer seemed to be as effective.

Some other attempts to eradicate insect pests by sexual sterilization were unsuccessful. But one of the most successful following the screw-worm experience was with the melon fly[40] on the South Pacific island of Rota. The flies were eradicated by a method of integrated control involving the use of a baited poison spray and the release of sterile insects. The oriental fruit fly[41] was also eliminated from Guam by the release of sterile

---

[38]*Cochliomyia hominivorax.*

[39]*A quiescent immature stage of the insect. Transformation from larva to adult takes place within the pupal case.*

[40]*Dacus cucurbitae,* one of the tephrited fruit flies.

[41]*Dacus dorsalis.*

insects. Experiments using similar techniques are being tried against the Mexican fruit fly[42] along the Mexico-California border, against the pink bollworm[43] attacking cotton in California, and against other pests.

Although the radiation sterilization method is perhaps the most spectacular to date of the unorthodox methods of pest control, other systems that are still largely in the experimental stage have potential for practical application. Some of those having various degrees of promise are described below:

- *Chemical sterilization* and release of males of the southern house mosquito[44] eliminated the indigenous population of the pests from an island off the coast of Florida. Chemicals that can be used for this purpose are called *chemosterilants*.

- *Hormones* are substances that are secreted by the glands of internal secretion, called *endocrine glands*. Insect hormones that control growth and development have been called "third generation insecticides" because of the prospect that such hormones, or synthetic materials that have similar action, can be used to interfere with the normal development of insects. In insects there are two interacting hormones, *ecdysone*, which stimulates normal development, and the *juvenile hormone*, which causes the insect to retain its immature form. Giant moth larvae can be produced by applying an excess of juvenile hormone, but the adult stage is never achieved, or if it is, the adult is abnormal and incapable of reproduction.

- *Pheromones* are substances that are excreted externally by organisms and have the function of affecting the behavior of other individuals. Sex attractants, trail markers, chemical warnings, repellents, or similar chemical signals are pheromones. The term is most commonly used to refer to those substances secreted by an animal to influence the behavior of other individuals of the same species. Many species of animals secrete sex attractants or stimulants. They are the aphrodisiacs of the animal world. In insects, females of some species secrete male attractants that are attractive in extremely small quantities. A trap containing a virgin female of the introduced pine sawfly[45] attracted more than 7,000 males during a 5-hour period, and after that, about 1,000 males per day for 5 days. Sex attractants have been used successfully in insect surveys. Knowledge of the chemical structure of attractants provides information that is useful in the development of synthetic pheromones.[46]

- *Genetic* control by the release of individuals having lethal genes is a possibility for bringing about the self-destruction of a population. Genes might be lethal only under certain conditions, causing the organisms to succumb, for example, to heat or cold.

- *Sonics* may have possibilities for mimicking the sounds of other organisms that are either attractive or repellent. The high-frequency shriek of insect-feeding bats has been suggested as a possibility for frightening insects away.

- *Irradiation* of wheat by a high-intensity gamma radiation source such as cobalt-60 has been demonstrated to be effective in killing insect eggs.

-----

[42]*Anastrepha ludens*.

[43]*Pectinophora gossypiella*.

[44]*Culex pipiens quinquefasciatus*.

[45]*Diprion similus*.

[46]M. Jacobson, *Insect Sex Attractants* (New York: Interscience Publishers, John Wiley & Sons, 1965).

- *Photoperiod* manipulations may be used to disrupt the life cycle of organisms that are sensitive to day length. Prolonging the length of the day with artificial light interrupts the development of the European corn borer and codling moth by preventing the larvae from going into diapause (inactive period).
- *Lures* (attractants) have possibilities that have been incompletely explored, especially for use in combination with toxicants and chemosterilants.

PROBLEM AREAS

The innovation of highly potent synthetic organic chemicals for pest control in the mid-1940s held forth the bright hope that it might be possible to eradicate, or at least bring under control, all of the more destructive pests that eat food crops, destroy physical structures, and carry diseases afflicting humans, livestock, and other domestic animals. But the massive use of several hundred kinds of pesticides that followed their introduction has had environmental effects surpassing any that were anticipated or thought possible. Many of the chemicals kill beneficial organisms as well as the target pests; some of them persist for so long in the global environment that they pose a semi-permanent hazard; some of those that decompose quickly are so toxic that they are dangerous to the users; some of them have made pest populations so resistant to chemicals that their utility is impaired or nullified; some of them are taken up in the food chain and concentrated through biomagnification to levels that are above safe standards for human food, or impair reproduction and other physiological functions of some forms of wildlife; and, finally, the prodigious amount of scientific investigation and experimentation needed to foresee and forestall these problems hampers and delays the development of more effective and safer products, and increases their cost.

The other side of the coin is that many of the hopes for pesticides have, in fact, been fulfilled. Malaria and other pest-borne diseases have been eradicated from some of the most disease-depressed areas, giving millions of people freedom from these debilitating diseases. By 1969, malaria cases in India had been reduced from a "normal" of 100 million cases per year to 300,000 cases, and the annual malaria death rate was reduced from 750,000 deaths to 1,500. Similar results were obtained in other parts of the world. Later, when DDT was abolished in some malarious areas, malaria flared up again because substitute materials were too costly, and it was necessary to reactivate the DDT program. Even today, in Africa alone, 500,000 children and infants die of malaria every year. It should not be assumed that all pests have become resistant to DDT. About one-half the species of malaria mosquitos can still be controlled with DDT. Many other species of insect pests are still susceptible to the insecticide, or have developed only partial resistance.

Food production has been greatly improved by control of plant diseases, insects, weeds, nematode worms, and rodents, increasing the food supply in a food-deficient world. Nearly all agricultural authorities contend that chemicals are necessary for food production in the modern world and that their sudden withdrawal would lead to economic chaos in the food-producing segments of developed countries and would deprive undeveloped countries of the opportunity to take advantage of other advances such as fertilization, irrigation, and the use of high-yield varieties of crops. Representatives of both the World Health Organization (WHO) and the Food and Agriculture Organization (FAO) of the United Nations adamantly defend the use of pesticides in their proper and useful place.

We have seen that biological control and other nonchemical methods are effective and safe. Why, then, bother with pesticides at all? Why not rely on biological control and other natural methods? It would surely be possible to do so, but only at great sacrifice in efficiency, increased deaths from disease, reduced nutrition, and higher costs for food. While the world still uses pesticides as its chief armamentarium against pests, improved methods are being developed steadily, integrated managed control is reducing the amounts and kinds of pesticides that are needed, and more effective, safer chemicals are being sought. It may never be possible to abolish pesticides completely, but continued research is the best hope for safe and economically sound answers to the pesticide problem—a global problem that calls for compassion for those in need as well as concern for wildlife and the integrity of the environment.

READINGS

DDT

CARSON, R., *Silent Spring*. Boston, Houghton Mifflin Co., 1962.

DAVIES, J. E. et al., "Pesticides in People." *Pesticides Monitoring Journal* 2(2): 80–85 (1968).

GUNTHER, F. A., "Did We Give Up Too Soon on DDT?" *Farm Chemicals*, January 1970, pp. 23,24,26,28.

HORSFALL, JAMES G., "A Socio-Economic Evaluation," in C.O. Chichester, ed., *Research in Pesticides*. New York, Academic Press, 1966, pp. 3–16. (Also in K. E. Maxwell, *Chemicals and Life*. Dickenson Publishing Co., Belmont, Calif., 1970.)

TAYLOR, T. G., "How an Eggshell is Made." *Scientific American* 222(3):89–95 (1970).

PEAKALL, D. B., "Pesticides and the Reproduction of Birds." *Scientific American* 222(4):72–78 (1970).

WIGGLESWORTH, V. B., "DDT and the Balance of Nature." *The Atlantic Monthly*, Dec. 1945, pp. 107–113.

WHITTEN, J. L., *That We May Live*. New York, D. Van Nostrand Co., 1966.

Control Methods

DE BACH, PAUL, ed., *Biological Control of Insect Pests and Weeds*. New York, Reinhold Publishing Corporation, 1964.

Insect-Pest Management and Control. Principles of Plant and Animals Pest Control, Vol. 3, 1969, Publ. 1965. Washington, D. C., National Academy of Sciences.

KNIPLING, E. F., "The Eradication of the Screw-worm Fly." *Scientific American* 203(4): 54–61 (Oct. 1960).

Report of the Secretary's Commission on Pesticides and their Relationship to Environmental Health, Parts I and II. Washington, D. C., U. S. Dept. of Health, Education and Welfare, 1969.

Scientific Aspects of Pest Control. A symposium arranged and conducted by the National Academy of Sciences National Research Council at Washington, D. C., Feb. 1–3, 1966, Washington, D. C., NAS–NRC, Pub. 1402.

WILLIAMS, C. M., "Third Generation Pesticides." *Scientific American* 217(1):13–17 (1967).

Integrated Pest Management. Pp. i–ix, 1–41. 1972. Council on Environmental Quality. U.S. Government Printing Office, Washington, D.C.

Pests and the Ecosystem

DELONG, DWIGHT M., *Man in a World of Insects*. Washington, D. C., 1963, Smithsonian Institute, Publ. 4556, from the Smithsonian Report for 1962, pp. 423–440.

HOWARD, L. O., *Fighting the Insects*. New York, The Macmillan Company, 1933.

NASH, T. A. M., *Africa's Bane. The tsetse Fly*. London, Collins, 1969.

WILLIAMS, G., *The Plague Killers*. New York, Charles Scribner's Sons, 1969.

ZINSSER, H., *Rats, Lice and History*. New York, Bantam Books, 1934.

FOOD QUALITY

People have come a long way in their eating habits since the human population subsisted on roots and berries or on warm animals bagged for their flesh, bone, and blood. Some of the most inspired creativity of civilized people has been applied to the culinary arts for the enhancement of gastronomic pleasure and aesthetic satisfaction in the performance of what is basically a mere necessity.

Today, in the advanced countries, food technology is a highly developed science—some would say overly developed. The food industry produces a vast array of mixed, seasoned, modified, and processed foods. People in the United States eat more than 125 billion dollars worth of food each year, the major portion of which contains additives. The biggest category of food additives is flavoring agents, with about 1,500 in use, of which more than 750 are synthetic. Spices are big business; the United States imports 250 million pounds annually. The annual market in the United States for iron compounds as flour additives alone is at least 15 million pounds. Additives include hundreds of materials, from amino acids to poppy seeds, designed to make food more attractive in various ways. Contaminants range from materials used at some stage of producing or processing the food, though not desired in the final product, to inadvertent filth from careless handling and deliberate adulterants.

FOOD ADDITIVES

What is an additive?

The FDA (Food and Drug Administration) of the U.S. Department of Health, Education and Welfare defines a food additive as "any substance that becomes part of food, or affects the characteristics of food, through direct or indirect use and with useful intention."

The legal definition of a food additive is more involved, and while not generally useful for our purposes, will help to explain some of the procedures and problems that are encountered. The Food Additives Amendment (1958) to the Federal Food, Drug and Cosmetic Act said that the term *additive* does not include pesticides and color additives because other laws cover them. The amendment also excluded from the meaning of the term all additives that had been sanctioned prior to enactment (1958) and additives which were "generally recognized as safe" by qualified people.

A more functional classification of foreign chemicals in food is one that distinguishes between: (1) those added directly to food with the purpose of intentionally modifying the characteristics of the food product by its presence; and (2) those which inadvertently gain entry into the food either from the environment or from treatment of the food during the growing or processing stage but are not desired in the final food product. We will refer to the substances intentionally incorporated in food as *additives,* and those in the second category as *contaminants*.

Origin of additives

Food additives are not exclusively the inventions of modern technology. When primitive people savored the taste of animal flesh singed with fire, they inadvertently discovered the delightful effects of new chemicals resulting from the action of heat and wood smoke. We know now that some of the substances produced by wood and charcoal smoke are toxic and carcinogenic; yet we continue the practice.

The Egyptians used food coloring 3,500 years ago. In 327 B.C., Alexander the Great brought from India a sweet reed called *Kand,* the original candy. Marco Polo, who visited the Orient during 1270 to 1295, told of sugar factories in southern China. Sometime later, Europeans learned to extract the sweet crystals that the Persians called *shakar*.

Ancient people placed a high value on common salt. It was used both as a seasoning and as a preservative. They fought wars over possession of salt deposits and many risked their lives in smuggling the condiment. The Book of Job says: "Can an unsavory food be eaten, if it is not seasoned with salt?" Caesar paid his soldiers partly in salt, called their *salarium,* from which comes our word "salary." The expression "not worth his salt" stems from the early practice of paying wages in salt.

Spices and leavening agents have been used for centuries. Defeated Rome paid a ransom to the Visigoths that included more than a ton of pepper. Condiments were highly prized. The travels of Marco Polo with his father and uncle to the Far East were largely the efforts of tradesmen looking for new routes for the spice trade. Christopher Columbus was looking for a trade route to the Spice Islands, now Indonesia, when he discovered America.

The Conquistadores inadvertently made a discovery more valuable than their stolen gold when they found the vanilla and cacao beans being used by people of the New World. The people that Cortez found in Mexico added lime to soften the corn for making cakes that the Spaniards called tortillas. The lime added calcium to their diet, a much needed mineral. In another strictly American development, Texas settlers are thought to have invented chili powder when they ground up dry peppers from Mexico.

Why additives are used

Substances are added to food to produce a wide variety of characteristics in the final food product. Additives are used in foods to enhance flavor, color, texture, appearance, or tactile sensation (for example, crunchiness), to improve the keeping qualities and make them safer for human consumption, to improve their nutritional value, and to make them more convenient to prepare and use. Substances are added for such diverse properties as leavening, curing, sweetening, emulsifying, clarifying, stabilizing, antifoaming, inhibition of oxidation, and mold prevention. Additives in modern use may be of natural origin, such as salt, spices, and vegetable gums, or they may be produced synthetically, such as the preservative calcium propionate, the flavoring benzyl alcohol, and the vitamin nicotinamide. Components are added to foods for at least 45 recognized purposes.

SAFETY STANDARDS

The standards for safety in food products have undergone a long history of evolution. Between 1883 and 1906, a few federal laws were in effect to protect consumers from adulterated or falsely labeled tea, dairy products, and imported food, and to provide for inspection of animals for disease before slaughtering. However, the first comprehensive law regulating adulteration was the *Federal Food and Drugs Act* enacted in 1906. Harvey W. Wiley, an indefatigable crusader for clean food, is credited with being the father of the 1906 law. It was replaced in 1938 by a new act, the *Federal Food, Drug and Cosmetic Act,* which made it illegal to add any unsafe substance to food except when it could not be avoided, and for such unavoidable cases, it provided for setting safe limits. An amendment called the *Food Additives Amendment* was enacted in 1958, requiring manufacturers to prove the safety of additives. No additive was to be allowed if it caused cancer in humans or animals. This provision became the cause of much contention between those charged with enforcing it and those who believed that the requirement was unrealistic.

The "GRAS" List

The Food Additives Amendment of 1958 did not adopt the same criteria for all substances that were added to food. Congress said that substances generally recognized as safe under the conditions of their intended use would be exempt from clearance and approval by the FDA. The Food Additives Amendment of 1958 also exempted "any substance used in accordance with a sanction or approval granted prior to the enactment . . ." Thus, a list of additives called the GRAS list (Generally Recognized As Safe) came into being. The choice of words was unfortunate. Many people made the interpretation that a GRAS substance was safer than one subjected to regulation by FDA. Not only was this not so, but with some substances the opposite was true. An additive could make the GRAS list merely by having been used for a long time without obviously causing adverse effects. (The original GRAS list contained some 600 food substances.) But under the 1958 law, only those substances that were already in use could be designated "generally recognized as safe." A new additive had to be subjected to rigorous laboratory study.

By mid-1970 it was recognized that many items needed further investigation. The primitive standards of acceptability formerly used were no longer acceptable. A comprehensive study was undertaken to determine whether individual items should continue to have GRAS status or

A sugar cane field and sugar refinery in Puerto Rico. People in the United States use an average of about 100 pounds of sugar per person per year.

whether they should be more rigidly regulated. It is expected that the GRAS list will become shorter, containing only items for which experimental evidence and many years of usage leave no question of safety.

SYNTHETIC SWEETENERS

Most people in the overfed parts of the world do not use sugar primarily for its food value but, instead, as a sweetener. There is a strong demand for nonnutritive sweeteners by those who want to indulge a sweet tooth without the fattening and other deleterious effects of excess sugar (see Figure 10.1). There is a need, also, for sweeteners by those such as diabetics who, for medical reasons, must avoid sugar in the diet. For both purposes, the need has been fulfilled largely by two synthetic chemicals, saccharin and cyclamate. (Common table sugar, or sucrose, is also a chemical, but one of natural origin.)

Saccharin was the first of the synthetic sweeteners (Figure 10.2a). Its sweetness was discovered by accident in 1879 by a young German

FIGURE 10.1 Sugar.

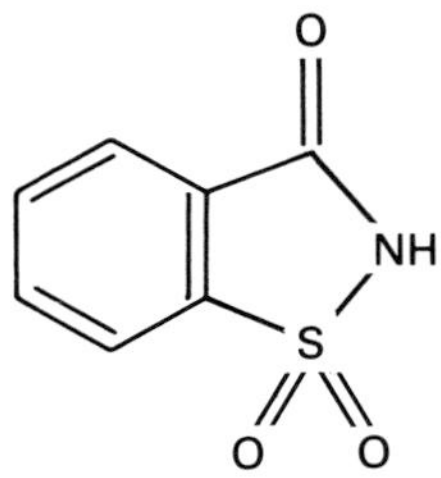

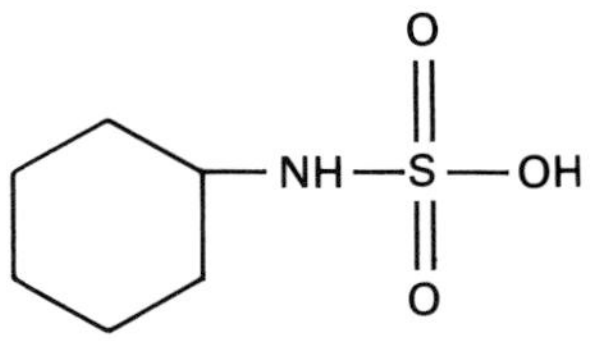

FIGURE 10.2 (a) Saccharin. (b) Cyclamic acid.

chemist, Constantin Fahlberg, in the laboratory at Johns Hopkins University. The substance turned out to be 200 to 500 times as sweet as sugar, depending on the concentrations compared, and is detectable at a concentration of 1 to 100,000. The sodium and calcium salts are the forms of saccharin that are used in foods and beverages.

In 1937, nearly sixty years after the discovery of saccharin, the sweetness of cyclamate was also discovered quite by accident by Michael Sveda, a graduate student working in the laboratory of Dr. Ludwig Audrieth at the University of Illinois. Cyclamate does not have the sweetening power of saccharin. Cyclamate is only 30 times sweeter than refined cane sugar and is detectable at a dilution of only 1 to 10,000. The advantages of the compound, therefore, were slow to be recognized. It was not used in food or beverages until thirteen years after its discovery and then only after the manufacturer (Abbott Laboratories) spent over a million dollars on research and development. Cyclamate is used as the sodium and calcium salts and as cyclamic acid (Figure 10.2b).

Flavor qualities   The flavors of saccharin and cyclamate are not identical to that of sugar, nor are they identical to each other. The sweetness of sugar is detected almost instantly, and its sensation disappears relatively cleanly and sharply. By contrast, the synthetic sweeteners build up to their sweetness intensity at a slower rate, and the sensation persists longer. This has certain undesirable carryover effects, such as aftertaste, taste fatigue, and tendency to cloy (satiate). The aftertaste is sometimes bitter, metallic, astringent, or drying. Cyclamate came into use when it was pointed out that these undesirable effects are less pronounced with cyclamate than with saccharin. It turned out that the two compounds were used to best advantage in combination, with about one-half the sweetness coming from each additive. A product sold under the brand name Sucaryl consisted, since 1955, of 10 parts cyclamate (sodium or calcium) and 1 part saccharin.

When the synthetics are used in food or beverages, the loss in flavor and other properties of sugar call for rebalancing such qualities as sweetness-tartness and texture. Sugar, in addition to imparting sweetness, also adds body, roundness, and mouthfeel, particularly in beverages. The synthetic sweeteners are often most effectively used when they can be combined with a reduced amount of sugar. The final criteria for the formulation of a food product containing the synthetic sweeteners are based on the results of taste evaluations called *organoleptic* tests.

Cyclamates Under Suspicion

When cyclamates were first used, they were, like saccharin, confined mainly to diet foods for diabetics and obese persons. In 1958, they were placed on the GRAS list, meaning that no restrictions were placed on their use. But as the popularity for low-calorie foods and drinks soared, disturbing reports began to come in from the laboratories of several research workers. The results of experiments gave indications that cyclamates were not as safe as they were once thought to be. By 1962 there was reason to believe that cyclamates were not, in fact, "generally recognized as safe." It was claimed by some people that many of the studies were financed by the sugar industry, but there is no concrete evidence that this was the case.

A new look at cyclamates

A review of the problem was made in 1968 by the National Academy of Sciences-National Research Council (NAS-NRC). Their report cited test results in which there was retardation in the growth of rats and pigs and evidence in other animals of damage to several organs including kidneys, liver, gastrointestinal tract, adrenals, and thyroid. However, the animals had been given massive doses—as high as 5 to 10 percent of their diets—putting in question the effects that could be expected from normal use of the product.

A popular diet cherry cola contained, in addition to flavorings, 0.21 percent cyclamate, 0.04 percent saccharin, and 0.05 percent benzoate of soda. Thus one 12-ounce can of diet cola contained 0.7 grams of cyclamate. Two cans of cola would exceed the maximum amount of cyclamate recommended by the FDA for a moderate-size child.

More research on cyclamates

In January of 1967, the FDA had begun to investigate the toxic effects of cyclamate and its metabolite[1] CHA (cyclohexylamine) in chick embryos. By September of 1969, they were reasonably certain that cyclamate and its breakdown product CHA had potential fetal effects. Saccharin and sucrose produced no comparable effects. In the meantime, a report from studies conducted at the University of Wisconsin indicated that cyclamates could produce bladder tumors in mice. This was confirmed by a two-year feeding study nearing completion. Cyclamates fed to rats at high doses produced malignant tumors in the bladders of a significant number of the treated animals. (The effect was later confirmed at much lower dosage rates.)

Cyclamates banned

The federal government invoked the Delaney clause of the Food, Drug and Cosmetic Act, which prohibits the sale of any food additive that can be shown to cause cancer in humans or animals. The low-calorie sweeteners consumed by Americans at the rate of 17 million pounds a year had been found to cause cancer in laboratory animals and would no longer be permitted in foods.

The FDA said that new scientific evidence showed that the safe maximum daily amount was only one-fifth of what was regarded as safe in its earlier action. It ordered all diet foods and drinks sweetened with cyclamate off the market by September 1, 1970, except as drugs for diabetics and obese persons under the care of a doctor. Meanwhile, the dispute

---

[1]Metabolites are the products of reactions resulting from the body's enzymatic attack on a substance, in this case a foreign material.

continued, with Abbott Laboratories seeking to market cyclamates again for general use. But tests on animals raised questions about a possible association between cyclamates and testicular atrophy and adverse effects on the cardiovascular system. As late as April 1975, the FDA stated that these matters, too, would require additional study.

The toxic agent   Part of the concern on the part of the FDA was the finding that cyclamate breaks down in the human body to cyclohexylamine, a substance known to cause chromosome damage in animals and the most probable cause of bladder cancer in rats fed high doses of the cyclamate. Cyclohexylamine was also found to cause chromosome breakage in human blood cells.

Saccharin in Question

The cyclamate findings stimulated an interest in possible side effects from the use of saccharin. Suspicions concerning its safety were aroused by a report in 1970 from a group working at the University of Wisconsin Medical School which said that saccharin produced bladder cancer when surgically implanted with cholesterol in mice. However, feeding studies with saccharin at levels up to 5 percent in the diet failed to produce tumors. It was evident that further studies were needed. Saccharin was suspended from the GRAS list in 1971 but was permitted interim use at the existing levels pending the outcome of research. A standard was adopted of 15 milligrams per kilogram (about 1 gram for a 154-pound adult) per day. This level provides a safety margin of 30 to 1 ($^1/_{30}$ the intake level at which adverse effects occur).[2] The FDA removed saccharin from the GRAS list in January, 1972, and issued an interim regulation placing restrictions on the use of the artificial sweetener until additional safety reviews could be completed. The use of saccharin was declared illegal except in products prominently labeled as special dietary foods.

In 1975, the National Academy of Sciences reported on the findings of a new review of the problem. The panel making the study concluded, "The results . . . thus far reported have not established conclusively whether saccharin is or is not carcinogenic when administered orally to test animals." The search for evidence that will clear saccharin of suspicion, or condemn it, continues.

Meanwhile, a third sweetener, aspartame, was approved for limited use by the FDA. It is only 200 times sweeter than sugar but, because it is a combination of two amino acids, phenylalanine and aspartic acid, it is therefore nutritive. The sweetener appears to be safe, but studies on its possible long-range effects are continuing.

FLAVOR ENHANCER: MONOSODIUM GLUTAMATE

Monosodium glutamate (MSG) is the best known of the modern flavor enhancers. It is known to many people by the brand names Accent®, Zest®, or Glutavene® (Figure 10.3).

---

[2]The usual standard is a safety margin of 100 to 1; the 30 to 1 was reported to have resulted from an error in calculations by the NAS-NRC Food Standards Committee, which originated the recommendation.

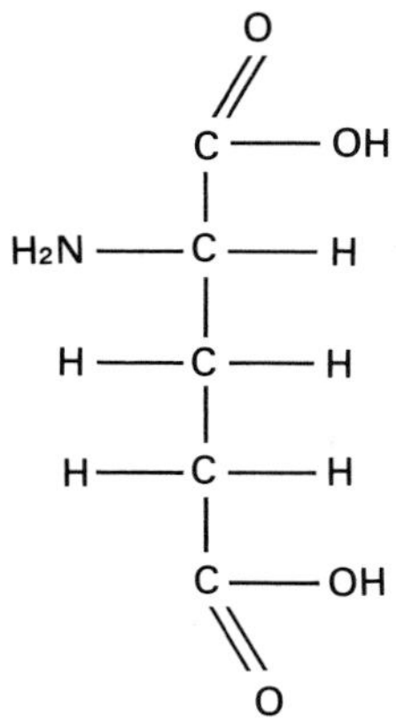

FIGURE 10.3 Glutamic acid.

MSG is used in many kinds of foods: prepared frozen foods that contain meat or fish, dry soup mixes, and various types of canned foods. It is deemed especially useful in proteinaceous food products and is used to intensify the flavor of commercial soups, chowders, canned meats, stews, meat pies, fish, and other seafoods and cheese spreads. The additive is thought to be especially useful for intensifying the flavor of high-protein foods and is an ingredient of some soy sauces used for that purpose. However, it does not improve the flavor of foods high in carbohydrates (sugars and starches), such as cereals, fruits, and candy.

Flavor   Flavor is a complex mixture of sensations produced by taste, odor, and other feelings that are transmitted by various kinds of sensory receptors in the tongue, lips, throat, nose, and mouth. The tongue has several thousand sensory receptors; the nose has millions. Heat, cold, effervescence, pungency, astringency, blandness, bite, tingling, numbness, freshness, touch, pressure, and texture are some of the sensations that contribute to flavor.

The basic sensations of taste for sweetness, sourness, saltiness, and bitterness are picked up by the chemical receptors in the tongue. They are concentrated in specific parts of the tongue: sweet sensors in the region of the tip, sour at the sides, salty around the edge of the tip and sides, and bitter at the rear.

The sense of odor, called the *olfactory* sense, is one of the most important aspects of flavor. Aromas contribute most, if not all, the flavor to some foods. Plug the back openings of the nostrils and you might have trouble distinguishing between an apple, a pear, and an onion if they were ground to the same consistency. While the taste receptors are sensitive to chemicals in solution, the olfactory receptors detect gaseous chemicals that become dissolved in the fluids of the specialized mucous membrane called the *olfactory epithelium,* high in the nasal cavity.

The sense of smell is one of the most delicately tuned sensory systems of the body. The olfactory receptors are especially well developed, along with the olfactory lobe of the brain, in some of the lower vertebrate animals. They are also extremely sensitive in many of the higher vertebrates. Even the human nose, which is abused and blunted, can detect $1/25,000,000$ of a milligram of skunk oil. We can detect an estimated 10,000 aromas.

The sensation of flavor may last microseconds or may persist for many minutes. The component taste and odor sensations of flavor are easily tired (particularly that of smell), but, paradoxically, there may be a prolonged sensation such as an aftertaste. The complexity of flavor is perhaps why the preparation of food has always been more an art than a science. Though flavor secrets have been handed down for centuries, we have only recently begun to gain an understanding of the components of flavor and of flavor perception. The use of new additives has contributed to our knowledge, though much remains that is mysterious about how they act.

Discovery of MSG

The Japanese have used seaweed[3] for centuries to flavor soups and other foods. Kikunae Ikeda, a chemist at the University of Tokyo, wondered what it was in seaweed that improved the flavor of food. His investigations led to the discovery of the unusual flavor-enhancing property of MSG with a large number of high-protein foods. Ikeda made his discovery in 1908 and shortly thereafter found a way to extract the amino acid from flour. He joined a Japanese chemical company, then called Suzuki & Company, and as a partner in the firm, began production of MSG in 1909. Today, world use of MSG is in excess of 200 million pounds annually.

Action of MSG

The manner in which MSG acts to intensify flavor is a mystery. The presence of sodium chloride (table salt) is required for the substance to impart an attractive flavor. The optimum concentration of MSG is between 0.2 and 0.5 percent of the food. At a concentration of 1 percent or more, it is apt to produce a sweetish taste. The additive has a taste of its own; many people perceive a slight salty-sweet flavor. Some believe that MSG acts primarily as a seasoner by combining its own flavor with those of the food. An early explanation of the action of MSG was that sodium monoglutamate, itself, has a meatlike taste (it also has a peptonelike odor). However, it was later discovered that the meaty flavor was caused by impurities in the crude product then available, which consisted of protein decomposition substances as contaminants. When the contaminants were removed, the flavor imparted by the added glutamate was also greatly reduced. Meat flavor is believed to be predominantly odor, and, since pure MSG is odorless, any enhancement from the pure product must be from a source other than olfactory. The most recent theory is that MSG stimulates the flow of saliva, thereby enhancing an appreciation for the food. The most widely held theory, however, is that MSG increases the sensitivity of the taste buds.

Safety of MSG in question

The FDA originally classified MSG as an artificial flavor, but since 1949, the substance has been regarded as an additive. Indications of impending trouble appeared as early as 1957, two years before MSG was placed on the GRAS list. At that time it was reported that MSG caused acute degenerative lesions in the inner retina of the eye of infant mice. The effect was confirmed again by different investigators ten years later. Injections of MSG in newborn mice were also shown to cause injury to the brain cells, including those in the region of the hypothalamus, a part of the brain important in regulating the body's hormone system. Similar brain damage was found in infant rhesus monkeys, but experiments with infant rats showed no damage in the subsequent adult brains, indicating

---

[3]Laminaria Japonica.

either a difference in species susceptibility or the possibility of a repair mechanism. Feeding MSG to infant macaque monkeys caused no detectable brain damage, but the same investigators confirmed again that feeding MSG to newborn mice caused brain injury. Moreover, adult rats, following treatment with MSG in infancy, had ovaries and adenohypophyses (anterior pituitary glands) that averaged about one-half the weight of those in untreated animals. John Olney at Washington University in St. Louis found gross injury to brain cells when animals were administered doses no greater than five times that which a human infant would be expected to receive in baby food. No damage was found from feeding large amounts of MSG to adult humans and animals, indicating a difference in age-susceptibility, possibly because of changes in the blood-brain barrier that permit significant amounts of MSG to reach the brain in newborn animals (and possibly in human infants).

Chinese restaurant syndrome

Another troublesome feature of MSG came to light in 1968 when it was discovered to be the cause of the "Chinese restaurant syndrome" (so-called because of the generous amounts of MSG presumed to be used in Chinese cooking). This condition is characterized by one or more of several symptoms, including severe headaches, burning sensations, facial pressure, and chest pain. It was found that there is great variation in the susceptibility of individuals and that the symptoms appear only if the meal is eaten on an empty stomach, an experience that is not unusual. The symptomatic response is related to the amount of MSG consumed. The trouble has also been called "Kwok's disease" because Dr. Robert Ho Man Kwok, who suffered severely from the Chinese restaurant syndrome, published what was perhaps the first detailed description of the symptoms.

MSG reviewed

Again, the FDA submitted their food-additive problem to the National Academy of Sciences–National Research Council and asked that a committee review the hazards from MSG. The NAS-NRC concluded that MSG is generally safe for use in foods and recommended that the FDA continue to permit its unrestricted use except for those foods "specifically designated for infants." The risk to infants was thought to be extremely small. But the reviewers could not find that MSG "confers any benefit" to the infant, even though baby food containing MSG may taste better to the mother. The committee also recommended that food containing MSG be clearly labeled to indicate its presence for the information of people who are sensitive to it and wish to avoid it. In view of the ubiquitous nature of MSG in food products and the fact that there is no control over the amount used, it is not easy to follow the reasoning that consumers could avoid it without rejecting restaurant foods and all prepared food products, and otherwise drastically changing their dietary habits.

NITRATES AND NITRITES

Nitrates and nitrites are among the most commonly used food additives, especially in smoked and cured meats such as corned beef, weiners, bologna, and similar products. Nitrates and nitrites are compounds that contain closely related radicals (groups of atoms) consisting of combinations of oxygen and nitrogen. Although both nitrates (NO_3^-) and nitrites

(NO_2^-) are used in foods, the nitrite form is preferred, despite the fact that the nitrite is the most toxic form of this pair.

Sodium nitrate is permitted in cured meat and certain smoked and cured fish to a limit of 500 ppm; the nitrite is permitted to 100 ppm. Nitrite may be used in smoked and cured tuna fish up to 10 ppm. Some countries (England, Wales, Norway, and Sweden) permit the use of nitrate and nitrite in certain types of cheese.

The nitrates and nitrites have demonstrable antimicrobial action, but their protective effect against spoilage is weak. The FDA holds the view that nitrate and nitrite are needed in bacon, bologna, frankfurters, and similar products to prevent spoilage by botulism bacteria, a danger the FDA considers to be more serious than the risk of nitrate-nitrite poisoning. But opponents contend that much of the preservative value is due to the fact that nitrates and nitrites are almost always used in combination with large amounts of salt. Another advantage to their use is that they impart a fresh coloration to the meat, which enhances its appearance. The bright color is due to the formation of nitric oxide compounds of hemoglobin or myoglobin, which have a stable red or pink color. For this reason the nitrate and nitrite additives are referred to as *color fixatives*.

How nitrate poisons

Poisoning of livestock from feed high in nitrates has been known since the last century. An investigation of deaths in a herd of cattle in Kansas in 1895 established that the poisoning was caused by their feeding on cornstalks having an unusually high concentration of potassium nitrate. The corn had been grown near a barn on land with an exceptionally high nitrogen content. It was noted that the blood of the poisoned animals was dark in color, but the meaning of this was not understood at the time.

One of the symptoms of nitrate poisoning is described as *cyanosis,* in which there is a bluish discoloration of the skin and mucous membranes. The red blood cells lose their capacity to carry oxygen, causing the blood to turn dark and resulting in death from asphyxiation. The failure of the red blood cells is due to the conversion of their hemoglobin to a closely related substance, methemoglobin, which cannot combine with oxygen. The conversion is brought about by the action of nitrite on the iron (Fe) in the hemoglobin molecule, which oxidizes it from divalent to trivalent iron. The condition produced by nitrate (or nitrite) poisoning is called *methemoglobinemia*.

$$\text{Hemoglobin (Fe}^{++}) \xrightarrow{\text{nitrite}} \text{Methemoglobin (Fe}^{+++})$$
$$\text{can combine with} \qquad\qquad\qquad \text{cannot combine with}$$
$$\text{oxygen} \qquad\qquad\qquad\qquad \text{oxygen}$$

For a nitrate to be highly toxic it must first be converted to nitrite. In cattle this is believed to be brought about by action of certain microorganisms prevalent in the rumen, the first pouch of the stomach.[4]

$$NO_3^- \xrightarrow[\text{reduction}]{\text{microbial}} NO_2^-$$
$$\text{nitrate} \qquad\qquad\qquad \text{nitrite}$$

---

[4]Members of the coliform group and the genus *Clostridium* are capable of reducing nitrate to nitrite.

Nitrate poisoning occurs in many animals, including cattle, sheep, horses, turkeys, and man. Cattle are especially susceptible. Nitrate poisoning is less common and usually less severe in adult humans than in cattle because the conditions in the human adult gut do not favor conversion to nitrite nitrogen from the nitrate form, which is readily eliminated from the body in the urine. Human infants, however, are highly susceptible to nitrate poisoning. This is believed to be caused by the lower gastric acidity which favors nitrate reduction. The makeup of the intestinal microflora might also be a factor as well as the high carbohydrate diet of infants which favors the microbial action. A case is on record of two children, two to three years of age, being poisoned from eating wieners containing 5,000 ppm nitrite—an unusually high concentration.

Nitrates in plants

Nitrates are normally present in many plants, often at levels that cause poisoning of livestock. Plants containing 1 percent nitrates are toxic to cattle and 0.5 percent is considered to be the maximum that can be tolerated. Excessive fertilization is often the cause of high nitrates in plants. Treatment with hormone-type weedkillers such as 2,4-D can also cause high nitrate levels.

The amounts of nitrate normally present in vegetables are not ordinarily dangerous to adult humans. Except under unusual circumstances, large amounts can be tolerated without adverse effects. However, if for some reason the nitrate is reduced to nitrite during storage of the food, the nitrite may reach levels that are poisonous to infants. Most of the reported cases have resulted from spinach purees in which the conversion may have taken place in frozen products. However, leafy vegetables are often fertilized heavily during growing and may contain high concentrations of nitrates initially.

Nitrates and smoking

The blood of heavy smokers has as much as 7 to 10 percent of the hemoglobin converted to carboxy-hemoglobin as a result of habitual inhalation of carbon monoxide. Carboxyhemoglobin is also incapable of transporting oxygen, so the effects of nitrite poisoning and carbon monoxide poisoning are at least additive. The high concentrations of carbon monoxide in the atmosphere of urban and industrialized areas adds appreciably to the burden of the oxygen-carrying function of the blood. There is also some nitric acid in the air from urban pollution, and nitrate salts are present in aerosol form. The contribution to nitrate poisoning is probably minor, but when added to the rest of the burden, may be significant.

Fetal effects

The effect of nitrates on the human fetus has not been fully studied. When pregnant cattle graze on land where the plants are high in nitrates, they may abort without showing any of the acute symptoms of poisoning. Abortion can be expected in pregnant animals that survive prolonged and severe poisoning. When fetuses are aborted in the absence of acute poisoning symptoms in the mother, lesions that would be expected from prolonged hypoxia (oxygen starvation) are found in the fetus. Microscopic examinations show thickening and blocking of the arterioles. More information is needed on the sensitivity of fetal and infant hemoglobin to the action of methemoglobin-forming agents.

Cytotoxic potential

A relatively recent cause for concern is the potential cellular toxic effect of nitrates and nitrites. Substances that adversely affect cells are said to by *cytotoxic*. Nitrous acid, which contains nitrite nitrogen, has been known to be mutagenic since 1953. Its action is believed to be through the formation of nitrosamines, the mutagenic, teratogenic, and carcinogenic properties of which are discussed in Chapters 12 and 13.

PROBLEMS WITH FAMILIAR ADDITIVES

We have seen that among three food additives of relatively recent origin, problems arose that required one of them (sodium cyclamate) to be banned and another (monosodium glutamate) to be restricted in its use. The third additive (saccharin) is under a cloud of suspicion. Examples of familiar additives that are toxic but in common use, now or previously, will illustrate additional problems associated with improving the attractiveness of food by the addition of chemicals.

Table Salt

Common table salt (sodium chloride) is probably the most widely used, and the most favorably accepted, of the chemical substances that are added solely for seasoning. Although the taste for salt is probably acquired and habit-forming in the same sense as tobacco or sweets, it adds to the joy of living for millions of people. Many users do not know that sodium in its salt form is toxic and that too much in the diet, even in quantities considered reasonable by most people, can sometimes produce subtle changes that are dangerous to health and, in the sense that it may shorten life, can be lethal.

Sodium is one of the more abundant elements in the environment. It is a component of all plants and animals and is present in all foods. Despite the fact that salt is concentrated during cooking, the seasoning effect of large amounts is so pleasing to the taste that extra salt is regularly added by cooks and diners alike. Medical people have long recognized that the drinking water in areas of exceptionally high salt content (1,700 ppm) is associated with a high incidence of hypertension (high blood pressure) and congestive heart failure, and that such water should be used only for cooking. Experiments with rats showed that levels of salt in their diet considered moderate by many people (2.8 to 5.6 percent salt, equivalent to ½ ounce per day in the ordinary human diet) produced hypertension and shortened the lives of the rats. Rats that received the highest levels (8.4 percent), which would have their human counterparts in areas where large quantities of salted fish or salted meat are eaten, had their life span reduced by 8 months—equivalent to 32 years for man. There were two other highly significant facts that came from the rat experiments. First, there was a high degree of individual variability in susceptibility to cardiovascular and other effects, and second, a degree of protection against the harmful effects of sodium was obtained by adding extra potassium to the diet.

Epidemiological studies in humans are incomplete, but the rat studies plus clinical observations suggest that man's favorite condiment is responsible for part of the cardiovascular death rate among people in our society.

Salt Substitute

Lithium chloride came into use about twenty years ago as a substitute for salt in the diets of people with heart disease who required a reduced sodium intake. Lithium chloride not only tastes like sodium chloride, it is ten times as salty. Unfortunately for those who used the substitute food additive, lithium is chronically toxic. Prolonged use affects several organ systems. One of the most damaging effects is severe kidney injury, especially if there is an insufficiency of sodium in the diet. Lithium also affects the nervous system. Several hundred known cases of central nervous system disturbance occurred before the effects were confirmed by toxicity tests on animals. Lithium chloride was placed in the category of a drug and removed from sale as a food. Today, lithium salts are increasingly used as drugs for treatment of manic-depressive psychosis.

Other Toxic Additives

There are many examples of food additives that were used extensively, and in some cases for long periods, before their injurious effects were recognized. Some of them have been abandoned.

- Coumarin is a flavoring substance obtained from the Tonka bean and is present in many other plants, including lavender, woodruff, and sweet clover. It was used in candy and liqueurs and in vanilla flavoring made from synthetic vanillin. Its use was discontinued in 1954 when it was found to cause severe liver damage in rats.
- "Butter yellow"[5] was one of the first synthetic dyes to be used for food coloring. It was banned after it was found to be a potent carcinogen.
- A food coloring designated FD&C Red No. 4 was originally on the FDA approved list of food colors but was banned in 1964 after it was found to damage the adrenal glands and urinary bladders of dogs to which the compound had been fed in high concentrations. It was later approved for use at lower concentrations, but only in maraschino cherries, deemed to be only a minor food item.
- Nitrogen trichloride, a gas, was used extensively for treating freshly milled flour in order to mature it quickly and produce desirable baking qualities. It was abandoned after the finding that feeding treated flour to dogs induced convulsions.
- Safrole is present, to the extent of about 75 percent, in an aromatic oil extracted from the root bark of sassafras,[6] now used in perfumery. It causes liver tumors in mice. Sassafras tea was once used as a spring tonic. Safrole is also a constituent of nutmeg and mace.
- Nutmeg has been reported to cause poisoning from ingestion of large quantities. It contain myristicin, a toxic substance closely related chemically to safrole. (Myristicin has also been found in cigarette smoke.)

---

[5] p-dimethylaminoazobenzene.

[6] *Sassafras albidum* (family Lauraceae).

- Thujone, from oil of wormwood *(Artemisia absinthium),* was used in France as a flavoring for absinthe, an intoxicating drink that was popular in the days of Toulouse-Lautrec. It produces convulsions and lesions of the cerebral cortex of the brain. France made it illegal in 1915, but it is still used in trace amounts for flavoring vermouth. (Perhaps dry martinis are the safest.)
- Red pepper *(Capsicum),* or cayenne pepper, contains capsaicin, an irritant at low concentrations and a vessicant (blistering agent) at high concentrations.
- Oil of bitter almond contains amygdalin, a substance that is converted to cyanide (prussic acid). Oils that are produced for food use are specially treated and labeled FFPA (free from prussic acid).
- When the seeds of bown mustard *(Brassica juncea)* are crushed and moistened, there takes place enzymatic production of allyl isothiocyanate, a potent irritant and toxicant to mice. The substance is also found in horseradish, broccolli, cabbage, and rocket salad *(Eruca sativa).*

FOOD ADDITIVES AND CANCER

Under the Delaney clause, a provision of the 1958 amendment to the Food, Drug and Cosmetic Act, any food additive that is shown to cause cancer in man or any animal at any dosage level cannot be sold for general use in foods or beverages. However, natural carcinogens are specifically exempted. In exempting carcinogens found in the environment, it seems as though the lawmakers were following the advice of poet John Dryden (1631–1700):

> Some Truths are not by Reason to be try'd,
> But we have sure Experience for our Guide.

The Delaney clause stirred up deep controversy. Dissidents pointed out that it dealt with only one type of disorder—cancer—and only one type of environmental exposure—food additives having a limited legal definition. There were other objections as well. It exempted natural materials that might cause cancer; and it placed no limit on concentrations even though some toxicologists believe that there are vast numbers of substances which will cause cancer, as well as other disorders, if the dose is raised high enough and the conditions made severe enough.

The Delaney clause is an example of the problems that arise from legislation establishing fixed standards based on scientific findings that are usually not fixed but are variable depending on a variety of circumstances, experimental conditions, and interpretations.

FOOD CONTAMINANTS

Food contaminants are unwanted foreign materials that gain entry either accidentally from the environment or by deliberate incorporation at some stage during growing, producing, processing, shipping, or storage for a specific temporary purpose. In contrast to additives, which are desired components of the finished product, contaminants serve no beneficial purpose at the consumer stage even though they may have been useful at some point during the supply process. All contaminants are presumed harmful unless proved otherwise.

Contaminants may consist of pesticides that are used to treat soils, food crops, or products in storage; antibiotics, hormones, and other chemicals added to livestock feed that remain in the meat at slaughter; veterinary drugs; spoilage due to molds or bacteria; filth from infestations of insects, rats, mice, and other vermin; pollutants from packaging materials and storage sources; or substances from deliberate adulteration with fraudulent intent.

Contaminants that remain on or in food from treatment with chemicals such as pesticides, antibiotics, and drugs are called *residues*. The amounts of residue allowed to remain on food products at the time of marketing following treatment with toxic materials are regulated by various laws and administrative regulations under several federal and stage governmental agencies. Since a detailed discussion of the laws and regulations for controlling residues is beyond the scope and purpose of this book, we will mention only those legal aspects that are pertinent to an understanding of the nature of toxic residues and the problems involved.

World problem

The British were the first to place limits on the amount of poisonous substances in food. Adulteration of bread was common enough in eighteenth century England to be the subject of at least three books and treatises. During the year 1900 in England, an alarming epidemic baffled physicians. Nearly 6,000 people became ill and at least 70 persons died in Manchester alone. It was finally determined that the illness was not a disease but mass poisoning caused by drinking beer that contained trace amounts of arsenic that came from an impurity in acid used for preparation of starch. (One opinion, however, was that the beer was poisoned with lead, picked up from the lead pipes used for tap beer in the pubs.) An outgrowth of the tragedy was that a Royal Commission on Arsenical Poisoning set a limit on the amount of arsenic that would be permitted in food. The maximum amount agreed upon was .01 grains of arsenic trioxide per pound of solid food (equal to 1.43 ppm).

But altho' Alum principally occasioned my Dislike to Bread, yet I never suspected that Lime, Chalk, Whiting, and burnt Bones were any of its constituent Parts: Certainly the most servile abandon'd Wretch could not be ignorant of the pernicious Effects arising from such odious Admixtures; the Practice of which would render him more detestable and dangerous than the lurking Assassin.*

*From An Essay On Bread, by H. Jackson, London, 1758.

Tolerances

Since the enactment of the Food and Drugs Act of 1906, and especially since the Food, Drug and Cosmetic Act of 1938, it has been illegal in the United States to sell any contaminated food in interstate commerce. However, from the beginning of attempts to insure the purity of foods, the lawmaking bodies, administrative agencies, scientific organizations, and general public were reconciled to the proposition that some contamination is inevitable. It has been generally believed that 100 percent purity is an unattainable goal except at prohibitive costs in money, manpower, and quantity of food available. High yields and productivity, low spoilage and waste, and improved quality and quantity were deemed to be more impor-

USDA photo.

Insect pests contaminate as well as destroy food products. Here the lesser grain borer and its larva have thoroughly riddled kernels of wheat.

tant than a small amount of food contamination. However, in recognition of the poisonous nature of many of the substances that might remain on or in food at mealtime, provisions were made for placing limits on the amounts that could be present. Such limits are called *tolerances*. A tolerance is defined as the maximum amount of a chemical residue or other pollutant that is legally permissible or "tolerated" on or in a food product.

PESTICIDE RESIDUES IN FOOD

Several hundred pesticide chemicals are used either directly on growing crops or on weeds, farm soil, or surrounding areas. Inevitably, food products contain traces of some of the pesticide chemicals. The 1954 amendment to the Food, Drug and Cosmetic Act empowered the FDA to establish tolerances for spray residues of all pesticide chemicals sold in interstate commerce, and later amendments strengthened the FDA's authority. Consequently, the FDA has published in the Federal Register and elsewhere tolerances for all pesticides on all food crops for which uses have been approved by the Environmental Protection Agency and the United States Departments of Agriculture; Health, Education and Welfare; and Interior. Each tolerance is based on extensive chemical analyses of the amounts of residue that remain after normal use, and on acute and chronic toxicity experiments on laboratory animals. Unfortunately, early tolerances were often derived from inadequate knowledge of chronic toxicity, and in several instances, the tolerances were little more than guesses based more on how much one could find on the food than on how much ought to be there. In several cases the tolerances have been drastically revised downward in the light of more complete information than was originally available. For example, the insecticide heptachlor, a persistent chlorinated hydrocarbon, was found to decompose after spraying on crops to a substance that is many times more poisonous than the original material. Use of the product was curtailed and eventually prohibited. With some other pesticides, it became necessary to modify or eliminate certain practices that left undesirable quantities of residues. An example was the use of DDT to treat dairy cows to control flies. Quantities of the chemical that fell on hay and other feed, or had been picked up by the cows from licking, were ingested by the cows and excreted in the milk. When DDT was found in milk, the use of the insecticide around dairy barns was prohibited, but for a while thereafter DDT continued to be dusted on alfalfa hay, some of which ended up as dairy feed. When milk and dairy cattle feed were eventually monitored more closely, the levels of DDT in milk subsided. However, DDT has become so generally distributed in the environment that the FDA found it necessary to change the tolerance in milk from zero (none permitted) to .05 ppm.

Today, the FDA conducts a surveillance program whereby several thousand samples of agricultural commodities are collected and analyzed each year. For example, during the period of July 1963 to June 1966, there were 49,044 samples collected and analyzed for dozens of pesticides. The ranges of residue levels are shown in Figure 10.4. About 3 percent of domestic random samples and 1.5 percent of imported samples were found to have residues that exceeded either the legal tolerances or analytical guidelines.

Not all of the pesticide residue problems have been solved. DDT and other organochlorine compounds have appeared in fish. DDT and other chlorinated hydrocarbons in the environment that result in residues in milk are of concern because of the potential effects on infants. DDT in human mother's milk is small but detectable and in some cases may exceed that in cow's milk. Most of the other organochlorine pesticides are more toxic than DDT and more hazardous to humans, but fortunately, their use is declining as newer and safer materials are being developed.

New chemicals inevitably pose potential hazards. The environmental effect of dioxin, an impurity in the weedkiller 2,4,5-T, is unknown (see Teratogenic Agents, Chapter 13). Parathion disappears from the environment but is too toxic to be used by anyone but an expert. Many herbicides (weedkillers) persist in the soil for months or years and their ultimate effect is largely unknown.

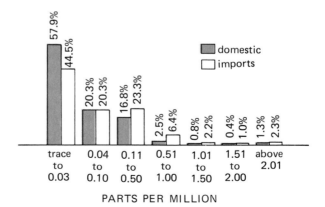

Redrawn from R. E. Duggan and K. Dawson, FDA Papers, June 1967.

FIGURE 10.4 Pesticide residues found in domestic and imported foods; percent containing residues at different levels.

ANTIBIOTICS

An antibiotic is a toxin produced by a microorganism, usually a fungus, with the capacity to destroy or inhibit other microorganisms. They are chemicals found widely in nature that have various degrees of toxicity, specificity,[7] adverse effects, and medical benefits when used therapeutically.

Antibiotics are generally believed to be one of the greatest discoveries of this century. They have become the leading weapon in the physician's arsenal for controlling infectious diseases and have saved countless lives. They have also been used extensively for other purposes such as animal disease control, the promotion of growth in livestock, treatment of plant diseases, and preservation of foods. The question arises: has man contaminated the environment by excessive and inadequately controlled use of the antibiotics?

Nearly 18 million pounds of antibiotics were produced in the United States in 1971, of which more than 10 million pounds were used in animal feed. Antibiotics are also used on food crops and other plants. Streptomycin is used on apples, pears, celery transplant beds, tobacco, peppers, tomatoes, hops, ornamentals, and seeds. Phytoactin®[8] is used only on western white pine, and Actidione®[9], a highly toxic fungicidal antibiotic, is used only on ornamental plants and turf.

There are various veterinary uses for antibiotics. They are used to treat cow udders for an inflammation called *mastitis*. Some of the antibiotic preparations were banned by the FDA because residues were found to persist in the milk for more than four days, the length of time permitted by the FDA.

Antibiotics as preservatives

One of the earlier uses for antibiotics in food production and processing was direct application on poultry and fish to prevent spoilage. In approving this use, the FDA in 1955 established a tolerance of 7 ppm for residues of the antibiotic chlortetracycline (Aureomycin®) in or on uncooked poultry. A year later a tolerance of 7 ppm was added for oxytetracycline (Terramycin®), and four years later, the FDA set tolerances for these antibiotics at 5ppm in or on fish, scallops, and shrimp, each in the fresh, uncooked, unfrozen form.

However, the use of antibiotics to prevent spoilage backfired. There emerged, in the words of the FDA, "resistant spoilage flora in poultry processing plants" and "adverse economics in the case of seafoods." The FDA concluded that the use of the antibiotics for food preservation "gives rise to selection of naturally resistant micro-organisms, induces the emergence of resistant strains, and results in food spoilage from increased mold and yeast development." The FDA scientists found that the "use of chlortetracycline and oxytetracycline on poultry and seafood may be a substitute for good manufacturing practice with consequent poor sanitation." Accordingly, the tolerances were revoked. In September of 1966 the FDA, in effect, banned the use of antibiotics for preservative

---

[7] The quality of affecting only certain organisms or tissues.

[8] Polyamidohygrostreptin.

[9] Cyclohexamide.

purposes when it revoked tolerances that had been previously established for chlortetracycline (Aureomycin®) and oxytetracycline (Terramycin®) in fish, shellfish, and poultry.

The development of antibiotics for animal feeding dates back to 1949, when it was accidentally discovered that fermentation residues fed to poultry and swine for their vitamin B_{12} content had growth-promoting benefits quite apart from the vitamin effects.[10] A large number of research projects followed and the concept of *infection level* was introduced to explain the acceleration of growth rates. Within ten years, results of feeding 17 different antibiotics were reported. The practice of using antibiotics commercially in the feed of meat animals expanded rapidly. The FDA approved and established acceptable tolerance levels in animal products for the following antibiotics: chlortetracycline, oxytetracycline, tylosin, bacitracin, penicillin, nystatin, lincomycin, monensin, and erythromycin.

Antibiotics are inconsistent in their growth-promoting effects. The best results are usually with animals that are being raised under unsanitary conditions or in pens that have been used for a long time and have become contaminated. This, together with the fact that animals raised under carefully controlled germ-free conditions do not respond to the antibiotics, indicates that the antibiotics are not "growth promoters" but germ controllers, possibly exerting their action on low-level infections, which occur when animals are continuously exposed to mildly pathogenic bacteria such as those which infest the gastrointestinal tract.

The growth response from feeding antibiotics is somewhat less than that generally claimed for hormone administration. In some cases the effect on growth occurs only when the animal is young. In cattle, the gain averages about 3 to 4 percent where benefits accrue, but research reports have been conflicting and indicate that the type of antibiotic, the dosage rate, and the proportions of roughage and feed concentrate are all factors. Apparently there is some increase in the rate and weight gain and in feed economy if the level of antibiotic is low enough to avoid upsetting the microflora balance of the rumen.[11]

Resistance

Bacteria are remarkably adaptable to changes in their environment. Among their capabilities is the increasingly evident resistance to antibiotics. People contribute to this adaptation when they introduce them orally, inject them, apply them topically (to the surface of the body), or feed them to animals. Microorganisms that are sensitive to the drugs will die, and those that are resistant will proliferate. The process is limited only by other natural antagonists in the environment. If the resistant organisms are pathogenic, the antibiotics that induced their ability to survive will be ineffective in treating the disease.

For many years penicillin was the standard treatment for staphylococ-

---

[10]The process of producing antibiotics such as penicillin and other products from fungus and other cultures of microorganisms is called *fermentation*.

[11]The rumen is the first pouch of the stomach of ruminants (animals that chew a cud, that is, return food to the mouth and rechew what has been swallowed).

cal infections. Then an alarming thing happened. Some hospitals could no longer keep "staph" under control with the antibiotic. Of 12,100 children admitted to the Children's Hospital Medical Center in Boston in 1967, the principal cause of illness of 17 percent was a bacterial disease. Many were permanently affected and 77 died. At first it was thought that overconfidence in penicillin had cause hospital workers to relax antiseptic discipline. This was a factor. But it was discovered that in addition, the *Staphylococcus* bacteria were simply not succumbing to the usual treatment. It was found that the bacteria produced a detoxifying enzyme, called *penicillinase*. The crisis led to more stringent antisepsis and the introduction of synthetic penicillins and other antibiotics which the enzyme could not as readily put out of action.

Animals and human disease

Several kinds of bacteria that have the ability to develop resistance are common to both man and animals. These include some of the microorganisms that are commonly found in the intestinal tract, called *gut microflora* or *enterobacteria*.[12] Some of them cause diarrhea and diseases of the urinary tract and are among those microorganisms sometimes picked up by American travelers in foreign countries.

The National Communicable Disease Center at Atlanta, Georgia, found that strains of *S. tiphimurium* isolated before 1948, when antibiotics were scarcely used on farms, were sensitive to tetracycline. But strains isolated in 1962, after antibiotics had been in use for several years, displayed resistance. There was resistance to tetracycline in 30 percent of the strains from poultry, 57 percent of the strains from hogs, and 94 percent of the strains from cattle. Similar findings were made by Dutch and British workers.

Antibiotics can remain in the body of animals or humans for long periods. Injectable preparations of streptomycin were shown to persist for at least 75 days in the kidneys of treated animals and for an unknown length of time thereafter. Theoretically, the chances of human diseases being caused by resistant bacteria from animal sources are very high. On the other hand, experience does not support this supposition. After more than 20 years of medicated feeds, there has been no calamity traceable to animal origins. The incidence of salmonellosis appears to be stable, mortality rates have not risen, and the virulence of *Salmonellae* is no greater. In hospitals there are indications of a decrease in multiresistant staphylococcus, but more and more resistant strains of bacteria are appearing, and many questions remain unanswered regarding the long-term ecological effects on people of continued exposure to antibiotics in the environment. The normal intestinal tract harbors more than 200 species of bacteria. It is estimated that more than 90 percent of them are anaerobic (independent of oxygen), and the effect of antibiotics on them has been studied relatively little.

Transferable resistance

One of the most worrisome facets of the resistance problem is the discovery that bacteria which have developed resistance can transfer their ability to resist antibiotics to susceptible bacteria. The transference takes place between individuals of two strains during a process called *conjugation,* wherein two bacterial cells come together and form a connection

---

[12]These include *Salmonellae, Shigellae,* and *Escherichia coli.*

through which genetic information is passed from one cell to the other. The spread of drug resistance that takes place in this manner is called *infectious resistance* or *transferable resistance*.

The implication of this finding is that if we expose ourselves to antibiotics indiscriminately, many of the microorganisms that normally inhabit our bodies—such as those in the gastrointestinal tract and elsewhere in the environment—may develop resistance that could be transferred to normally nonresistant pathogenic organisms to which we are occasionally exposed.

When resistance to several antibiotics is transmitted to other strains or species, it is called *infectious multiple resistance*. Some bacteria are resistant to as many as nine different antibiotics. Thus far, the only bacteria known to be involved in *infectious* drug resistance are those classified as gram-negative entero-bacteria. These are bacteria that are found in the lower intestine of man and animals. They cause several diseases, including salmonellosis, typhoid fever, and bacillary dysentery. Evidence indicates that bacterial strains retain their resistance only as long as they are subjected to constant antibiotic exposure, called *antibiotic pressure,* and that when antibiotics are removed from their environment, they gradually regress, that is, they lose their drug resistance.[13] However, strains of bacteria may retain their acquired resistance for a long time. In one case, bacteria from a pig were resistant to an antibiotic for seven months after the drug had been withdrawn.

Possibly even more far-reaching is the finding that bacteria can transfer other genetic information, such as the ability to bring about certain enzymatic reactions. In one case, a bacterial strain that was usually associated with chickens changed to a type very close to the one found in cattle. This Jekyll-Hyde behavior of bacteria makes it almost meaningless to classify bacteria as either human or animal, and with continued use of antibiotics and other drugs, may necessitate an endless and frantic search for new anti-bacterials that have the capacity to somehow break through the infectious multiple resistance cycle.

SEX HORMONES IN MEAT PRODUCTION

In August 1970 Sweden informed the United States that it would no longer permit the importation of meat from animals that had been fattened with hormones. Diethylstilbestrol (DES)[14] is the hormone most widely used for this purpose in the United States, where about three-fourths of the beef cattle were given the hormone in mixture with their feed. At least twenty countries have prohibited the use of the hormone for fattening livestock because small amounts have been found to cause cancer in laboratory animals and the results are supported by clinical evidence that it is carcinogenic in humans.

DES is a synthetic female sex hormone. Its effect is similar to that of estradiol,[15] the most potent of the three main female steroid hormones

---

[13]These include *Salmonella, Shigella,* and *E. coli.*

[14]Also called stilbestrol.

[15]Estradiol-17 B.

USDA photo

In addition to producing new breeds and types for greater meat production, livestock breeders have added hormones, antibiotics, enzymes, arsenicals, and other chemicals to improve the efficiency of livestock rations.

and the major secretory product of the ovary. The female sex hormones as a group are called *estrogens* and their effect is described as estrogenic.

Effects of DES in animals

Diethylstibestrol was one of the first synthetic, nonsteroid estrogens discovered and is still the most biologically active, having the equivalent potency of estradiol. DES, unlike the natural steroid estrogens, can be administered by mouth, and for that reason, as well as its potency, was readily adaptable for use in medicine.

Effects of natural estrogens

The estrogens are, in large part, responsible for the changes that take place in girls at puberty, and they play a role in bringing about that intangible quality called femininity. Estrogens cause the maturation and development of the vagina, uterus, and fallopian tubes and cause, along with secretions of the pituitary gland, enlargement of the breasts and the formation of fatty tissue. Indirectly, and in a way not well understood, they contribute to the shapes of the bones and skeleton, the contour of the body, and softening of the skin. Pigmentation of the skin of the nipples and the growth of hair in the armpits and pubic region are also the effects of estrogen. A cyclic intensity of estrogenic activity is superimposed on the feminizing effects and influences some of the features of the normal menstrual cycle, particularly at puberty and again at menopause. There are indications that some aspects of aging are associated with a decline in estrogen activity. During pregnancy the estrogens are secreted by the placenta and at that time human urine is an abundant source of these natural female hormones.

The horse and other animals of the genus *Equus* are remarkable estrogen factories. The mare, when pregnant, excretes more than 100 milligrams daily in the urine, but this is still less than that produced by some stallions, who despite their clearly evident virility, deposit more estrogen in the environment than almost any other creature.

The estrogens are unique among the animal hormones in that a large number of chemical compounds have estrogenic activity. They are found abundantly in the natural environment, from the flowers of plants to the mud of the Red Sea. Estrogenic compounds are known to occur in concentrations that cause disorders in animals. When sheep in Australia were allowed to graze on so-called subterranean clover,[16] the results were disturbed estrous cycles, infertility, and abortion or birth difficulties if they became pregnant. The defects were traced to the presence of large amounts of a weak estrogenic chemical called genistein that is produced by the plant. Genistein is found in several plants of different families.

The estrogenic substances found in plants are weak compared to the animal estrogens. Estradiol is 1,000 times more potent than coumestrol, a botanical estrogenic substance found in ladino clover; and coumestrol itself is 30 times more active than genistein, Legumes show higher contents of estrogenic substances than most other groups of plants. Coumestrol has also been isolated from alfalfa, one of the most widely used hay crops for livestock, particularly dairy feed. Have strains of alfalfa been unknowingly selected for their growth-promoting properties? Should they be? Is coumestrol being excreted in milk? These questions are difficult to answer and perhaps cannot be answered at present.

Estrogenic activity has been detected in a large number of food plants, including vegetables, cereals, and edible oils. Estrogens are found in pollen, which may account for the reported estrogenicity of honey. The natural estrogenic compounds in foods are generally of such weak potency that the probability of their causing adverse effects from the normal consumption of food by humans appears to be remote.

The effect of estrogens on poultry fattening was reported during World War II, but even though FDA approval was granted for stilbestrol implants as early as 1947, the idea was not extensively commercialized for several years. A renewed interest developed in the use of stilbestrol and other hormones in cattle, sheep, pigs, and poultry in the early 1950s, and by 1954 the oral administration of DES to beef cattle came into general use. Shortly thereafter, approval was granted for the use of DES pellets for implantation in the ears of beef cattle. By 1959, both oral administration and pellet implantation were being used for sheep.

When DES is fed to livestock, the effect is similar to that of castration. Castration of animals became a common practice following the discovery that if male animals had their testicles removed, they were gentler and easier to handle. The practice increased in comparatively recent times because castration produces beef that is fatter and more tender than meat obtained from bulls. The testicles produce a steroid hormone, testosterone, which has comparable effects on growth and development of the male skeleton, muscles, and organs and influences on sexual drive and sex characteristics that the estrogens have in the female. From the standpoint of beef production, castration especially affects the parts of the

---

[16]*Trifolium subterraneum.*

body that develop late. For example, the untrimmed loin is heavier and the proportion of hindquarter to forequarter is larger in castrates.

In some European countries meat from bull carcasses is just as acceptable as from steers and often brings a premium price.[17] Testosterone does, in fact, cause animals to gain weight. In one experiment, steers treated with testosterone gained about one-half pound more per day. But testosterone must be injected frequently, so this is impractical; however, Russian scientists have developed a technique for partial castration that leaves the testosterone-secreting portion of the testicle intact for increased meat production.

The benefits from DES to the consumer have been questioned. American biologists showed that cattle which were double-implanted with DES made slightly larger gains in weight, but steers that received no hormone produced the fattest (not the heaviest) carcasses with the highest yield grade.

Experiments on the effects of DES in beef cattle, either by implantation or incorporated in the feed, show that the treated animals gain weight slightly more rapidly and require slightly less feed per pound of weight gain, but that the carcass grade is slightly inferior. There is, on the average, about a 10 percent increase in weight, though not necessarily in meat. Evidence is not conclusive that there is a real technological gain.

An unpredictable feature of stilbestrol is the effect in different species of animals. Whereas an implant of 12 to 15 milligrams is required for chickens, the same amount in a lamb that is several times larger is excessive. In 600-pound steers, more than a hundred times larger than a chicken, 24 to 36 milligrams are adequate. Why do the effective dosage rates in different species not have the same ratios to body size? The answer is not known; but the question raises a corollary: what is the effect in humans of different age, sex, body weight, and physical condition?

Hormones in chicken An early use of DES was for chemical castration of chickens in poultry production. The method is called *chemical caponization. Capon* is the term used to describe a castrated cock, especially one being fattened for the table. The procedure was to implant a pellet containing 15 to 30 milligrams of DES subcutaneously (beneath the skin) in the neck region of the young bird. Approval for the practice was originally granted on the supposition that any remaining hormone residue would be discarded with the neck. However, it was found that portions of the neck are often cooked along with the remainder of the bird and excessively high residues often remained. Deep concern was aroused following reports that when mink raisers fed discarded chicken necks to their animals, the mink became temporarily sterile.

The effect of hormones on poultry is variable. About half the reports indicate some improvement, while half show no effect. Weight gain, when it occurs, is almost entirely due to greater deposition of fat.

On December 10, 1959, the FDA announced that it was banning the use of stilbestrol for use in poultry and prohibiting the sale of stilbestrol-treated birds. The FDA did not, at that time, restrict the use of estrogenic

---

[17]A steer is a male bovine animal castrated before maturity; however, the term at one time was also used to refer to any male cattle raised for beef.

hormones in large animals because of the belief that residues of the hormones were in smaller amounts than in poultry and because there was no proof (then) that meat from implanted animals, even when eaten in far greater than normal amounts, were carcinogenic. Since then, small amounts of stilbestrol have been found to cause cancer in laboratory animals.

Evidence that DES may have serious effects at the cellular level is indicated by the inhibition of cleavage during cell division in sea urchin eggs and mitotic (cell division) abnormalities in cultures of rabbit fibroblasts. DES and its phosphate salt have been used therapeutically for carcinomas of the prostate.

A rare vaginal cancer found in seven young women aged 15 to 22 was associated with treating their mothers at the time of their pregnancy with the synthetic hormone. DES was once thought to be beneficial during the first three months of pregnancy to control bleeding and counteract a history of miscarriages.

Residues in beef   About 40 million cattle are slaughtered each year in the United States. In view of the profound effect of steroid hormones in the human system, the FDA formerly required that feed-pen operators withdraw hormone-treated feed 48 hours before marketing. This was changed in October 1971 to seven days, which, according to the FDA, was long enough for the hormone to be excreted by the animal below levels that would present a hazard to consumers. The difficulty of this procedure is that there is no way to effectively enforce the time limitation on feeding hormones. It is an expensive procedure to make the switch, and in a highly competitive business subject to sharp price fluctuations, the beef producer can be under severe economic pressures that may make it difficult for him to face up to what may seem to be an unrealistic and prejudicial requirement.

Under the Delaney clause of the Food, Drug and Cosmetic Act, no food may contain a residue that causes cancer in laboratory animals. Because of this and other toxic properties, no DES residue in meat is permitted. However, government tests repeatedly turned up DES residues over a period of several years. Since there was no feasible way to prevent DES residues from showing up in beef liver, the FDA finally ordered a total ban, effective January 1, 1973, on the use of animal feed containing DES.[18] Later, the courts ruled the ban illegal because no hearings had been held, but cattlemen remained reluctant to resume the use of DES on a large scale until the issue could be resolved.

PROBLEM AREAS

Eating is one of the most satisfying responses to the basic survival instincts that motivate all higher animals. In humans, there is also an almost irresistible compulsion to experiment and to look for new discoveries. The age-old search for new, exotic, and flavorful foods is one of the best examples of this. One of the first heroes must have been the daring

---

[18]DES implants, which are equally effective and require smaller amounts of the synthetic estrogen, were unaffected by the ruling.

cavewife who sprinkled the salty white crust from a dry lake on her barbecued cutlet, or dipped a sparerib in sea water before singeing it over the aromatic flame of a driftwood fire. Few people would deny her the plaudits of all those who have savored the delights of what may have been the first food additive. It is not surprising that similar innovations would extend far beyond the early use of salt and common spices.

On the other hand, taboos against tainted food are found in primitive and civilized societies alike. We are taught since childhood to heed them carefully. Mother cries sharply, "Don't touch that—it might be poison!" The child drops the suspicious-looking object promptly. It is not necessary that the object actually be poisonous to get this response. It is enough that it *might be poisonous*. It is well recognized by nearly all adults that synthetic chemicals can have injurious as well as beneficial effects. Since many chemicals are used in food preparation, we have established a system of legal taboos, administered in the United States by the Food and Drug Administration. The FDA does not always accurately reflect the taboos that society has set for itself, and a series of oversights, misjudgments, and mistakes can hardly have a soothing effect on a public that has a deep-seated distrust of chemical tampering with food.

The FDA does not promise absolute purity or absolute freedom from risk. On the contrary, society only asks, through its legislators, that the FDA guarantee reasonably pure and reasonably safe foods. Thus, the FDA often makes a compromise between what it thinks the public would like to have and what it thinks is an unacceptable impurity. Take, for example, bacon, or frankfurters. These, as well as some of the other preserved meats, have been treated with liberal amounts of the known poisons nitrates and nitrites. The FDA cannot imagine that people would want to forgo these tasty foods. Bacon and eggs, or a hot dog on a bun, are supposed to be as American as apple pie, although there is no evidence that any of them originated in America. The FDA has ruled that the nitrate-nitrite additives, though they involve some hazard, are necessary to preserve the foods against botulism. In effect, the FDA says, "Poisoning from botulism is worse than poisoning from nitrates and nitrites, and since people are going to eat these foods anyway, we will see to it that they are exposed to the lesser of the two evils." However, in October, 1975, the U.S. Department of Agriculture recommended that the FDA reduce the permissible level of nitrite in meat and poultry products such as frankfurters, bologna, and luncheon meats to 156 parts per million. The residual level in such products as cooked sausages would drop from 200 to 100 parts per million. Further testing was needed to determine the safe level of nitrite in bacon.

A different kind of problem surrounds the use of substances in foods that might cause cancer, birth defects, or mutations. Since the law applies specifically to only one of these—cancer—most of the controversial issues have centered around carcinogens (agents that cause cancer). But should the restriction be extended to include other disorders? Should it be extended to include nitrates and nitrites, which do not cause cancer themselves but, under some conditions, change chemically to substances called *nitrosamines*, which can cause cancer? Biologists do not even have a definition for cancer that everyone can agree on. Some say that any substance that causes a tumor in experimental animals is also apt to cause cancer in people; others say that the tumor should be called a cancer only if it persists after withdrawing further feeding of the offending chemical. Still others wonder if it is a mistake for the lawmakers to exempt cancer-causing agents that occur naturally in foods, of which there are several well documented examples.

The poet-philosopher Lucretius said, "What is food to one may be fierce poison to others." More than two thousand years later, the problem is still with us.

READINGS

Food Additives

Anon., *Additives in Our Food*. FDA Publ. No. 43, October 1968.

Anon., *Criteria Proposed for Classing Substances as GRAS or as Regulated Food Additives*. FDA Papers, Feb. 1971, p. 33.

Anon., *Food Additives. What They Are and How They Are Used*. Washington, D.C., Manufacturing Chemists' Association, Inc., 1961, 1971.

Toxicants Occurring Naturally in Foods, Second Edition. Washington, D.C., National Academy of Sciences-National Research Council, 1973.

FURIA, T. M., *Handbook of Food Additives*. Cleveland, Ohio, The Chemical Rubber Co., 1968.

MODELL, W., "Mass Drug Catastrophes and the Roles of Science and Technology," *Science* 156:346–351 (1967).

Food Contaminants

Anon., *Salmonella, the Ubiquitous Bug*. FDA Papers, 1(1): 13–19. (Feb. 1967).

DUGGAN, R. E. and K. DAWSON, *Pesticides: A Report on Residues in Food*. FDA Papers, June 1967, pp. 4–8. (Also in K. E. Maxwell, *Chemicals and Life*. Belmont, Calif., Dickenson Publishing Co., 1970, pp. 281–291.)

HAROLD, L. C. and R. A. BALDWIN, *Ecological Effects of Antibiotics*. FDA Papers, Feb. 1967, pp. 20–24.

The Use of Drugs in Animal Feeds. Proceedings of a Symposium. Washington, D.C., National Academy of Sciences, 1969. Publ. 1969.

VAN HOUWELING, C. D., *Drugs in Animal Feed? A Question Without an Answer*. FDA Papers 1(7):11–11 (1967).

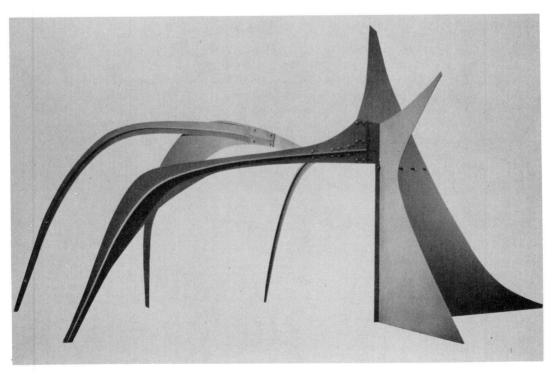

Cancer, "The Crab," by Alexander Calder. Photo courtesy The Museum of Fine Arts, Houston.

THE MICROSPHERE: LIFE INSIDE THE CELL

The cell is the structural building block of plant and animal life. But it is also much more. Some living cells are complete animals or plants in themselves. Enclosed within their single cells are all the structures necessary for nutrition, growth, reproduction, and behavior that make up the characteristics of living organisms. It is not surprising, then, that in the more complex multicellular animals and plants, the cells are the sites of action for many of the important functions of life.

THE INTERNAL ENVIRONMENT

The *internal environment* is an expression coined in 1878 by the French physiologist Claude Bernard. He referred to the exchanges that take place between the lymph, the blood, and the cells and their effect on the equilibrium that is important to life. He said, "The constancy of the internal environment is the necessary condition of the free life."

The cells of the body are surrounded by a fluid that supplements the circulatory systems of the blood and the lymph. This fluid is called *tissue fluid*. It serves as an intermediary to bring the cells in contact with the constituents of the blood plasma and the lymph whereby there is constant exchange of the chemicals of life. Said Claude Bernard, "All the vital mechanisms . . . have only one object, that of preserving constant conditions of life in the internal environment."

Modern scientists have increasingly looked upon the internal and external environments as parts of an integrated whole. Many of the effects of the external environment take place primarily in the internal environment immediately surrounding and within the cells, even though these effects may be visibly manifested only in the gross reactions or behavior of the organism. Bernard was the first to show the specific manner in which a toxin from the external environment acts upon the body. He showed that the poisonous action of carbon monoxide was in its ability to replace oxygen in the molecules of hemoglobin and that death or injury resulted from oxygen starvation. His reference to the importance of the internal environment may be interpreted as a warning against tampering with external conditions in ways that upset the equilibrium of the internal environment.

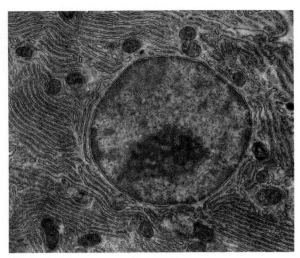

Photo courtesy of Dr. Donald W. Fawcett, Dept. of Anatomy, Harvard Medical School.

Portion of a cell from the pancreas of a bat, photographed with an electron microscope at X29,000 magnification. The large spherical object is the nucleus containing a smaller object, the nucleolus. The nucleus contains the chromosomes, rich in DNA.

CELL STRUCTURES

A typical cell is surrounded by a cell membrane, or in plants, by a cell wall, enclosing a gelatinous material called *cytoplasm*. It is possible to observe within the cytoplasm several kinds of structures. Knowledge of their nature and functions has been greatly aided by biochemical studies and electron microscopy. While the ordinary lens microscope is at best capable of magnifying an object 2,000 times, the electron microscope can magnify up to 200,000 times on photographic film and 500,000 times on a fluorescent screen. Using special electronic gear it can magnify to about 2,000,000 times. The electron microscope uses streams of electrons instead of photons of light.

The chromosomes   Usually the most prominent body inside a living cell is the *nucleus*, itself enclosed in a nuclear membrane. Within the nucleus is the genetic material that appears at certain times during cell growth and multiplication in the form of distinct, often elongated, bodies called *chromosomes*.[1]

The chromosomal material in cells was discovered by a German anatomist, Walther Flemming, who published his findings in 1882 in a book entitled *Cell Substance, Nucleus, and Cell Division*. Because the nuclear material had an affinity for the dyes with which he was working, he called it *chromatin*, from the Greek word for color. He was able to follow the substance through the various stages of cell division and to observe that under some circumstances it formed into distinct bodies, which came to be called *chromosomes*. Flemming had no inkling of the genetic significance of his findings. Gregor Mendel had published his classic studies on inheritance in pea plants twelve years earlier, but few

---

[1]*Chroma* (Gr.) = color; *soma* (Gr.) = body; hence "colored bodies."

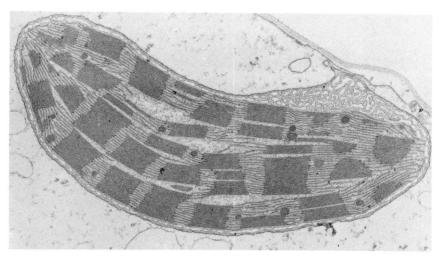

Photo courtesy of Dr. L. K. Shumway, Depts. of Genetics and Botany, Washington State University.

Thin section of corn leaf tissue showing a chloroplast, at X32,000 magnification. Chlorophyll, a green pigment within the membrane system of the chloroplast, is the site of the light reactions of photosynthesis.

people knew about it and no one understood its importance. It was not until twenty years later, when Mendel's work was rediscovered, that Flemming's studies of chromosomes were seen to have a bearing on Mendel's rules of inheritance.

INHERITANCE: MENDEL'S DISCOVERIES

Gregor Mendel was an Austrian monk who had an interest in botany and a flair for mathematics. He was the first person to work out the principles of inheritance on a scientific basis. Mendel described his discovery in 1866 in the *Transactions of the Brünn Natural History Society*. He worked with pairs of complementary traits, such as tallness and shortness. By carefully pollinating pea plants of different heights, flower color, and other qualities, he was able to show that the pairs of traits were inherited by the progeny in mathematically precise patterns.

Mendel noticed that when he crossed tall plants with dwarf plants, tall plants were more apt to appear in the progeny than dwarf plants. But he found that the ability to produce seeds that would grow into dwarf plants remained latent in many of the tall plants. For example, when he cross-pollinated dwarf plants with true-breeding tall plants, the hybrid seeds sprouted only tall plants. And when seeds from self-pollinated dwarf peas were planted, only dwarf pea plants grew. However, when he self-pollinated the tall hybrids, a fourth of the seeds grew into true-breeding dwarf peas, a fourth produced true-breeding tall peas, and one-half developed into tall plants that were not true-breeding, having unexpressed characteristics for dwarfness. This is a type of inheritance in which tallness is *dominant* and dwarfness is *recessive*.

Mendel decided that the factors governing each pair of traits were contributed equally by both parents and that these factors remained distinct in the offspring.

Genes

The unseen factors supposed to be responsible for the Mendelian pairs of characteristics came to be known as *genes*, from the Greek word *genos* meaning "descent." The idea of the gene as the vehicle for paired traits is useful in explaining a great deal about inheritance. The genes are thought of as units arranged linearly along the chromosomes. Since the chromosomes in the body cells are in pairs, the genes must also be in pairs. Each gene, then, is responsible for one of a pair of traits such as tallness and shortness or red flower and white flower. In the fruit fly *Drosophila*, one pair of genes is for body color. If both genes of the pair are for gray color, the fly will be gray. If both genes are for black color, the fly will be black. But if the chromosomes of a fly contain one gene responsible for gray color and one gene responsible for black color, the fly will be gray. The gene for gray body color is said to be *dominant*. There are many variations of this scheme for the action of genes, the study of which is part of the science called *genetics*. But exactly what takes place when the factors in genes are transmitted from one generation to the next did not become evident until new discoveries were made about the structure of the chromosomes.

The double helix

Chromosomes consist of double strands of nuclear material (see Figure 11.1). Each of the strands contains a backbonelike support consisting of two kinds of components: *sugar* units alternating with *phosphate* units. To each of the sugar units there is attached another kind of unit called a *nitrogen base*. The three kinds of structures make up a larger unit in the strand called a *nucleotide*, consisting of a sugar, a phosphate, and a nitrogen base. The entire strand of nucleotides is called a *polynucleotide*, or nucleic acid. The two strands of polynucleotides that make up the chromosome are held together loosely with a zipperlike arrangement in which the bases of one strand are fastened to the bases of the other strand by means of *hydrogen bonds*. The holding force of the hydrogen bonds is relatively weak, permitting the strands to come apart when the time comes for replication.

Thus the double strand of nucleotides has a ladderlike structure with the sugar-phosphate units forming the rails and the nitrogen bases—connected through the hydrogen bonds—forming the rungs. The structure does not, however, lie straight but is twisted in a way that suggests a spiral staircase. For this reason, it has been called a *double helix*.[2] The proposed structural arrangement is sometimes called the Watson-Crick model because of the work of two colleagues at Cambridge University, F.H.C. Crick, an English biochemist, and J. D. Watson, an American biologist, who worked out the probable structure of the DNA molecule on the basis of crucial x-ray diffraction studies by a British physicist, M.H.F. Wilkins. The three scientists shared the 1962 Nobel Prize for their discoveries.

---

[2] A single helix is a structure such as a coiled spring with a longitudinal core, or a wire that has been wound around a rod (not a spiral, such as that of a watch spring).

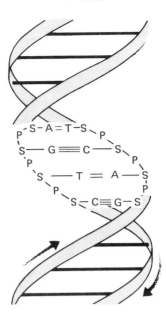

FIGURE 11.1 X-ray diffraction studies revealed that DNA is a double helix, resembling a spiral stairway with the rails being formed by alternating sugar-phosphate groups and the steps consisting of pairs of nitrogen bases. The two halves of the helix are held together by the hydrogen bonds of the base-pairs.

DNA: THE MASTER MOLECULE

The entire double helix forms a large molecule having a molecular weight of several million, its exact size depending on the species and the particular chromosome. The polynucleotide of the chromosome is named after the kind of sugar it contains, deoxyribose, and is therefore called deoxyribonucleic acid, or simply DNA.

DNA is uniquely important. It has the dual responsibility for instructing the cell in the manufacture of biological substances needed for growth and development as well as for passing all of the inheritable traits of the organism from one generation to the next. Thus DNA is the control center for both the individual organism and its society, since it oversees both the orderly functioning of the organism and the integrity of the species.

The code of life   How does DNA store its instructions and how are its commands given? The information is in the arrangement of the nitrogen bases that are attached to the polynucleotide strands in the DNA. The message is spelled out by the order in which the different kinds of nitrogen bases are arranged. The nitrogen bases form "words" much as letters of the alphabet spell out words. The exact sequence of the nitrogen bases along the strands of the DNA has been called the *genetic code*. Thus the blueprint for the organism, its structures and its functions, and for its progeny as well, is written in the genetic code.

One might think that the information in the genetic code would resist change and would remain constant within the species from one generation to the next or, at least, for the life span of the individual organism. This is generally so, but not always. An occasional miscue, often induced by environmental effects, causes changes in the code, so that new instructions are contained in the DNA.

Since the discovery of nucleic acid in 1869 by a young German chemist named Freidrich Miescher, studies by a number of investigators over the years showed that there is a significant pattern to the nitrogen bases in the DNA molecule that is constant in all living things, whether in man, worm, moss, sponge, or virus. The total units of adenine (a purine) in the molecule is about equal to the units of thymine (a pyrimidine); and the number of units of guanine (a purine) is equal to the units of cytosine (a pyrimidine).

Along the zipperlike connection of the two strands, adenine (A) is always paired with thymine (T); and guanine (G) is always paired with cytosine (G). (See Figure 11.2.) Throughout the length of the DNA molecule, the connections between the two strands are made up of these complementary pairs. During the process of replication, the two strands of the DNA molecule come apart along the zipperlike connections and new nitrogen bases attach themselves to those of the original strands. New adenines become attached to the old thymines, new cytosines become fixed on the old guanines, and so on, until a new strand of complementary nucleotides is formed on the old strand. The new molecule is ready for a new cell or a new organism.

Since the genes are the paired factors for the inheritable characteristics, it follows that there is a relationship between the genes and the

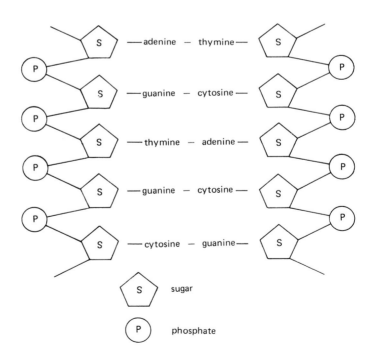

FIGURE 11.2 Structure of a section of the DNA double strand.

nucleotides that make up the DNA of the chromosomes. It is beyond the scope of this book to describe in detail the molecular mechanism of inheritance; a brief description will suffice to show the relationship between inheritance and the effects of extraneous influences on the chromosomes.

THE VARIETY OF TRAITS

The human body, which begins as a single fertilized egg cell (*ovum*), grows into a mature adult with an enormous number of cells, variously estimated at from 100 billion to as many as 50 trillion. Each of the 100 billion or more cells—with a few exceptions—contains within its nucleus a microscopic replica of the total genetic material that came from the male and female parent sex cells.

The genetic makeup of a person is subject to the vagaries of chance almost beyond imagination. One cannot predict which partner of a given pair of genes will end up in any one sperm or egg cell during their formation. Nor can anyone tell in advance what combination of the 10,000 or so human gene partners will pair up at the time of fertilization.

A child can have any one of 100 trillion gene combinations. The variety of possible combinations is usually enough to explain the seemingly endless variety of characteristics seen among humans, or among the individuals of any other species of plant or animal. But there are other reasons as well why differences show up, sometimes unexpectedly. These differences are often caused by changes within the chromosomes or in the DNA itself. Environmental conditions are directly responsible for many of the changes. They are profoundly important to the welfare of the individual and the future of the species because many of the effects are injurious, some of them lethal, and a few are beneficial, being the vehicle for improved adaptation and evolution of the species. Environmental factors that cause injurious biological effects such as illness, abnormalities, death, or population decline, often exert their first effects unseen, inside the cells. In these cases it is vitally important to know what is happening so that corrective action can be taken before excessive damage is done to the organism or the population. Such situations are dealt with in Chapters 12, 13, and 14.

READINGS

AUERBACH, C. *The Science of Genetics*. New York, Harper and Row, 1961.

ASIMOV, I., *The Genetic Code*. New York, The New American Library, 1962.

The Molecular Basis of Life. San Francisco, W. H. Freeman and Co., 1968.

WATSON, J. D., *The Double Helix*. New York, The New American Library. 1968.

GENETIC INJURY: MUTATIONS

Inheritance that broke the rules

The world did not know about Mendel's discoveries until after his death. The principles of inheritance were rediscovered nearly forty years later, almost simultaneously by three botanists working independently and unknown to each other. One of them, Hugo De Vries of the Netherlands, made an important additional discovery. While working with the evening primrose, he noticed that occasionally a new variety, which differed markedly from its parents, would appear and that it would perpetuate its new characteristics in future generations. De Vries had discovered the regular occurrence of what later came to be recognized as aberrations in the chromosome content of the cells.

Farmers and husbandrymen had long before been familiar with occasional freaks, called "sports," that arise spontaneously among plants and animals. Sometimes the sports had desirable qualities and were propagated. New varieties of plants and new strains of animals often show up this way. These occurrences are sometimes caused by the appearance of new genes caused by changes in the molecular makeup of the DNA. Such alterations are called *mutations*.

Broadly speaking, a mutation is an inheritable change in the genetic material of a cell or an organism. Almost every cell in the body contains an exact copy of all the genetic material originally present in the fertilized egg cell. The most important exceptions are the germ cells (sperm and ova) which contain only half as much. Germ cells are the gametes (male and female sex cells) and the cells that give rise to gametes.

Mutations in the somatic (body) cells will die out with the death of the organism. But mutations that occur in the germ cells, or during their formation, can be passed on to future generations. Thus a mutation in a germ cell can have vastly more far-reaching effects. For example, if there were no mutations, evolution would not take place and all life on earth would be exactly alike.

Evolution

The history of life on earth is a parade of living types extending over a period of three billion years. The broad sweep of evolution, from the most primitive single-celled organisms to the most complex of the present-day forms, is recorded by the fossilized remains of animals and plants in sedimentary rocks. Each variation was molded by spontaneous mutation. Many of the mutated forms were unable to adapt to the dynamic changes in climate, food supply, and enemies. Others were culled out because characteristics that were useful under one set of conditions became useless or dangerous under a new set of conditions.

Every plant and every creature is an elaborate composite of genetic characteristics: a piece added here, a color deleted there, during millions of years of evolution. The individuals of each species that had the gene pattern which adapted them best to their environment were the ones that survived in the competitive struggle up the evolutionary ladder. They were more fit to win the contest for food, sexual partners, resistance against enemies, and protection against the elements. Each characteristic is the product of countless generations of natural selection. Though the odds that a random change in one or more of the genes would cause an individual to be more favorably adapted to its environment are slim, a mutation occasionally occurs that is beneficial.

Suppose, for example, that a tribe of hunters had lived for thousands of years in a densely wooded region. The success of any individual would depend on keen eyesight with which to spot game and aim an arrow true to the mark. Nearsightedness would be a disadvantage and might be disastrous if his affliction caused him to be more vulnerable to the attacks of predatory animals and human enemies. Clearly, a successful though extremely nearsighted person from our modern society would not fare very well, and might not be able to survive at all, in a primitive hunting society. Nor would a man born into the hunting society be successful if his chromosomes contained a mutant gene for nearsightedness. This is evolution in action, by survival of the "fit." Charles Darwin called it *natural selection*.

CHARACTERISTICS OF MUTATIONS

Mutations that take place under ordinary circumstances in nature are called *spontaneous mutations*. Many of these are of evolutionary significance. They occur as the result of more or less minor imperfections in the complicated mechanism of chromosome division, separation, and replication before or during the formation of the gametes in the male and female sexual organs. The behavior of molecules within the cells is precisely programmed by the presence of just the right amount of hundreds of enzymes operating under dozens of stimulating and restraining influences. But even nature itself can seldom bring off an operation as complex as the production of sperm and ova, called *gametogenesis*, without occasionally making a mistake. Any one of several unexpected circumstances, such as a fever caused by an invading organism or its toxins, or a localized condition of stress from other sources, might be enough for a molecule to be twisted the wrong way or cause it to be held up so that it would have to wedge itself in the wrong place in line. Many such miscues go unexplained.

We have also seen that the nitrogen bases are arranged along each strand of the DNA molecule, seemingly at random. But in reality, they are formed according to a precisely programmed sequence. The order in which the nucleotide units are arranged along the strand is critically important, for this will determine in the end whether the cells will develop into a variety of plant that is tall or short, whether an egg cell will become a mouse or a man, or whether a person's eyes will be blue or brown. The nucleotide units, which are common to all life, call the signals for the incredibly complicated "game plan" involving millions of varieties of characteristics in living organisms. When there is an imperfection in the DNA, the signal is apt to be garbled. The result is a mutation.

Injurious mutations

Experiments with lower organisms have shown that the overwhelming majority of mutations are harmful to the organism. Some mutations are so deleterious that no progeny can be produced. The fertilized ovum may be so severely affected that it cannot develop or, if it does develop, the fetus may die. Even if birth occurs, the offspring may die in infancy or never reach maturity. A mutation that is severe enough to cause death before the mutant gene can be passed on to the progeny is called a *lethal mutation*.

Some mutations are not severe enough to be immediately lethal, in which case they may cause harmful characteristics to persist from generation to generation. The total of such deleterious genes in a population is called the *genetic load*. If a particular mutant gene is removed from the population by natural selection as fast as it appears, the condition is one of *genetic equilibrium* and it will be found with the same frequency of occurrence generation after generation. There is little that man can do to prevent most of the mutations that result from natural forces in the environment. But anything that man does to *increase* the genetic load has the potential for dire consequences to individuals and to the population as a whole.

Frequency of mutations

Sudden changes in the genetic content of cells occur with surprising frequency. Thomas Hunt Morgan, an American geneticist, worked out a method for pinpointing lethal mutations anywhere along one of the four pairs of chromosomes of the fruit fly. He found that a lethal gene would be expected to appear along the length of a particular chromosome about once every 200 times that a chromosome underwent replication. Thus for every 200 sex cells (sperm or eggs) produced by a fruit fly, one of them would have a lethal gene somewhere on that chromosome. Since there are at least 500 genes along the chromosome, each capable of undergoing lethal mutation, the chance that a particular gene would be lethal was 1 out of 200×500, or 1 out of 100,000 replications. This is now taken by many geneticists as the average mutation rate in higher organisms.

Humans have 46 chromosomes (23 pairs) (see Table 12.1)—until 1956, it was thought that there were 48 chromosomes. Modern techniques, including the use of chemicals to stop cell division, made it possible to learn many important facts about chromosomes and their behavior. It is estimated that humans have at least 10,000 different genes. Suppose that the rate of appearance of a certain deleterious gene in man is 1 out of 100,000. Then the chance of at least one of the genes in a sex cell being strongly deleterious is 10,000 out of 100,000 or 1 in 10. However, there are probably four times as many mutations that are weakly damaging. So the chances that a sperm or an ovum would have at least one weakly deleterious mutation would be $4 + 1$ in 10, or 1 in 2. Since these mutant genes are not evenly distributed among the cells, the odds that some of the cells would have none at all are favorable. Even so, it can be assumed that nearly half the sperm and ova produced by humans carry at least one mutant gene somewhere in their 23 chromosomes.

Though most of the mutations are highly deleterious, many of them

TABLE 12.1   Chromosome Numbers in the Body
Cells of Several Species of Animals and Plants
Compared to Man

| | Chromosome | |
Organism	Pairs	Total
man	23	46
mouse	20	40
fruit fly*	4	8
guinea pig	4	8
round worm†	2	4
crayfish	hundreds	
onion	4	8
pea plant	7	14
broad bean‡	6	12
rye	7	14
cabbage	9	18

*Drosophila melanogaster
†Ascaris
‡Vicia faba

lethal, geneticists and cell biologists are more concerned about the less drastic changes in the genetic material. Alterations that do not kill the cell or the organism can be carried from one generation to the next, bringing about important changes in the population.

It is not always possible to say that a given mutation is good or bad. The varieties of minor changes that have taken place in the nucleotides of the DNA molecules during millions of years of the struggle for survival made evolution possible. Today, they make up the *gene pool* that will bring about further changes in the characteristics of the human species through natural selection. It appears that we are accumulating mutations faster than they are being eliminated.

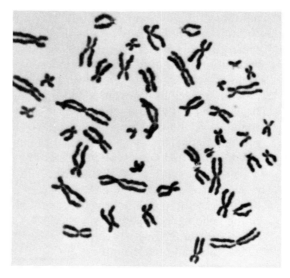

Photo courtesy of Carolina Biological Supply Company.

Somatic chromosomes of a normal human female.

Unfortunately, we increase the genetic load of undesirable genes by caring for those who are disordered physically and mentally and by trying to preserve their lives from infancy to old age, though morally we have no choice. Thus, the rate at which mildly deleterious genes are removed from the population by natural selection is interfered with. The frequency of defects in newborn babies runs at least 10 to 15 percent. According to Barton Childs of Johns Hopkins University, between 4 and 20 percent of all conceptions, and ½ percent of all children born, have chromosome abnormalities. As many as 20 percent of the pediatric patients at the University hospital had chronic afflictions in which genetic defects were involved.

KINDS OF MUTATIONS

Changes in the genetic material can take place in several ways. Modifications can occur either in the physical structure of the chromosomal makeup of the cell or in the chemical makeup of the DNA. The two main types, then, are chromosome alterations and DNA alterations, sometimes called *chromosomal mutations* and *genetic mutations*.

Chromosomal mutations

Gross changes in the physical nature or arrangement of the chromosomes can occur in several ways. Failure of the chromosomes to separate during cell division is called *nondisjunction*. The result may be a new cell with one or two extra chromosomes. In Down's syndrome, sometimes called mongolism, a congenital disease characterized by a form of mental deficiency and certain physical abnormalities, the cells may contain an extra chromosome (3 chromosomes of type 21 instead of the normal 2), or one of the chromosomes may be enlarged. The disorder is believed to be associated with excessive cellular production of materials under the instructions of the extra DNA. Women over the age of 38 have about 1 chance in 50 of having a child with Down's syndrome.

One form of chromosome-number mutation involves an abnormal number of the chromosomes that play a role in the determination of sex. In humans and in many groups of animals as well as plants, sex is determined by chromosomes called X and Y (see Figure 12.1).

If during cell division at the time of sperm formation in a man there is nondisjunction of the sex chromosomes, one of the resulting cells will contain no sex chromosomes and the other will contain both, or XY (see Figure 12.2). If a sperm containing XY fertilizes a normal ovum having X, the cells of the offspring will contain XXY, or double the normal complement of chromosomal material for femaleness, which in humans results in what is called male-type intersex. The effects of aberrations in the sex chromosomes differ greatly in mice, fruit flies, and humans.

Genic (DNA) mutations

We have seen that DNA replicates itself, faithfully transmitting its heritage to the progeny in accordance with the instructions programmed in the genetic code. However, the DNA is not infallible; sometimes mistakes are made. Changes can and do occur in the chemical structure of the DNA molecule before, during, or after the process of replication. If the mistake occurs during the formation of the gamete, and if the gamete makes a union that results in a fertile egg, the altered information carried in the DNA produces new characteristics in surviving progeny. Changes

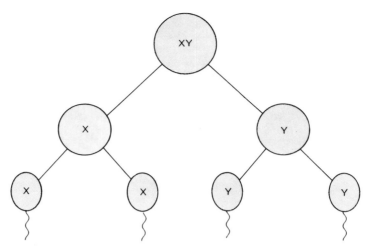

FIGURE 12.1 Transmission of the sex chromosomes (X and Y) during spermatogenesis (formation of the sperm in the testes).

involving one or a few nucleotide bases are called *point mutations*. These are not visible under the microscope and often are not sufficiently damaging to kill the cells; thus, they may persist in future generations.

ENZYMES AND MUTATIONS

Living cells are in a state of remarkable activity. Within the cells, myriads of chemical reactions take place with split-second precision. Though the reactions differ in nature and complexity, they must proceed rapidly and evenly despite the limited range of temperature and other conditions. The efficiency required could never be achieved without some means of promoting and controlling the millions of reactions that are involved in the transfer of molecules through membranes, synthesis of new compounds, splitting of molecules that are no longer needed, and the transport of

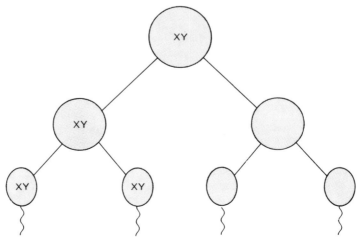

FIGURE 12.2 Fate of the sex chromosomes (X and Y) when nondisjunction occurs during spermatogenesis.

electrons, protons, and molecular pieces from one place to another. These and similar reactions are controlled by specialized proteins called *enzymes*, sometimes called biological catalysts. A *catalyst* is an agent that controls the rate of chemical reaction.

A human cell is estimated to contain as many as 100,000 different kinds of enzymes which supervise up to 2,000 chemical reactions. Several hundred enzymes have been identified, and many of them have been isolated and crystallized. Because several steps are often involved in changing one chemical into another, as many as ten or more enzymes may be needed for the process. Since enzymes are not consumed in the reactions but are either left intact or regenerated, only very small amounts of the numerous kinds of enzymes are needed to carry out the functions of the living organism.

Enzymes and genes

Until 1941, the way in which genes exerted their influence on the cells, and thus on the organism itself, was a mystery. In 1941, George Beadle, a geneticist, and Edward Tatum, a biochemist, began a series of experiments with the red bread mold *Neurospora crassa* in which they found that some mutants had lost their ability to form certain amino acids, the building blocks in the structure of proteins. Beadle and Tatum concluded that the role of the gene was to supervise the synthesis of a particular type of protein—an enzyme, the function of which is to promote specific chemical reactions within the organism. When a mutation occurred, the enzyme was lacking and so was the ability to complete the steps for the manufacture of certain organic compounds essential to the organism. This finding was the basis for their "one gene—one enzyme" hypothesis. Since enzymes are proteins, we might also call the concept "one gene–one protein" in recognition of the action of a gene (or group of nucleotides) in supervising the production of proteins generally. For their work, Beadle and Tatum shared the 1958 Nobel Prize in medicine and physiology.

More than 3,000 human hereditary disorders caused by mutant genes have been identified. The specific causes of the mutations are unknown, although several congenital diseases are known to be related to enzyme malfunctions.

- A type of cerebral palsy called the Lesch-Nyhan syndrome involves a deficiency in an enzyme (PRT-ase) that regulates uric acid metabolism. Excessive production of uric acid causes brain damage and early death.
- Alkaptonuria is an inherited disorder in which a faulty gene cannot form the enzyme that enables the substance alkapton to be converted to carbon dioxide and water. The disease was recognized as early as 1902 as being caused by a defective recessive gene.
- Albinism is caused by an abnormal recessive gene that prevents the formation of an enzyme system that converts the amino acid tyrosine to the black pigment melanin. About one person in twenty thousand is affected; there is little or no color in the skin, hair, or the iris of the eyes.
- Tay-Sachs is an enzyme deficiency disease, one of a half dozen involving accumulation of lipids (fats) in the brain.
- One of the most widely known genetic enzyme deficiency diseases is phenylketonuria (PKU), which causes mental retardation and other defects. The body is unable to form the enzyme that oxidizes phenylalanine to tyrosine. The damage is thought to result from the accumulation of

phenylalanine in the body. About one child in ten or twenty thousand is born with phenylketonuria.

- Other genetic disorders include hemophilia, sickle cell anemia, agammaglobulinemia, and galactosemia.

MUTAGENIC AGENTS

Mutagenic agents are chemical or physical agents that cause mutations. A slight increase in the mutation rate might not be noticed, so the causative agent could go undetected, though causing great harm over a period of many generations. Genetic defects are held responsible for the anomalies in as many as 10 to 20 percent of all conceptions, and are believed to cause the failure of a high proportion of human conceptions to produce a living child. It is estimated that congenital abnormalities, including minor birth defects, occur in 5 to 15 percent of the children that are born. About 0.5 percent of all births have serious chromosome abnormalities.

Radiation We have seen that mutations often appear spontaneously, that is, from natural causes. This was recognized as early as 1886 when De Vries made his observations on aberrant forms of evening primroses. The causes themselves, however, remained a mystery. The first clue came from studies on the common fruit fly *Drosophila*. The american geneticist Hermann Muller found that by raising the temperature, he could increase the mutation rate. He decided that changes in the molecular structure of the chromosomes must be involved in mutations and that heat hastened molecular alterations. His most dramatic experiments, however, were with x-rays, which greatly increased mutations. Later it was found that mutations can also be induced by other forms of radiation. Muller's breakthrough came in 1926. For this and other work, he was awarded the Nobel Prize in medicine and physiology.

Chemical mutagens We now know that unaccountable mistakes in the replication of DNA may be from any number of natural causes: cosmic rays; solar radiation; emissions from radioactive rocks, water, soil, and food plants; or man-made causes such as mutagenic chemicals in food, drugs, or pollutants.

The first evidence of direct chemical interference with the mechanics of heredity was the discovery in 1937 by an American botanist, Albert Blakeslee, that the alkaloid colchicine obtained from the autumn crocus[1] could produce mutations in plants.[2] The alkaloid does not exert its effect by modifying the structure of the DNA, as did Muller's x-rays, but by interrupting nuclear division without obstructing the replication of the chromosomes. As a result, the chromosomes double in number without the cell itself dividing. In this way it is possible to produce plants whose cells contain multiple numbers of chromosomes (see Figure 12.3). The condition is called *polyploidy*. New strains of plants have been produced artificially by this technique.

Soon other chemicals were found to have mutagenic action, often by altering the chemical structure of DNA. Nitrous acid (HNO_2) caused

---

[1] *Colchicum autumnale*, an ancient medicinal plant, still used for treatment of gout.

[2] Although Blakeslee is usually given credit for the discovery of the chromosomal effects of colchicine, R. J. Ludford, who worked with animal cells *in vitro* and *in vivo*, was probably the first to recognize the effects of colchicine on spindle function and arrested mitosis (Arch. Exp. Zellforsch. 18:411–441, 1936).

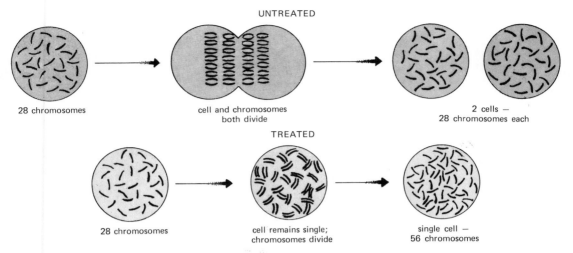

UNTREATED

28 chromosomes   cell and chromosomes
                        both divide

2 cells —
28 chromosomes each

TREATED

28 chromosomes   cell remains single;
                        chromosomes divide

single cell —
56 chromosomes

Redrawn from *Imprint on Living: A Report on Progress*, Agricultural Research Service, USDA, 1969.

FIGURE 12.3 Colchicine—an ancient antigout drug still in use as a medicine—has potent mitotic (nuclear division) effects on plants. Here treatment with the chemical produces cells with double the normal number of chromosomes. Colchicine has been useful in experiments designed to develop new varieties of plants.

mutations in *Aspergillus niger*, a fungus commonly found on food products, rotting fruit, and dead vegetation. Sodium nitrate and sodium nitrite have been used for at least thirty years as preservatives and color enhancers in such meat products as bologna, frankfurters, smoked ham and bacon, Vienna sausage, and smoked fish. The fact that the nitrite and under some conditions the nitrate are converted to nitrous acid in the stomach leads to the suspicion that this food additive may be dangerous. One possibility is that nitrous acid and nitrites are converted to *nitrosamines*, which are known cancer-causing agents. (See Chapter 13.)

In the early 1940s, the chemical warfare agent mustard gas (dichlorodiethyl sulfide) was found to be a strong mutagen (see Figure 12.4). The discovery was an outgrowth of an attempt at the University of Edinburgh to find an antidote for the poison gas. The investigators noted that the injury caused by mustard gas closely resembled the burns caused by x-rays, already known to be mutagenic. When geneticist Charlotte Auerbach and pharmacologist John Robson tested the war gas on fruit flies, it did, indeed, prove to be a powerful mutagen. This important discovery was made in 1941 but was not reported until 1946, after World War II. Subsequently, other sulfur mustards and nitrogen mustards were shown to be mutagens.

$$S \Big\langle \begin{array}{l} CH_2-CH_2-Cl \\ CH_2-CH_2-Cl \end{array}$$

mustard gas

$$H_2N-\overset{\overset{\textstyle O}{\|}}{C}-O-CH_2-CH_3$$

urethane

FIGURE 12.4 Mustard gas and urethane have long been known to be strong mutagens.

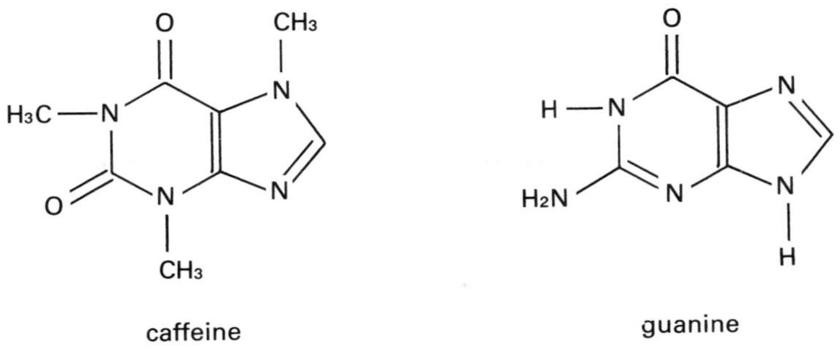

FIGURE 12.5 Molecular configurations of the purines, caffeine and guanine.

Other mutagens were discovered at a rapid rate. Urethane (ethyl carbamate) caused chromosome aberrations in the evening primrose and other flowering plants. Mutations in fruit flies were obtained from treatment with formaldehyde and various other chemicals. Today, a large number of substances are known to be mutagens. They include drugs, industrial chemicals, solvents, food additives, pesticides, radioactive minerals, and some natural food products.

Many of the mutagens are substances that we commonly use or regularly come in contact with. Caffeine has been known since the late 1940s to be capable of inducing mutations in bacteria and fungi and to cause chromosome aberrations in onion root tips. Later, caffeine was found to be mutagenic in fruit flies, in which it causes serious chromosome aberrations including extra X or Y chromosomes. Finally, it was shown that caffeine causes chromosome breaks in cultured human cancer cells and in human leucocytes (white blood cells). However, tests on mice have thus far failed to establish mutagenic effects even at doses much higher than those normally taken by humans.

The possibility that caffeine is mutagenic in man is of great interest in view of the widespread consumption of the substance in coffee, tea, cola drinks, chocolate, and in combination with pain-killing drugs and a variety of other medicines. It seems to be more than coincidence that caffeine is a purine—as are adenine and guanine, components of DNA (see Figure 12.5). An attractive theory is that caffeine acts by substituting for adenine or guanine in the DNA molecule, but this explanation is not widely accepted. A more plausible explanation based on experimental findings is that caffeine can combine with DNA and change its physical properties.[3] One view based on experimental evidence is that it may inhibit the natural repair mechanism of DNA that is damaged by other mutagenic chemicals or radiation. But there is no evidence that caffeine in the doses normally used has any significant effect on human DNA.

Whether LSD[4] is a mutagen or has other cytotoxic effects is a much debated topic. There is some evidence that it causes chromosome abnormalities. Exposure of laboratory animals to high concentrations of LSD has caused mutations and chromosome aberrations. On the other hand, much of the evidence is negative. Chromosome damage, when found, is

---

[3]Caffeine, theobromine, and theophylline are members of a group called methylated oxypurines (xanthines); a large number of alkylated oxypurines have cytological effects.

[4]Lysergic acid diethylamide.

usually from concentrations and duration of exposure that would not be experienced by humans using ordinary dosages. It appears tentatively that LSD taken in moderate doses does not have any detectable mutagenic effects in humans.

Large numbers of drugs and other chemicals cause alterations in the normal chromosome number or structure in laboratory samples either *in vivo* (within the test organism) or *in vitro* (outside the organism).

Many of the mutagens are of greatest interest because of their specific effects on growth, development, and organization of somatic (body) cells. When only a few somatic cells are affected, the organism may not suffer greatly; but when large numbers of cells are affected, forming abnormal tissue, the results may be uncontrollable growth of the tissues (carcinogenesis). These effects are discussed in Chapter 13.

PROBLEM AREAS

Mutations and the future of mankind

Throughout millions of years of evolution, man and other organisms have acquired and retained the ability to detoxify foreign chemicals and to repair injured tissues. But during a mere half century, thousands of new chemicals have been synthesized that do not occur in nature. Many of them are potent biologically. Many of them persist in the environment. Others are used regularly in medicines and foods or are released to the environment, so that large numbers of people are constantly exposed to their effects. An increasingly large number of these foreign substances are known to be mutagenic or to cause chromosome aberrations in laboratory tests. In some cases it is the metabolites (substances resulting from the organism's effort to detoxify the foreign chemical) that are injurious. Thus some chemicals that appear harmless are, in fact, dangerous.

Muller's experiments with the effects of x-rays on fruit flies convinced him that nearly all mutations are harmful and that only rarely does a mutation benefit either the individual or the species. If mutagenic agents that increase the rate of mutation are introduced into the environment, they can cause great harm and perhaps even threaten the survival of the human race. Muller warned repeatedly against needless x-ray therapy and diagnosis in medical practice and advocated shielding the gonads during exposure to x-rays. He brought attention to the risk of nuclear bomb testing and the inevitability of an increase in the mutation rate from radioactive fallout.

Some biologists believe that mutagenic chemicals may be as dangerous as radiation. Chemicals, like high-energy radiation, can cause changes in the genetic material by point mutations (those occurring at specific gene locations) as well as by chromosome breaks and abnormalities.

One view is that effects of chemicals, drugs, and high-energy radiation are less liable to cause a genetic disruption on the level of a catastrophe, such as that from a lethal toxin, than to result in more subtle, long-range influences on the vitality of the population. In the words of E. A. Carlson of the State University of New York, Stony Brook, the cumulative effect will be "expressed as lowered resistance to disease, lowered life span, increased infertility and general physiological weakness."[5]

---

[5]M. Harris, "Mutagenicity of Chemicals and Drugs," *Science*, 171:51–52 (1971).

READINGS

ASIMOV, I and DOBZHANSKY, T., *The Genetic Effects of Radiation*. Oak Ridge, Tennessee, U.S. Atomic Energy Commission, 1966.

HARRIS, M., "Mutagenicity of Chemicals and Drugs," *Science* 171:51–52 (1971).

Hermann Muller, FDA and Chemical Mutagens. FDA Papers, July–Aug. 1969, pp. 15–21.

HOOK, E. B., "Monitoring Human Birth Defects and Mutations to Detect Environmental Effects," *Science* 172:1363–6 (1971).

MOORE, J. A., *Heredity and Development*, Second Edition. New York, Oxford University Press, 1972.

OLSON, E. C., *The Evolution of Life*. New York, The New American Library, 1965.

VOLPE, E. P., *Human Heredity and Birth Defects*. Indianapolis, Pegasus, A Div. of Bobbs-Merrill Co., 1971.

VOLPE, E. P., *Understanding Evolution*. Dubuque, Iowa, William C. Brown Co., Publishers, 1967.

SOMATIC INJURY: TERATOLOGY AND ONCOLOGY

Somatic is from the Greek word *soma*, meaning body, hence somatic injury is bodily injury. The adult human body consists of a complex community of some 100 billion or more cells. When the organism is at its theoretical best, each cell performs its function perfectly in a way that is integrated with the common cause. But living organisms are not machines, and cells of similar characteristics do not all function alike. We find great differences among cells of apparently identical function. A wide range of biological variability is normal. It is only when the differences are so great that the organism obviously suffers that we say that the cells are "abnormal."

Cells can, and frequently do, stray from their appointed tasks. When the deviation is small, the disturbance to the normal functioning of the organism will be so minor that it may be insignificant. But if the cells of a tissue or organ get so far out of line that they disrupt the normal functioning of the complex community of cells that make up the body, then they cause disease or deformities.

Malfunctions that have environmental origins are sometimes associated with physical imperfections in the development of tissues. When the effect is seen in abnormal development of the embryo, the process is called *teratogenesis*. Another type of malfunction is seen as a group of cellular diseases called *cancer*, commonly found in many kinds of plants and animals, including humans. The two types of malfunctions are not necessarily related, although in some cases they may be caused by the same or similar agents. The agents that cause embryonic abnormalities are called *teratogens*, and those that cause cancer are called *carcinogens*. We will discuss them separately.

TERATOLOGY

Teratology is the science that deals with congenital malformations,[1] or abnormal development of the embryo and fetus, and the causes of deformities. The word is from the Greek *tera* meaning "monster." There

---

[1]In broad terminology a congenital disorder is a defect that exists at the time of birth. It is strictly defined as a defect acquired during development in the uterus, as distinguished from heredity.

was little interest in the study of birth defects except as a descriptive science until a disaster occurred in the early 1960s involving the birth of several thousand deformed children caused by a teratogenic drug called thalidomide.

The incident shocked both the public and the scientific community and stimulated an accelerated interest in the causes of embryonic and fetal abnormalities. Teratology is now thought of as a new science involving the research efforts of embryologists, pathologists, toxicologists, pharmacologists, chemists, pediatricians, and geneticists.

We now recognize that malformations occur in 3 to 5 percent of the infant population and that 5 to 7 percent of all children born alive will have birth defects that require medical attention before two years of age. There is increasing evidence that environmental intrusions are involved in many such cases.

The teratological awakening

The thalidomide episode demonstrated the harm that can be done to the fetus by seemingly harmless chemicals. Thalidomide was a sedative drug considered by many people to be so mild that it was sold in some countries as an over-the-counter sleeping tablet and used by the young and the old. It was used by many pregnant women. The drug was manufactured and sold widely in Europe and other parts of the world where it was sold either by prescription or as a proprietary (nonprescription) medicine. Early experiments on laboratory animals indicated that it was almost free of injurious effects, and medical experience confirmed that it was as safe to adult humans. Those who attempted suicide with it survived large doses.

Under the trade name Cantergan, the drug became by 1960 the most widely used sleeping tablet in West Germany, where it was available without a prescription. Several other countries began marketing the drug.

About that time, reports of an increase in strange malformations in newborn babies aroused the concern of physicians. The combination of deformities was considered to be exceedingly rare. One aspect of the disorder was characteristic of phocomelia,[2] meaning seal-like limb, in which there is a hand or foot but no arm or leg. However, it often affected all four limbs which were deformed, reduced, or absent.

At first the anomaly was thought to be a new and puzzling clinical disorder. Investigators reported that they could find no hereditary evidence in the families, no blood type incompatibility, and no chromosome aberrations in the children. Radiation of either the parental gonads or the embryo can cause malformations, but none of the mothers had been x-rayed in the early months of pregnancy. Limb defects had been produced in the offspring of rats fed on a deficient diet, but the mothers of the deformed children seemed to be adequately nourished. Two alert physicians, first W. Lenz of Hamburg, Germany, then W. G. McBride, an Australian physician, traced the deformities to the use of thalidomide by pregnant women. They noted that the drug had disastrous effects if the woman had taken it early in pregnancy. Nearly all of the deformities were associated with the use of the sedative within the first few weeks of pregnancy.

---

[2]*Phoke* (Gr.) = seal; *melos* (Gr.) = limb.

In the meantime, thalidomide had been proposed for sale in the United States under the trade name Kevadon. Free samples were distributed to more than 1,200 investigators and nearly 20,000 patients were treated. But a physician in the Food and Drug Administration, Dr. Frances O. Kelsey, who had been given the application to process, demanded further evidence of the drug's safety. When word of the calamitous results in Europe and elsewhere came out, her precautionary hunch was confirmed. The FDA rejected the application and the drug was never sold in the United States.

Teratogenic effects

Many of the biological effects of drugs and other chemicals have been studied, but most of the attention has been given to animals and adult humans. Suprisingly little is known about the effects of drugs, environmental pollutants, and other chemicals on the embryo and fetus, though these are the most sensitive stages in the life of the mammalian organism. Nutrients, hormones, and other essential materials cross the placenta from the mother's blood stream to the blood of the fetus. There is evidence that foreign substances also cross the placental barrier. For example, it is well known that the fetus can become addicted to narcotics. Other drugs, as well, can have adverse effects if taken during pregnancy. The effects of toxins on the embryo are sometimes so different from their effects after birth that physiologically the unborn organism may be thought of as a different species.

Teratogenic effects do not necessarily produce deformed offspring. Resorption (disappearance) of the embryo and spontaneous abortion are frequent consequences, although these may have other causes also. Reduction in fetal weight may be a teratogenic effect that would be difficult to recognize. It has been postulated that any toxicant capable of causing death of the embryo will also have teratogenic action at certain times and concentrations.

TERATOGENIC MECHANISM

The first few weeks are critical in the life of the embryo. In the human fetus, tiny arms and legs appear within 42 days after conception. It is at that time, or shortly before, that damage to the fetus is most apt to occur. During that period, for example, thalidomide is intensely teratogenic. During the most critical period, a single, moderate therapeutic dose of the drug will produce deformity in nearly 100 percent of the fetuses.

The conceptus

In humans and other mammals[3] the terms *embryo* and *fetus* are generally used to refer to different stages in the development of the unborn product of conception, called the *conceptus*. The embryo is considered to be the product of conception when the organs and definitive structures are being formed during the period called differentiation. This stage is also called *embryogenesis* or *organogenesis*. The completion of embryogenesis is not sharply defined, for the growth of the conceptus is continuous. It is generally considered to be completed by about the be-

---

[3]Mammals are animals that suckle their young, such as cats, kangaroos, whales, mice, and man.

ginning of the third month in humans. Development of the fetus is considered to follow embryogenesis. Fetal development is also sometimes called the *maturation* stage of conceptus.

When and how teratogens act

Teratogenic effects nearly always occur in the embryonic stage, during organogenesis. There is relatively little effect before differentiation begins during the first week following pregnancy in humans. The cells of the embryo are then uncommitted and can suffer some damage without seriously affecting their future. After organogenesis is complete, the developing fetus is also less sensitive to exogenous substances.

The way in which chemicals have such drastic effects on the fetus is not well understood. Thalidomide produced no identifiable changes in the chromosomes of the deformed children, although the possibility of molecular changes has not been ruled out. In animal studies, thalidomide administered to the male can affect the progeny, showing that the drug has the capability for mutagenic action.

Embryonic and fetal metabolism

Surprisingly little is known about substances that cross the placenta and their concentrations in the mother's blood and in the fetus. Teratogenic agents must have a direct action on either the embryo or the mother, or on the placental membranes which are the site of exchange between the mother and embryo. Metabolically the fetus is a different organism than either an infant or an adult. We know little about its tolerance for foreign substances or for changes in the mother's metabolism that may have only minor effects on her adult physiology.

NUTRITION AND TERATOGENIC EFFECTS

People have long realized that the nutrition of the mother is reflected in the child and that an adequate diet is important. But the magnitude of nutritional effects was not recognized until recently. A deficiency of certain nutrients or an excess of certain nutrients can produce fetal malformations in experimental animals. In humans as well as animals, both the quantity and quality of offspring are adversely affected by malnutrition. Experimentation with animals is causing us to revise our concepts concerning the effects of the mother's diet on the welfare of the offspring.

- Inadequate protein in the diet can cause death of the fetus, spontaneous abortion, or reduced size. Excess of protein can also be harmful.
- Excess cholesterol causes cleft palate in cats.
- Specific amino acids are requirements for rapid growth and for development of certain fetal organs. An excess of phenylalanine, an essential amino acid, prevents normal development of the nervous system in humans, believed to be similar to the effects of the congenital disease phenylketonuria.
- Vitamins: Vitamin A is teratogenic when either deficient or present in excess in the diet of animals. There are also documented cases of teratogenic effects from overdoses in humans. Vitamin B_6 deficiency in rats causes small, anemic young and several kinds of deformations.

DRUGS AND TERATOGENIC EFFECTS

The thalidomide episode was a warning to pregnant women of the dangers inherent in the habitual use of medicines, even those that are seemingly mild. Women often take extensive medication from the beginning of pregnancy until the birth of the child. They are also exposed to other toxic substances in the air, water, food, and various parts of the household and public environments. The presence of such a vast array of potentially injurious substances in the external and internal environments of the body makes the recognition of damaging effects from specific toxicants extremely difficult. It has been estimated that an increase of as much as 5 percent in fetal malformations from the use of a drug would have little chance of being detected.

It was well known before the thalidomide episode that some drugs are teratogens. Insulin, cortisone, steroids, and folic acid antagonists are examples of those that were known or believed to be capable of producing malformations. In some cases the evidence was obtained experimentally with animals, and in other cases from therapeutic experience with the drugs in humans. *Side Effects of Drugs*[4] cites more than 30 drugs and classes of drugs for which there is evidence of teratogenic effects. They include antibiotics, corticosteroids, sex hormones, aspirin, monoamine oxidase inhibitors, EDTA, meclazine hydrochloride, triton, triparanol, antivitamins, morphine, heroin, sulfa drugs, the antimalarials quinine and atabrine, and vaccines if given during pregnancy.

TERATOGENS OTHER THAN MEDICINAL DRUGS AND NUTRIENTS

Many causes other than medicines and nutrients have been found to have teratogenic effects in laboratory animals and humans. Some of them are as follows:

- LSD can be teratogenic in mice. In humans, 4 out of 14 therapeutically aborted fetuses of LSD users had brain and skull defects that are rarely found. However, both human and animal studies have been inconsistent. The question of LSD teratogenicity has not been resolved.
- Nicotine injected into mice during the mid-term of pregnancy caused skeletal malformations. Nicotine also has a lethal effect on mouse embryos.
- Lead and cadmium produce fetal abnormalities in hamsters. The increasing environmental burden of the heavy metals indicates a need for more attention to their potential teratogenic effects.
- Trypan blue and other azo dyes are teratogenic in chicks.
- Carbon monoxide is harmful to both mother and fetus because it decreases the capacity of the blood to supply oxygen to the tissues. Oxygen deficiency, called *hypoxia*, increases the teratogenic action of some drugs. This effect is called *synergism*. The increase in exposure of the population to carbon monoxide from cigarette smoking and combustion sources has adverse effects of unknown magnitude. There may be many undetected cases of teratological synergism, including the effect of x-rays and other forms of radiation on the action of various toxic substances.
- Trimethyl phosphate is used as a gasoline additive, an industrial chemical reagent, and a solvent and flame retardant for paints and plastics, and was proposed as a food additive for stabilizing egg whites. It is mutagenic in mice. Evaluation of potential human hazards is needed.

---

[4]L. Meyler and A. Herxheimer, *Side Effects of Drugs*, vol. 6 (Baltimore, Md.: The Williams & Wilkins Company; and Amsterdam: Excerpta Medica Foundation, 1968).

- The weedkiller 2,4,5-T (trichlorophenoxyacetic acid) caused fetal abnormalities in laboratory animals. Further investigations showed that the teratogenic effects were caused by dioxin,[5] an impurity in the weedkiller (see Chapter 9).

- Insecticides: Carbaryl (Sevin) is teratogenic in dogs; Kelthane produced malformations in mice; and certain organophosphorus insecticides caused abnormalities in chick embryos.

- Several naturally occurring plant poisons of a class called alkaloids cause fetal abnormalities in laboratory animals and domestic livestock. Nicotine is in this category. Lupine[6] plants cause "crooked calf disease"; loco weed plants[7] produce skeletal deformities in lambs; and *Veratrum* plants[8] produce skull deformities in lambs. Range plants belonging to several genera are teratogenic.

- Several human diseases are highly teratogenic: Syphilis is a damaging teratogenic bacterial disease, causing serious fetal injury. Most of the serious teratogenic disease pathogens are viruses. Rubella, also known as "German measles," can cause damage to any one of several organs, including the nervous system, if the mother is infected with the virus during the first three months of pregnancy. An accumulation of evidence also puts guilt on influenza, infectious hepatitis, the coxsackie virus, and poliomyelitis. There is some suspicion that the Echo virus, mumps, measles, and smallpox are teratogenic agents.

- Antibodies of various types can be harmful. If an Rh negative mother has an Rh positive child, the result may be a "blue baby" with severe anemia and jaundice that is often fatal. The disorder is called erythroblastosis fetalis.

ONCOLOGY[9]: CANCER AND CAUSES

Few aspects of the environment are more ominously destructive to the organism, to human health, and, indeed, to society than the physical, chemical, and biological agents that cause cells to go on a rampage of uncontrolled growth. Such unrestrained behavior is characteristic of cells in tumors. The abnormal tissues of tumors are called *neoplasms*, meaning new forms, and the cells that form them are described as *neoplastic*.[10]

A neoplasm, or cancer,[11] is a disorder of the cells. However, it is not a single disease but many diseases with a common characteristic— abnormal cellular behavior and growth pattern. In humans, as many as 86 different types of "cancer" have been identified.

Cancer is one of the oldest human diseases. But its cause and cure

---

[5]2,3,6,7-tetrachlorodibenzodioxine.

[6]Genus *Lupinus*.

[7]Certain species of the genera *Astragalus* and *Oxytropis*.

[8]*Vetratrum californicum*, commonly called false hellebore, corn lily, or skunk cabbage, is one of the several toxic species.

[9]Oncology is from the Greek word *onkos*, meaning protuberance or bulk; hence oncology is the study of swellings, especially tumors.

[10]*Neos* (Gr.) = new; *plasma* (Gr.) = form.

[11]From the Greek *karkinos* and the Latin *canceris* meaning "crab."

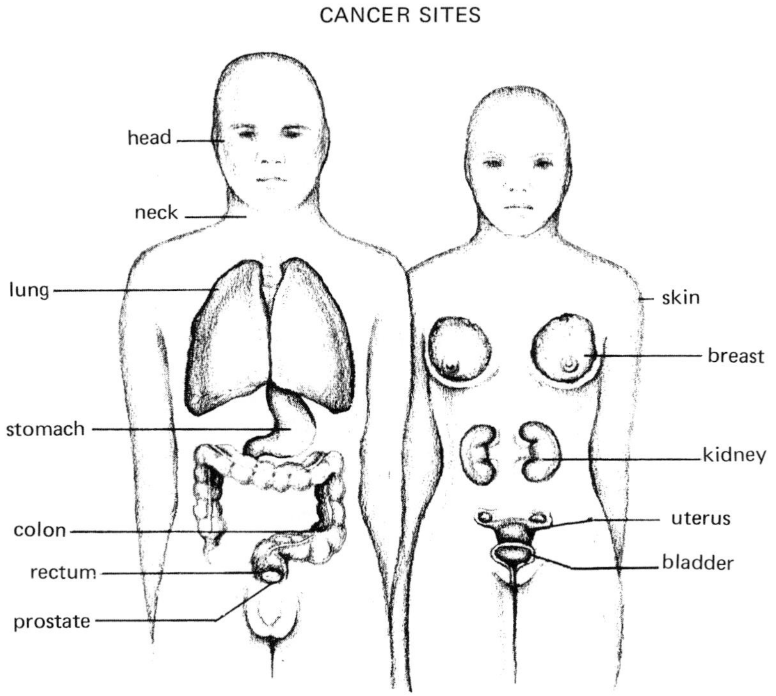

CANCER SITES

Redrawn from *Progress Against Cancer*, 1969, U.S. Dept. of Health, Education and Welfare.

FIGURE 13.1 Sites of the major forms of "solid" tumors that affect man. Generalized cancers such as leukemias and lymphomas are not shown.

remain, even today, one of the most baffling problems of medical science. It does not have one cause, but many: chemical, physical, environmental, and cultural (see Figure 13.1). Though many agents have been implicated, the variety of contributing factors in the inducement of cancer in man, animals, and plants weaves an incoherent pattern of puzzling complexity.

THE CANCER CELL

The cancer cell is so disordered in its behavior that it reproduces in a pattern of uncontrolled growth, sometimes at an accelerated rate. Many scientists believe that the cell's escape from the regulatory authority of the organism is the result of something that has gone wrong with the cell's genetic machinery—the DNA and RNA—so that the signal-calling mechanism is disarranged. According to this theory, vital changes in the nucleic acids foul the channels of communication within the cell and cause the transmission of garbled messages for the production of amino acids, enzymes, and other proteins. For this reason, such a cell is said to be the result of a somatic mutation. The Greek word *soma* means "body"; thus a somatic mutation is a mutation that occurs in a body cell in contrast to a hereditary mutation of the germ cells. A cell resulting from a somatic mutation is sometimes called a *mutant cell*. However, this concept is applicable only in the broadest sense, inasmuch as the misleading instructions given to the cell for its growth and behavior may be derived from the addition of foreign nucleic acids to the normal complement of DNA and RNA by virus invasions. Some of the disturbance may even be caused by substances that are carried in the extra-nuclear cytoplasm.

ENVIRONMENT, CULTURE, AND CANCER

Iran is a part of the world in which there is a high rate of esophageal cancer. Along the coast of the Caspian sea there are three regions with different climates, soils, plants, and life habits—and three different patterns of cancer. One region is well-watered and heavily forested; the main crops are rice and tea. Here, the incidence of cancer of the esophagus is lowest, about 6 per 100,000. A second region is intermediate between the two extremes; the rainfall is low to moderate and the main crop is cotton. Here the incidence of esophageal cancer is more than double that in the wet area, about 15 per 100,000. Still another region, slightly inland, is the dry, semidesert of Gonbad Gorgan; it is an area of poor productivity, populated by camels and hut-dwelling nomads. Here in the desert of Gonbad the cancer rate is the highest, 50 to 70 cases per 100,000 people.

What is it in three regions of the same country that could cause three different patterns of cancer of the esophagus? One thing that aroused suspicion was that the people of the desert area of Gonbad habitually drink large quantities of scalding hot tea. Exposing the lining of the esophagus to the fiery treatment might in time develop cancer. It was a plausible explanation that justified further study. But the theory did not hold up. People in nearby areas also drink their tea scorching hot and they do not have an exceptionally high rate of cancer. The evidence now points to low humidity and high salinity (salt content) of the soil as contributing environmental agents in the regions of low rainfall.

There are similar patterns in Kenya in East Africa and in Central Asia. As on the Iranean plateau, saline soil types coincide with a high rate of esophageal cancer. In Russia, there is a disease dividing-line between the saline soils of the coastal area, the nonsaline soils of the forest, and the black soil called *chernozem* north and west of the Caspian sea.

Whether conditions in Asia and Africa are causal or merely coincidental are not known for certain. One possible relationship of saline soils to cancer is that acid soils make trace elements such as iron, zinc, manganese, copper, and boron more available to plants. Although the trace elements may be present in large concentrations in saline (and alkaline) soils, they are not as readily available to plants. Soils in areas of high rainfall are usually more acidic due to leaching but also because the abundance of dead vegetation provides the soil with acidic materials. Is diet, then, a contributing factor along with humidity and possibly the irritating effect of saline dust? The riddle of the relative importance of the various environmental conditions on the incidence of cancer remains to be solved (see Figure 13.2).

A disease known as Burkitt's lymphoma accounts for about half of all cases of cancer in African children (see Figure 13.3). The disease most frequently affects the jaw but it also involves other organs such as ovaries and kidneys. Burkitt's tumor is found in a wide geographic belt extending

Courtesy of C. S. Muir and Dr. Janez Kmet, International Agency for Research on Cancer, Lyon, France. From *Science* 175: 846–853, February 25, 1972. Copyright 1972 by the American Association for the Advancement of Science.

FIGURE 13.2 Esophageal cancer belt in Asia. Studies to solve the riddle of its cause include the full spectrum of environmental features: climate, soil, flora, fauna; disease patterns in man, animals, and plants; social and economic structure; diet, habits, and customs.

across Central Africa. It occurs almost exclusively below 5,000 feet elevation, at temperatures above 60° F, and where the rainfall is more than 20 inches annually. This low, humid area in Africa is a vast breeding ground for man-biting mosquitos. Some of them are known virus carriers. A finding of interest is that a spontaneous lymphosarcoma of African clawed frogs is associated with a virus that resembles the viruslike particles found in Burkitt's tumor. Some investigators believe that mosquitos may act as vectors (carriers) in transmitting the virus from frogs to humans.

Skin cancer has been associated in several regions with arsenic intake. In Taiwan, a causal relationship was established between the rate of skin cancer among villagers and the arsenic in their drinking water. There was clearly a dose-effect correlation between the amounts of arsenic in the well water used in different communities and the frequency of the disease. In India, mouth cancer accounts for about one-third of all cancer in males. It is believed to be related to betel chewing, common in India and throughout the East Indies. The practice consists of taking "pan," a mixture of betel nuts[12] coated with slaked lime, usually wrapped in a bulath leaf,[13] to which tobacco and other ingredients are sometimes added. Pan increases salivation but curbs the appetite.

Diet and cancer   An epidemiological[14] study in Iceland from 1951 to 1960 showed that

---

[12]*Areca catechu* L. (Palmaceae).

[13]*Piper betle* L. (Piperaceae), a vine related to pepper.

[14]Epidemiology is the study of factors that influence the frequency and distribution of diseases.

Redrawn from *Progress Against Cancer*, 1969, U.S. Dept. of Health, Education and Welfare.

FIGURE 13.3 Burkitt's tumor, a cancer primarily of African children, occurs in a geographic belt of low, humid areas across central Africa.

the incidence of stomach cancer among farmers was more than 2½ times as high as among white-collar workers, who had the lowest rate among several occupations studied. In between were laborers, seamen, and craftsmen (see Figure 13.4). But the high rate of stomach cancer among farmers is not truly an occupational exposure. It was found that the farmers consumed more home-smoked meats and singed foods which are high in polycyclic aromatic hydrocarbons (PAH). In comparison to commercially sold foods, home-smoked preparations contain 20 times as much benzpyrene[15] one of the most potent carcinogens known. While stomach cancer in Iceland is related to occupation and rural-urban living, the presence of PAHs in the diet is the specific cause of differences in mortality.

Occupation and cancer

There was a high incidence of scrotal cancer among chimney sweeps in eighteenth-century England. Soot was the culprit. Even earlier, an occupational association with cancer was described by Agricola, a German physician-mineralogist. Agricola was an assumed name. His real name was Bauer (meaning "farmer" in German). His writing on medicines and minerals told of an ailment resembling lung cancer affecting miners of the fifteenth century.

High risk occupations exist in many industries. In most cases preventive measures can be taken, either by protecting the workers from exposure to dangerous conditions or by substituting safer materials. Because carcinogenic (cancer-causing) agents are numerous, there may be many

---

[15]3, 4-benz(a)pyrene.

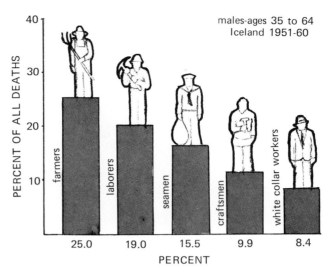

Redrawn from *Progress Against Cancer,* 1969, U.S. Dept. of Health, Education and Welfare.

FIGURE 13.4 Relationship between stomach cancer and occupation in Iceland, 1951 to 1960. Farmers consume more home-smoked and singed foods than other groups and, consequently, more polycyclic aromatic hydrocarbons. Such foods contain up to 20 times as much penzpyrene, a potent carcinogen, as commercially prepared food products.

cases of injury from undetermined sources. But many occupational relationships have been recognized:

- Among several occupations studied in Iceland, stomach cancer is highest among farmers and lowest among white-collar workers. The rate has been correlated with diet. (See Diet and Cancer).
- Bladder cancer among aniline dye workers was caused by exposure to 2-naphthylamine and several related chemicals.
- Osteosarcomas (bone cancer) occurred in radium watch dial painters, caused by wetting the brushes with the mouth.
- Uranium mining workers have a high incidence of lung cancer from inhaling uranium dust and radon gas unless special precautions are taken. The high death rate from lung disease in uranium miners was reported as early as 1879 by a German pathologist.
- Early laboratory workers with radioactive substances suffered burns and probably cancers. (Madame Curie, Nobel Laureate who pioneered work on radium and other isotopes, died of leukemia.)
- Radiologists who were exposed to high levels of x-rays developed burns and cancer of the hands until the hazard was recognized and precautionary procedures were instituted. Special protective measures are essential for those engaged in work involving exposure to radiation, such as testing x-ray tubes, x-ray diagnosis, x-ray therapeutics, and work with other radiation sources such as radioactive isotopes of high intensity or prolonged radiation (see Chapter 14).
- Lung cancer is prevalent in workers who inhale radioactive ores, chromates, asbestos, and iron.
- At least 20 cases of a form of liver cancer called angiosarcoma have been linked to vinyl chloride, a chemical used in the manufacture of plastics.

CAUSATIVE AGENTS

An agent that induces cancer is called a *carcinogen*, and its action is said to be *carcinogenic*.[16] Known carcinogenic agents include natural and synthetic chemicals, radiation, and viruses. Hormonal changes and immunilogic responses are also known to be involved, although how they relate to the other factors is obscure. Chemical carcinogens in the environment are probably the most important causes of cancer in man. Radiation and viruses are thought to account for 5 to 7 percent of human cancer. In southern latitudes, the high intensity of ultraviolet radiation increases the incidence of skin cancer.

Chemicals   The first success in identifying a cause of cancer goes back to 1775 when an alert English physician, Sir Percival Pott, noticed that chimney sweeps frequently had cancer of the scrotum. He said that the cause was an accumulation of soot in the crotch and other crevices of the skin. He recommended frequent bathing.

This incredibly important information lay dormant for 140 years until two Japanese doctors, Yamagiwa and Ichikawa, decided in 1915 to experiment with a related material, coal tar. When painted on rabbits' ears, the coal tar caused cancer. Later, a team of British scientists followed up the lead by isolating a pure chemical substance from coal tar, called benzpyrene.[17] They found this to be the main causative agent of the type of cancers produced on rabbits' ears by Yamagiwa and Ichikawa.

Benzpyrene remains today one of the most potent carcinogenic agents known. It is present in minute quantities in such diverse elements as tobacco smoke, charcoal broiled steaks, smokestack effluent, and automobile exhaust. Benzpyrene is a compound that belongs to a large group called polycyclic hydrocarbons, many of which induce cancers of the skin or subcutaneous tissues in laboratory animals. The suspicion is growing that benzpyrene and its relatives may be among the most widely distributed killers in the environment. Some of the other carcinogens are as follows.

- Nitrosamines are strongly carcinogenic. They are of special interest because of the common occurrence of secondary amines which are readily converted to nitrosamines:

$$R_2NH \quad + \quad HONO \quad \rightarrow \quad R_2NNO \quad + \quad H_2O$$
secondary amine   nitrous acid   nitrosamine   water

There is suspicion that sodium nitrite used as a food preservative may tend to produce nitrosamines by this reaction.

- A weedkiller, aminotriazole, was used to kill weeds in Oregon cranberry bogs until trace amounts were found on the ripe berries. The U.S. Food and Drug Administration promptly banned its use in 1959 when laboratory experiments showed that the chemical caused liver tumors. The action will be

---

[16]These terms are commonly accepted; however, inasmuch as a *carcinoma* is a particular type of cancer, some writers prefer to use *oncogen* and *oncogenic*, relating to the broader term *oncoma* for tumor.

[17]3,4-benz(a)pyrene.

remembered as the "cranberry scare" because the action came shortly before Thanksgiving Day, when people were shopping for their traditional turkey dinners.

- Tars, which are crude mixtures of organic materials, contain many toxic substances that are irritating to the skin. Only some of those that cause inflammations are carcinogenic. However, tars are more active in producing cancers than the isolated carcinogens. The joint effect of carcinogenic and noncarcinogenic agents is a case of synergism, whereby the action of one or more substances enhances the injurious effects of another. Two materials so involved in causing cancer have been called *cocarcinogens*.

- The U. S. Food and Drug Administration prohibited the use of an Azo dye, Butter Yellow, because the coloring agent was found to cause cancer in laboratory animals. (Azo dyes have the ability to bind with nucleic acids and proteins.) A remarkable feature of this finding was that subsequent experiments did not produce the same results until it was discovered that the original investigator, Riojun Kinosito of Osaka University, had fed his animals on a diet of polished rice, which is deficient in riboflavin (vitamin B_2). The vitamin deficiency triggered the carcinogenic action of the dye apparently because the vitamin is essential for the production of detoxifying enzymes.

Air contamination

Lung cancer kills more people in the United States than any other form of cancer, claiming about 55,000 lives in 1968; it is the leading cause of cancer death in men. Almost 40 percent of all cancer deaths of men in the United Kingdom are due to lung cancer.

Air pollution from industrial sources is an important, though apparently a relatively minor, cause of the current high incidence of lung cancer. Organic fractions (the benzene soluble portion) from particulates taken from urban air are carcinogenic. The complete composition of the extracts is not known, though one of the ingredients is benzpyrene, a known carcinogen of high potency.

Nitrogen oxides are suspected of being implicated in lung cancer because they are easily converted to nitrous acid and nitrosamines, compounds of established carcinogenic activity. Ozone also may be involved. Exposure of mice in the laboratory to as little as .05 ppm, not far from a normal concentration, increased the lethal effects of x-ray irradiation. PANs (peroxyacyl nitrates), another family of nitrogen-containing atmospheric pollutants, are believed to be carcinogenic. Asbestos, heavy metal dusts, finely divided silica, and soot also appear to increase the tendency to develop cancer. Automotive exhausts contain benzpyrene and related polycyclic hydrocarbons as well as other substances of probable carcinogenic properties. Tetraethyl lead experimentally produced lymphoma neoplasms in guinea pigs.

Cigarette smoking

Lung cancer is largely a preventable disease because the principal cause is known. It is caused mainly by cigarette smoking (see Figures 13.5 and 13.6). Although association between smoking and cancer was firmly demonstrated twenty years ago, and was suspected long before, the relationship has been widely accepted only within recent years. The 1964 report of the Surgeon General of the United States, based on epidemiological studies, strongly implicated cigarettes as the major cause of lung cancer. Compared to nonsmokers, male cigarette smokers who

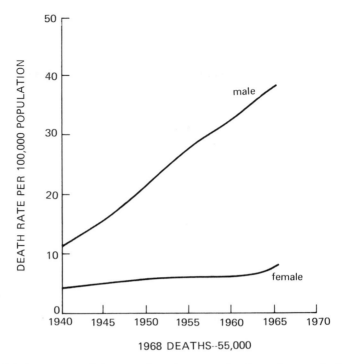

1968 DEATHS--55,000

FIGURE 13.5 Lung cancer claims more lives than any other form of cancer in the United States. Since cigarette smoking is one of the main causes, lung cancer is largely preventable.

use less than one pack a day run 10 times the risk, and those who smoke more than one pack a day run 30 times the risk of getting lung cancer.

At present, more men than women smoke cigarettes (see Figure 13.7). But a recent U. S. Public Health Service report revealed that as many teenage females smoke as teenage males. Thus, one can predict a striking rise in lung cancer in women beginning about 1995 and that eventually lung cancer will be just as common in women as in men.

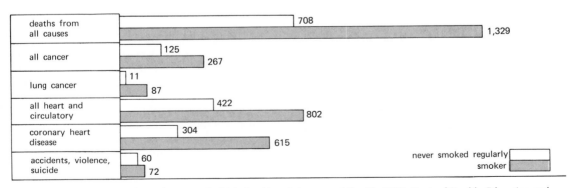

Redrawn from J. L. Hedrick, *Smoking, Tobacco and Health*, 1969, Dept. of Health, Education and Welfare.

FIGURE 13.6 Death rates of cigarette smokers versus nonsmokers (males, age 45–64; rates per 100,000 person-years). These diseases are related to smoking.

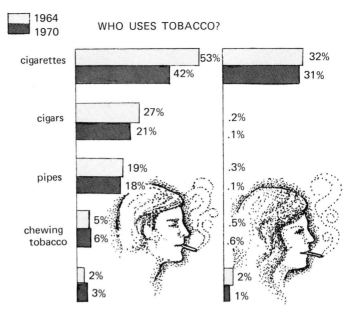

From the Economic Research Service, USDA, 1973.

FIGURE 13.7   Tobacco use by men and women, 17 and over, in the United States, 1964 and 1970.

The composition of cigarette smoke is complex. One of the ingredients is the ubiquitous benzpyrene. There is evidence that the content of "tars" is related to the cancer-inducing properties of the smoke. This aspect requires further research and study.

Toxic foods   The mouth is a major pathway for contact with the environment. Food that we eat may become contaminated with bacteria and fungi, some of which produce known carcinogens. Alkaloids and other toxic substances are common in herbs and foods.

- Safrol, a flavoring found in oil of sassafras, mace, and nutmeg is tumorigenic in laboratory animals. Oil of sassafras was used as a flavoring in foods until outlawed in 1958. Safrol is a minor constituent of several other spices.

- An ordinary mold, *Aspergillus flavus*, produces toxins that are highly hepatotoxic and carcinogenic. The discovery was made following the loss of several thousand turkeys in England during 1960 from an epidemic that was called "turkey X disease" before it was found that the epidemic was caused by feeding the turkeys a batch of moldy peanuts from Brazil. The fungus produces substances called *aflatoxins*, called aflatoxin B and G because they produce blue and green fluorescence under ultraviolet light. They are among the most potent carcinogens known, producing their effects in microgram quantities.[18] In the laboratory the fungal toxin is carcinogenic to such widely diverse species as trout, ducks, and rats. The fungus grows on cereals in storage such as peanuts, beans, cottonseed, soybeans, pistachio nuts, Brazil nuts, wheat, rice, and corn. Prevention consists of avoiding damp storage conditions and inspection of stored cereals before use.

---

[18] 1 microgram = about 1/28,000,000 of an ounce.

- Selenium is essential to the nutrition of some mammals and birds at concentrations of .01 to 0.1 ppm in the diet. Higher concentrations are hepatotoxic (causing liver injury) and hepatocarcinogenic. The element is taken up by plants from soil and incorporated in the protein.

Radiation   Skin cancer has long been associated with the amount of exposure to sunshine. People who work outdoors such as farmers and seafaring people have a higher rate of skin cancer than those who are less exposed to the sun. Among the white population of the United States, the incidence of skin cancer is higher in the South and West than in the North. Skin cancer occurs more frequently in light-complexioned people than in those who are heavily pigmented. Ultraviolet radiation increases the incidence of skin cancer. The principal culprit is that portion of sunlight in the ultraviolet range having a wavelength of about 3,000 Ångstrom units, just below the visible spectrum.

Carcinogenic effects of more highly energetic radiation, such as x-rays and radioactive chemicals, was not suspected by early workers. The injurious effects became evident only when radiologists and laboratory workers developed burns, ulcers, dry skin, and finally cancer on parts of the body subjected to high exposure, mainly on the hands. One pioneer worker, Jean Clunet of France, applied radium to the skin of a rat; the rat developed cancer. For many years, his results and the growing evidence among people who were exposed to radiation aroused little concern among either laboratory scientists or the medical profession. However, ionizing radiation is now recognized as one of the most potentially dangerous carcinogenic agents. This topic is treated more fully in Chapter 14.

Viruses   The first inkling that biological agents might be associated with cancer came in 1908 with the finding by two Danes, Ellermann and Bang, that a virus produced leukemia in chickens. Their work attracted little notice among cancer workers because at the time it was not realized that leukemias were neoplastic in nature. However, in 1911 a virus was found to be involved in the inducement of solid tumors in animals. Peyton Rous, at the Rockefeller Institute for Medical Research, was engaged in the study of a sarcoma (tumor of the connective tissue) in chickens. He made a cell-free extract from the tumor and injected it into disease-free hens of the same strain. They developed cancer identical to the original tumors. For this discovery, Rous was later awarded a share of the Nobel Prize in medicine and physiology.

Viruses are ultramicroscopic parasites that grow and reproduce only inside living cells. They have long been thought of as being on the borderline between life and nonlife, though in most respects they appear to be degenerate cells rather than precursors of more advanced organisms. Some kinds of viruses can be isolated in the laboratory and purified in the form of crystals. In this state, isolated from their native cellular habitat, they are as lifeless as a grain of salt. Reinoculated into a healthy host, they find the environment in which they can again reproduce and continue their parasitic life.

Most viruses are too small to be seen through the ordinary microscope. Usually magnifications of 100,000 times or more are needed to detect them visually. Powerful electron microscopes revealed that viruses have various sizes and shapes: spherical, rod-shaped, doughnutlike, flattened, or plate-like. They are much less complex than plant or animal cells. Biochemical studies show that most viruses consist of a core of DNA— some of them contain RNA instead—surrounded by a thin coat of protein. Thus, viruses consist mainly of nucleic acids, the molecules that carry the coded instructions for the functioning of cells and for the transmission of the genetic information.

Herpes-like virus bodies were found in cells from Burkitt's lymphoma patients in 1964 and, since then, have been reproduced in laboratory cultures of the human lymphoid cells. Similar herpes-like virus particles have been found in cell cultures of white blood cells from patients with a type of cancer called myelogenous leukemia. It is hoped that techniques for the large-scale production and study of viruses that may cause human and animal cancers will lead to an understanding of the role of viruses in causing cancer and eventually to the development of immune serums or other cures.

PROBLEM AREAS

Research is constantly bringing to light more known carcinogens. Since it is impossible to avoid exposure to many of them, several questions arise. Are the effects of all the known carcinogens to which we are exposed cumulative? Are they synergistic? What can we do to reduce them, or is that really necessary? Many substances that prove to be carcinogenic at very high dosages in laboratory animals may be only slightly dangerous or not dangerous at all in concentrations to which people are normally exposed. Therefore, there has been criticism that we are crying "wolf" in some cases where little hazard actually exists.

Can cancer be cured?

Neoplastic cells kill about 300,000 people in the United States each year, nearly one a minute. Cancer is the number two scourge of our people. If the present rate of cancer continues, 50 million Americans now living will be victimized by it; nearly one in five will succumb to its afflictions. Awareness of the environmental causes of cancer can be helpful in blunting the onslaught of this disease. Also, the treatments that are now available could greatly reduce the loss of life if people would act early enough. Some physicians believe that the cure rate could be doubled if patients could be reached a year earlier.

A major obstacle to curing cancer is the aura of gloom that surrounds the disease. Cancer has replaced the black death of earlier times as the pestilence of our age. In the minds of many, no one ever recovers from an attack of cancer. Actually, some forms have very hopeful prognosis if detected at an early stage. People who believe that nothing can be done to help them are not inclined to rush to a doctor only to be told that they have what they are convinced is a terrifyingly incurable disease. Knowledge of the truth will do much to allay the fears and mistaken beliefs that cause hundreds of thousands of people to forfeit their best chance of a cure—that of going to a doctor quickly.

READINGS

The Health Consequences of Smoking. Public Health Service Publ. No. 1696:1–199. Washington, D.C., U.S. Dept. Health, Education and Welfare, 1967.

HOAR, R. M., "Hormones in Reproduction and Teratology in Guinea Pigs," *Science* 158:529 (1967).

McCUTCHEON, R. S., "Teratology," in *Essays in Toxicology*, Vol. 1 (Frank R. Blood, ed.). New York, Academic Press, 1969.

SHIMKIN, M. B., *Science and Cancer.* Public Health Service. Washington, D.C., U.S. Dept. Health, Education and Welfare, 1969.

TAUSSIG, N. B., "The Thalidomide Syndrome," *Scientific American* 207 (2): 29–35 (August 1962).

WILLARD, N., "The Riddle of Many Regions," *World Health*, February–March, 1970, pp. 14–19.

CHAPTER 14   RADIATION

Radiation permeates the universe, the solar system, and the earth. The surface of the planet would be bombarded with radiation from the sun so intense that life would be in danger if it were not for the atmosphere that surrounds the earth. The atmosphere screens out much of the sun's radiation, including most of that which would be lethal to life. Cosmic rays, shot from outer space at high velocity, strike the earth continually and penetrate deeply into the surface. Naturally occurring radioactive elements are present in the rocks, water, and air and in all living organisms.

Some of the incoming radiation is trapped by the earth's magnetic field. The far-reaching portion of the atmosphere, called the *magnetosphere,* contains an area of high-energy radiation called the *Van Allen region*. It is roughly doughnut shaped, extending from about 500 miles above the earth at the equator to an altitude of about 40,000 miles. The Van Allen region was at first thought to present a major obstacle to space travel, but further studies showed that astronauts would pass through the zones of high radiation quickly and that precautions could be taken to prevent excessive exposure.

We have recently learned how to produce radiation in many forms, and our exposure to the normal level of environmental radiation that remained relatively constant for millions of years has been greatly increased. We are now bombarded with man-made radiation from x-ray machines, radioactive fallout from nuclear explosions, effluent from nuclear power plants, radioactive materials and other radiation sources in research laboratories, industrial plants and hospitals, wastes from reprocessing nuclear fuels, and from mining and processing radioactive products.

Extremely small amounts of radiation can be profoundly injurious, although different forms of radiation have different biological effects. Visible light and infrared heat rays are kinds of radiation that are generally beneficial, but some of the more potent forms of radiation, for example x-rays, destroy parts of living cells and tissues. Because these highly energetic forms of radiation tend to split substances, including matter, into ions, they are called *ionizing radiation*. The various forms of ionizing radiation differ greatly in their penetrating power and in the intensity of their injurious effects.

KINDS OF RADIATION

Radiation is of two main types. Sunlight—the form of radiation people are most familiar with—is an example of the type called *electromagnetic radiation*. It covers a broad spectrum of radiant energy and a wide range of physical and biological effects. The other type, called *particulate radiation*, consists of parts of atoms traveling at high speed and often with tremendous energy. Both electromagnetic and particulate radiation have great biological importance and account for some of the most serious environmental pollution problems.

ELECTROMAGNETIC RADIATION

Electromagnetic radiation includes a broad spectrum of energy having the nature of light. Radio waves, infrared rays, visible light rays, ultraviolet rays, x-rays, and gamma rays are all part of the electromagnetic spectrum (see Figure 14.1). All the different kinds of electromagnetic radiation are nothing more than light rays of different wave length and frequency. The fact that our eyes are sensitive only to visible light, actually a very small part of the light spectrum, is enough to remind us how different the world would look if, through some accident of nature, we could also see x-rays or radio waves.

The energy of electromagnetic radiation is transmitted in packets called *photons*. All photons, of whatever part of the spectrum, travel at a speed of 186,282[1] miles per second, the speed of light. During their flight through space, they seem to vibrate, so that the rays have a wavelike quality.

The energy of a photon is described as being inversely proportional to its wavelength. This means that the shorter the wavelength, the more energy there is in the photon. Therefore, as the wavelength of the electromagnetic spectrum decreases, the radiation is said to become more energetic.[2] For example, the visible rays of sunlight are more energetic than radio rays of longer wavelength, and ultraviolet light is more energetic than visible light. X-rays having a wavelength of 0.1 nanometer (1 Å) possess 3,000 times as much energy as ultraviolet rays having a wavelength of 300 nanometers.[3] As the wavelength decreases and the energy of the radiation increases, the potentialities to disrupt living tissue become very great.

---

[1]The velocity of light in a vacuum.

[2]The energy of a photon is called a *quantum*. *Quantum* = quantity, amount; *quantus* (L) = how much.

[3]Metric measurements used for expressing small units are:

Measurement	Abbreviation	Fraction of a meter (1 meter = 39.37 inches)
centimeter	cm	.01
millimeter	mm	.001
micrometer [formerly *micron* (μ)]	μm	.000001
nanometer [formerly *millimicron* (mμ)]	nm	.000000001
ångstrom	Å	.00000000001
picometer [formerly *micromicron* ($\mu\mu$m)]	pm	.000000000001

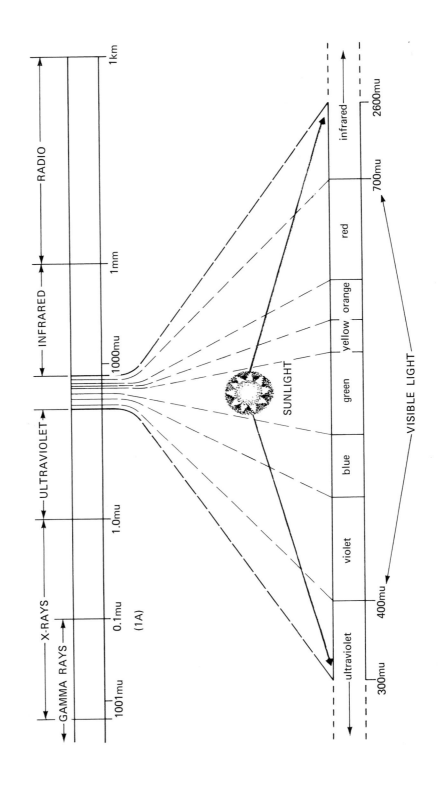

FIGURE 14.1 The electromagnetic spectrum.

Microwaves

Microwaves are radio waves in or near the extremely high frequency, or shorter wavelength, range. Hence, they are electromagnetic radiations in the more energetic portion of the radio wave spectrum.[4] Microwave energy is much too low to disrupt living tissues by ionization. Instead, the energy is absorbed as *oscillation* energy and is converted to heat. This makes it possible to use microwaves for cooking. But the same radiations are capable of heat injury to tissues in the body when exposed to the radiations, thus making it necessary to carefully maintain the door seals of microwave ovens. While microwaves can cause bodily damage with heat, most of our discussion will concern the more energetic radiations that can disrupt tissues, including cellular structures, by ionization, or splitting, of molecules.

Ultraviolet rays

Ultraviolet light is electromagnetic radiation just beyond the shortest wavelength of violet light to which the eye is sensitive. It has profound physical and biological effects. Though shorter in wavelength than the most energetic rays of visible light, ultraviolet light rays are longer than the average for x-rays, overlapping the shorter and more energetic x-rays.

Ultraviolet light is produced in abundance by the sun and can be generated artificially by the carbon arc or the mercury vapor arc. Ultraviolet radiation penetrates human skin only superficially, probably to a maximum depth of 1 millimeter, which, however, is enough to cause serious injury. Ordinary glass about 2 millimeters thick filters out (absorbs) ultraviolet rays shorter than 310 nanometers. It is fortunate that the air, smoke, and dust filter out much of the ultraviolet radiation from the sun, especially the shorter wavelengths, for otherwise the damaging effects would drastically change life on earth. The shorter wavelength ultraviolet radiation near the x-ray region can cause damaging ionization.

X-rays

X-rays include a broad spectrum of wavelengths. Therefore, the energy, penetrating power, and disruptive force of x-rays differ greatly depending on the wavelength. An "average" x-ray has a wavelength of approximately 0.1 nanometer, a distance about equal to the diameter of a molecule of water.

The discovery of x-rays was made by the German physicist Wilhelm K. Roentgen in 1895. He was working on the luminescence of chemicals produced by cathode rays when he noticed a strange phenomenon occurring some distance from the tube (see Figure 14.2). A sheet of paper coated with a luminescent substance (barium platinocyanide) was glowing. Since the cathode tube was blocked off by cardboard, it was not at first apparent that the rays from the cathode tube could cause the luminescence. Roentgen reasoned that there must be some kind of penetrating radiation that was invisible to the eye. He experimented and found that the radiation could penetrate thin layers of metal. Because he had no idea of what the strange radiations could be, he called them by the traditional symbol for the unknown, "x" rays. The name continued to be used after the nature of x-rays became known. Roentgen received the Nobel Prize in 1901, the first year of its award.

Gamma rays

Gamma rays are similar to x-rays, with which their wavelengths overlap, but because most gamma rays are of shorter wavelength, they gener-

---

[4]Technically, microwaves are frequencies designated as those in the region from 100 MHz to 300,000 MHz (in wavelength, roughly from about 3 mm to 200 cm).

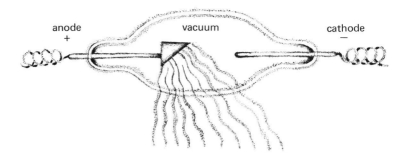

FIGURE 14.2 Diagrammatic representation of an x-ray tube. Electrons emanating from the cathode strike the target at high speed. Part of their kinetic energy is transformed into the energy of x-radiation.

ally have higher energy and are more penetrating than x-rays. Short wavelength x-rays, sometimes called *hard* x-rays, may be identical to gamma rays at the longer wavelength end of their spectrum. Gamma rays have wavelengths of about 0.1 nanometer and shorter. Regardless of the wavelength, the term *gamma rays* is used only in reference to electromagnetic radiation that is emitted from the atomic nucleus of a radioactive element. For example, some atoms of radium, uranium, and plutonium give off gamma rays.

PARTICULATE RADIATION

Particulate radiation is produced when one or more of the various components of the atom, for example, an electron, a proton, or a neutron, are emitted from a radioactive element. Such particles may be ejected from the atom with great force, sometimes along with gamma rays as the atom gives up its energy, or in technical terms, as it decays to a less energetic

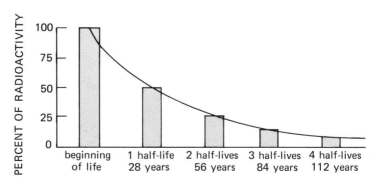

Redrawn from R. L. Mead and William R. Corliss, *Power From Radioisotopes,* rev. 1966, U.S. Atomic Energy Commission.

FIGURE 14.3 The time it takes for one-half the nuclei of a pure sample of radioactive substance to decay is called its *half-life*. The half-lives of different radionuclides differ enormously. For example, nitrogen-12 has a half-life of $^{12}/_{1000}$ second, while the half-life of iodine-129 is 16 million years. Above is the radioactive decay pattern of strontium-90.

Redrawn from N. A. Frigerio, *Your Body and Radiation,* 1967. U.S. Atomic Energy Commission.

FIGURE 14.4 Some radioactive substances emit beta particles, electrons traveling at high speed. When a charged particle such as this speeds along through tissue, it collides with other atoms and parts of molecules, knocking off free radicals and ions. In doing so, it ricochets easily and takes off in another direction. Above is the track of a beta particle.

form of matter. The time it takes one-half the nuclei of the atoms in a sample of radioactive element to decay is called the half-life (see Figure 14.3). But whether the radiation from nuclear disintegration is electromagnetic or particulate, the emanations are so energetic and forceful that they can do great damage to living cells and tissues.

Beta particles   Beta radiation consists of high-speed electrons called *beta particles.* They are emitted by many of the radioactive substances during their decay to more stable states. An electron is exceedingly small; it weighs only 1/1,835 as much as a proton. When an electron is ejected at high speed from the atom of a radioactive element, it ionizes substances it collides with in its path. Thus, beta particles damage tissue. But they are less penetrating than x-rays or gamma rays (see Figure 14.4). Carbon-14, phosphorus-32, hydrogen-3 (tritium), and strontium-90 are examples of beta emitters (see Figure 14.5).

Alpha particles   Some radioactive isotopes emit fast-moving particles, called *alpha particles,* containing 2 protons and 2 neutrons. This is equivalent in makeup to a helium nucleus. Since alpha particles contain no electrons, they are positively charged (see Figure 14.6). Alpha particles are less penetrating than either beta particles or x-rays and gamma rays (see Figure 14.7).

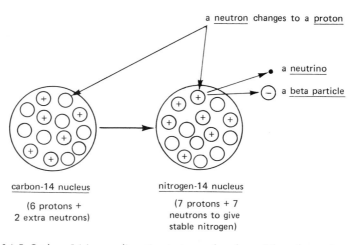

a neutron changes to a proton

a neutrino

a beta particle

carbon-14 nucleus

(6 protons + 2 extra neutrons)

nitrogen-14 nucleus

(7 protons + 7 neutrons to give stable nitrogen)

FIGURE 14.5 Carbon-14 is a radioactive isotope of carbon. When the nucleus of a carbon-14 atom decays, a neutrino and a beta particle are emitted and a neutron changes to a proton. (This results in the transmutation of the element to stable nitrogen.)

FIGURE 14.6 An alpha particle contains two protons and two neutrons, the equivalent of a highly energetic helium nucleus.

They can be stopped by a sheet of paper. Alpha radiation, therefore, is dangerous only when the emitter is in direct contact with tissue, such as when inhaled or ingested. But in such cases the alpha particles are readily absorbed by tissue, so the danger is very great. Radium-226, uranium-238, and plutonium-239 are examples of radionuclides that emit alpha particles.

Protons   Protons are 1,835 times heavier than electrons. While the enrgetic electron (beta particle) drives into tissue like a tiny particle of sand, the proton lumbers along like a rock, knocking off pieces of atoms and molecules as it goes. The proton does not penetrate as far as an electron of the same energy, but it causes more disruption in a small area. It is said to leave a heavier track. The effect is similar to that of an alpha particle.

Energetic neutrons   Neutrons have approximately the same mass as protons but have no charge. They are obtained from either a neutron beam or from the atomic disintegration of one of the radioactive isotopes, for example, uranium-235. Fast neutrons are four or five times more lethal to mice than x-radiation. Neutrons shot from uranium-235 was the basis for the *chain reaction* of uranium in Fermi's famous "pile" under the bleachers of the stadium at the University of Chicago in 1942, where experiments on the first atomic bomb were conducted.

Cosmic rays   Cosmic rays are energetic subatomic particles from the sun and outer space. They strike the earth at high velocities, some of them penetrating several thousand feet of solid rock. Cosmic rays are part of the background radiation that affects all living organisms. The dosage from cosmic rays at sea level is estimated to be equal to about 0.001 roentgen per day. The amount of radiation that can be tolerated by humans without detectable injury is about 0.01 roentgen per day. Because the difference between background cosmic radiation and detectable human tolerance is small (about ten-fold), cosmic rays probably have significant effects on

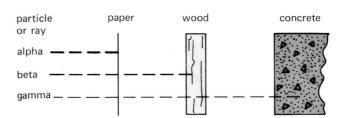

Redrawn from E. W. Phelan, *Radioisotopes in Medicine*, 1967, U.S. Atomic Energy Commission.

FIGURE 14.7 Penetrating ability of alpha, beta, and gamma radiation. Gamma rays penetrate into concrete and it takes a thick wall to stop them, but a block of wood stops beta particles, and it takes only a sheet of paper to stop alpha particles.

cells and tissue, no doubt causing mutations and other chromosomal aberrations. It is estimated that cosmic rays account for about one-fourth of the mutations caused by radiation.

BIOLOGICAL EFFECTS OF IONIZING RADIATION

Radioactive substances are among the most toxic materials known. The magnitude of their injurious effects is in a range that might be said figuratively to represent a "quantum jump" in comparison to ordinary organic chemical poisons. Radium is 25,000 times more lethal than arsenic. Pure radium costs about $40,000 per pound, but 40 cents worth is lethal.

The biological importance of radiation became known when in 1895 Wilhelm Roentgen placed his hand between the newly discovered x-ray tube and a flourescent screen and noted that the bones cast a deeper shadow than the flesh. The following year, the French physicist Antoine Becquerel discovered natural radioactivity in a uranium compound,[5] a discovery for which he shared the Nobel Prize with the Curies in 1903. An accident with one of the earliest known radioactive substances gave Becquerel the first indication of the biological potency of the newly found radiation when he burned himself with a vial of radium that he was carrying in his pocket.

But the most tragic early evidence of the potency of radiation was the death of Marie Curie, although the lesson seemed to go unheeded for many years. Marie Curie, with the help of her scientist husband, Pierre Curie, found that uranium ore contained a substance as yet unidentified that was several hundred times as radioactive as the uranium with which she was working. They succeeded in isolating from the ore a radioactive element which they named *polonium* after Marie's native country Poland. But polonium accounted for only a fraction of the radioactivity, and in 1898 they identified *radium*, an intensely radioactive substance. Pierre, upon learning of Becquerel's experience with radium, performed an experiment that deliberately produced a burn on his own arm.

Marie Curie became famous. Although denied membership in the snobbish French Academy by one vote because she was a woman, she shared the 1903 Nobel Prize with her husband and Becquerel. In 1911 she again received the Nobel Prize, this time for the discovery of two new elements. Marie Curie died of leukemia, a cancer of the formative white blood cells of the body, almost certainly brought on by the high levels of radiation to which she was exposed during her work. Irene Joliot-Curie, the scientist daughter of the husband and wife team, also died of leukemia apparently from radiation exposure.

When x-rays were a relatively new discovery, their usefulness in medical diagnosis and other applications was quickly recognized. However, the dangerous nature of x-radiation and its potential for serious, even fatal, injury were largely unknown. Indiscriminate use of x-ray machines caused numerous cases of injury and illness. Miners suffered toxic effects from radium and thorium; and the careless use of radioactive materials,

---

[5]Potassium uranyl sulfate.

most notably radium, caused serious ailments and premature deaths among workers. The potency of x-rays and radionuclides is so great that the enormously expanded use of ionizing radiation in its many forms in research, medicine, and industry is potentially hazardous to people working in areas of high radiation and in the environment at large.

UNITS OF RADIATION

Ionizing radiation, except in rare circumstances, cannot be seen, heard, tasted, smelled, or felt. The intensity of radiation, therefore, cannot be measured directly. It is measured, instead, by the degree of molecular or atomic disruption (ionization) when the radiation passes through a specified medium. The amount of radiation may be expressed in terms of (a) ionization produced in a given quantity of air (roentgens), (b) energy absorbed by tissue (rads), or (c) degree of biological effect estimated to occur in the body (rems).

Roentgen   The *roentgen* (R)[6] is a unit of radiation that is useful only for x-rays and gamma rays and is the standard measure of radiation used by radiologists. The designation is named after the German physicist Wilhelm Roentgen, who discovered x-rays.

Rad   Since the roentgen was designed to be used only with x-rays and gamma rays, it was necessary to devise a unit that could be used to specifically define the dosage from alpha and beta particles and fast neutrons as well as other kinds of radiation. The unit selected, called the *rad* (radiation absorbed dose), is now the most widely used designation for radiation dosage. For practical purposes, the rad is nearly the equivalent of the roentgen.

Rem   The *rem* was intended to compensate for the properties of the wide varieties of radiation encountered by relating them to the effect of electromagnetic radiation. The rem is an estimate of the amount of radiation *of any type* which produces the same biological injury in man as that resulting from the absorption of a given amount of x-radiation or gamma radiation (see Figure 14.8).

TYPES OF BIOLOGICAL EFFECTS

Ionizing radiation differs from ordinary environmental contaminants in several ways. One of the most important is that some of the radioactive elements continue to emit radiation for extremely long periods of time—in some cases for thousands of years—comprising a special kind of hazard to man and to the environment. The increasing quantity of radioactive waste products poses problems that are unique and require greater knowledge than is now available concerning both physical and biological aspects of the long-term effects (see Half-life).

The injurious effects of radiation are seen mainly in two forms. One is direct damage to the tissues and organs of the exposed organism, called

---

[6]Roentgen (pronounced *runt'-gen,* with a hard g) was formerly abbreviated r.

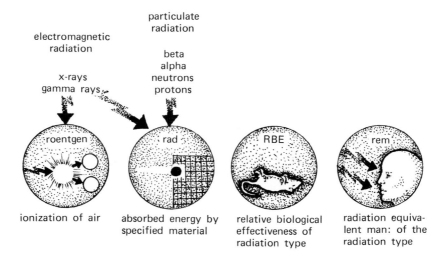

FIGURE 14.8 Units for expressing "doses" of ionizing radiation. RBE (relative biological effectiveness) is the biological activity of a radiation type in relation to the biological activity of x-rays. RBE times rad equals the equivalent radiation effect in man or mammal.

somatic effects, from the Greek *soma*, meaning "body." The other is the indirect damage to subsequent generations through the disruptive action of radiation on the reproductive tissues of the parent, called *genetic effects*, from the Greek word *genesis*, meaning "descent." Genetic effects are caused by the action of radiation on the molecular or physical structure of the chromosomes. Even in the production of somatic injury, the focal point of the ionizing radiation may be its disruptive action on the DNA makeup of the chromosomes of the somatic cells. Remember that DNA is the vital substance that has the dual function of guiding the somatic cellular activities of the individual organism as well as perpetuating racial characteristics in the progeny.

Somatic Effects

The first evidence of the damaging effects of radiation came from unfortunate experiences of people who, because of their occupations, either used or came in contact with sources of high-intensity radiation and suffered chronic exposure. Radiologists, uranium mine workmen, and painters of radium dials were among those who suffered disastrous effects. Patients who received large doses of x-rays and those who were given natural radionuclides such as radium and thorium for therapeutic purposes were also afflicted.[7] Additional evidence of the degree and kind of damage from radiation was forthcoming from studies of the Nagasaki and Hiroshima survivors, from the unforeseen exposure of the Marshall Island people following a weapons test in 1954, and from accidental exposures among laboratory and industrial workers. Finally, laboratory animal

---

[7]Radium and thorium are no longer recommended for internal use.

TABLE 14.1   Estimated Doses for Varying Degrees of Injury to Man*

Accumulated Dose (R) or Dose Rate	Period of Time	Effect
500 R/day	2 days	Mortality close to 100%
100 R/day	Until death	Mean survival time approximately fifteen days; 100% mortality in 30 days
60 R/day	10 days	Morbidity and mortality high with crippling disabilities
30 R/day	10 days	Disability, moderate
10 R/day	365 days	Some deaths
3 R/day	Few months	No drop in efficiency
0.5 R/day	Many months	No large scale drop in life span

*Based on 250 kvp x-rays. Corrections should be made for higher energy radiations. e.g. 1000 kvp x- or gamma radiation would have a relative biological effectiveness of approximately 70% of 250 kvp x-rays.
From *Radiological Health Handbook*, U.S. Public Health Service, U.S. Dept. Health, Education and Welfare, 1960. Reprinted by permission.

experimentation has been a rich source of information on the effects of ionizing radiation, dating as far back as 1927, when Hermann Muller demonstrated that x-rays can cause drastic changes in fruit flies. Many other investigators have since then shown damaging effects to animals and plants.

High radiation exposure

A lethal dose of penetrating ionizing radiation to humans (about 500 rads,[8] see Table 14.1) causes an acute form of *radiation sickness*. Within a few hours the patient becomes nauseated and experiences vomiting and fatigue. In the typical radiation syndrome, the symptoms last a day or two after which the patient feels fairly normal for a time. However, his blood is not up to par. There is a gradual decrease in the number of red and white blood cells over a period of several weeks until it is obvious that anemia has developed. The blood has lost much of its capacity to carry oxygen and the patient feels weak. Because of the drop in white cells, he is more susceptible to infection. There is also a decrease in the number of blood platelets, essential for clotting. As a result, the patient begins to bleed in various parts of the body. There is hermorrhaging from the nose, gums, and intestines. Blood clots blotch the skin and mucous membranes. If there are wounds, they heal poorly. The victim declines in vitality until he dies from anemia, infection, and hemorrhage.

Individuals that do not die within a few weeks are still not in the clear. There is a sharp decrease in life expectancy. Typically in mammals, such as mice, there is an interval of low mortality with a peak at about 3 months following exposure, after which the death rate increases until 9 months after exposure. Delayed deaths may occur for several months thereafter as the result of progressive deterioration from the acute effects of the exposure (Figure 14.9). During pregnancy, the fetus is more susceptible to radiation injury than adults, the effects depending on the number of days from conception (Figure 14.10).

---

[8]This is equal to about one-sixth of the energy that would be released if a pound of TNT were exploded inside the body.

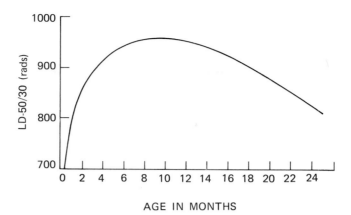

Redrawn from N.A. Frigerio, *Your Body and Radiation*, 1967 (revised), U. S. Atomic Energy Commission.

FIGURE 14.9 Effect of gamma rays form cobalt-60 on mice of different ages. The amount of radiation required to produce the death of 50 percent of the mice within 30 days (LD-50/30) increases to maturity, showing decreased sensitivity; but sensitivity then increases with advancing age.

Parts of the body differ in their sensitivity to radiation injury. The most sensitive tissues from acute dosages are the intestines, the lymph nodes, and the organs responsible for regenerating blood cells—the spleen and bone marrow. The intestinal wall partially breaks down and becomes inoperative in its normal function as the main barrier to infection from the bacteria that inhabit the gut. The reduced number of leucocytes (white blood cells) that normally attack invading microorganisms are unable to cope with bacterial attack, and the body succumbs to overwhelming infection aggravated by the failure of the blood to clot because there are fewer platelets. There is also partial asphyxiation from anemia caused by the disappearance of the red blood cells.

A further effect of ionizing radiation is that it destroys the body's *immune response*. One of the body's defenses against extraneous substances, especially invading organisms, is a biochemical defense consisting of the rapid production of *antibodies* that neutralize the intruders. Thus, in addition to lowering resistance by the destruction of white blood cells, radiation impairs the ability of the body to combat infection with antibodies, leaving it vulnerable to further invading microorganisms.

Fractionated dose

The biological effect of radiation is sometimes decreased when the total radiation dose is divided into several exposure periods over a period of time. For example, the susceptibility of patients whose radiation was fractionated over an 8-day period was significantly less than those who received the same amount of radiation in a single dose. In a typical experiment using fractioned exposures, twice as much radiation was required to produce a characteristic response.

Low-penetrating radiation

The effects of the low-penetrating types of particulate radiation are typically less severe than from penetrating radiation. This was demonstrated by a mishap that followed a test explosion at Bikini Atoll in 1954. Radioactive material was thrown up from a 15-megaton thermonuclear device. Unpredicted winds at high altitudes swept the nuclear debris

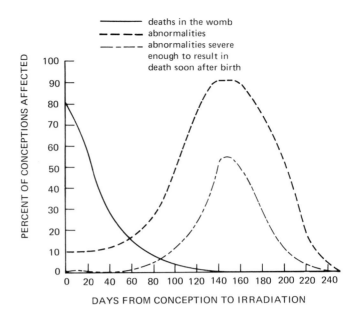

Redrawn from N. A. Frigerio, *Your Body and Radiation*, 1967 (revised), U. S. Atomic Energy Commission.

FIGURE 14.10 The probable effects of x-rays during pregnancy (250 rads of 250-kilowatt peak x-rays).

off its expected course and deposited radioactive material on Rongelap, an inhabited island about 100 miles east of Bikini. Radioactive debris also fell on a Japanese fishing vessel, the *Lucky Dragon*, that was operating in the area.

The natives of Rongelap suffered exposure that required medical attention and prolonged observation. They were evacuated from the atoll and it was not safe enough for them to be returned until 1957. The Rongelap people received a heavy dose of beta radiation from radioactive material that settled on their skins.[9] For a few days they were bothered with considerable itching and burning but there was no erythema (redness of the skin). Then in about 3 weeks, patches of darkened skin and raised areas appeared, especially on the scalp, and there was some loss of hair. Within a few months the injuries completely healed. With highly penetrating radiation such as photons of x-rays and gamma rays, much less exposure would have been required to cause serious injury.

Delayed somatic effects   If the animal or human patient survives the radiation sickness from acute exposure, the possibility of illness or death remains from one or more of several profound effects on the cells and tissues. The foremost delayed somatic effect is the increased chance that cells will become cancerous, but the appearance of cancer may be delayed for months or years. The type of cancer most commonly associated with radiation, and the first to appear, is leukemia, a cancer of the blood-forming tissues in which the white cells increase in number and become malignant.

---

[9]Over 2,000 rads.

Leukemia may show up about a year after irradiation, or even later. The chance of eventually developing one of the many other forms of radiation-induced cancers is estimated to be about five times that of leukemia.

There is also the increased probability that the victim will be afflicted later in life with anemia, various cardiovascular disorders, or eye cataracts, as well as premature aging and reduced longevity. If the organism is young, there will be stunting or other impairment in growth. Fertility is reduced or temporarily impaired. Exposure of the female ovaries to a dose of 300 rads or the male testes to 1,000 rads causes permanent sterility.

Delayed effects showed up among the residents of Rongelap Island, which received fallout from the Bikini test of 1954. The sixty-four residents of Rongelap were exposed to both gamma and beta radiation.[10] The dose was not lethal but caused nausea and vomiting, and there was a reduction in blood cell elements that lasted for several months. Besides the skin exposure that resulted in "beta burns" and loss of hair, there was an intake of radioactive iodine with contaminated food and water. The iodine was deposited in the thyroid glands and this proved to be the most serious aspect of the exposure. Healing of the skin and regrowth of hair, as well as nearly normal recovery of the formed blood elements, were complete by the end of the first year. The effects on the thyroid glands were longer lasting.

Observers noted abnormalities of the thyroid glands ten to fifteen years after exposure. It is highly probable that the effects were the result of deposition of radioactive iodine in the thyroid at the time of fallout. Ninety percent of the thyroid abnormalities were in children who were under ten years of age at the time of exposure. Some of the children were retarded in growth and development, an effect that was related to the thyroid injury.[11]

Life span and aging

Experimental work on animals shows that radiation shortens the life expectancy of rats, mice, dogs and other animals. The causes of reduced life span are not confined to cancer but extend generally over the normal spectrum of diseases. From such data, it has been estimated that at low levels of radiation exposure, the life expectancy of humans may be reduced by as much as 5 days per roentgen. This means that, during a normal human life span, an average exposure to man-made radiation of six rem (or roentgen) would cut the life span by about 1 month. The effect is usually proportional to exposure, so the life-shortening effects of higher exposure rates, such as from excessive diagnostic or therapeutic x-rays, would be proportionately greater.

Other delayed somatic effects of radiation include cataracts, sterility or impairment of reproduction depending on exposure rate, and accelerated aging. These have all been demonstrated in laboratory animals. The cancer-causing effects of radiation have been extensively documented for

---

[10]175 rads of gamma and 2,000 rads of beta radiation.

[11]It was estimated that the dose to the thyroid glands of the children was 700 to 1,400 roentgens from radioactive iodine, in addition to 175 roentgens of gamma radiation from deposition of radionuclides outside the body.

both laboratory animals and humans. Some radiation workers believe that diagnostic radiation of pregnant women may increase the risk of cancer to the child, although experiments with fetal mice show them to be surprisingly resistant to radiation before birth.

Genetic Effects

When Hermann Muller x-rayed his fruit flies and found that the mutation rate was thereby greatly increased over that which occurred from the effects of the natural environment (see Chapter 12), his discovery made it possible for geneticists to study large numbers of mutations in the laboratory. But beyond that, it removed much of the cloud of mystery that had surrounded mutations. It showed, Muller reasoned, that mutations were the result of changes within the genes brought about by conditions in the environment to which the flies were subjected. Furthermore, these changes could be initiated by human intervention. Muller set the stage for the subsequent discovery by Charlotte Auerbach that not just radiation but ordinary chemicals as well could cause gene mutations. Twenty years after his breakthrough discovery in 1926, Muller was finally awarded the Nobel Prize in medicine and physiology. Later, other geneticists found that mutations could be caused by different forms of radiation.

Mutation rates   Mutations in the human population are difficult to study because of the long human life span, the comparatively small number of offspring, and the fact that human mating and reproductive habits are almost impossible to study under anything approaching controlled conditions. Geneticists have turned to small organisms that reproduce rapidly and can be grown in the laboratory under carefully controlled conditions.

The common fruit fly, *Drosophila melanogaster*, has only four pairs of chromosomes. The chromosomes can be identified and mapped easily and the locations of the genes can be determined. This insect has served science admirably in divulging much that we know about mutations, as well as inheritance in general, from the time the American geneticist Thomas Hunt Morgan began his experiments with the insects in 1907. Morgan and his students found that it was possible to pinpoint on a chromosome the spot at which a gene for a particular characteristic was located. Soon they were drawing up "chromosome maps" for their fruit flies. It was on this foundation that Hermann Muller, a student of Morgan's, was able to speed up the rate of natural mutation in fruit flies in ways (especially with x-rays) that enabled him to observe the deleterious effects of mutations.

There is little that we can do at present to protect ourselves from natural radiation, but it is important to know what the effect is going to be when we expose ourselves to *additional* ionizing radiation.

Effect of background radiation   It was said earlier that various types of radiation, generally of low intensity, are part of the natural environment. This is called *background radiation*. The earth and all its plant and animal inhabitants are bombarded daily by high-energy particles from the sun and outer space. Rocks and soil contain radioactive isotopes of many of the elements in variable but usually small amounts.

Radioactive elements are picked up by water and are taken into plants along with mineral nutrients, thereby gaining entry into the food chain.

Thus, food itself varies in radioactive content, and even the tissues of our bodies inevitably contain sources of radiation in varying degrees depending upon the concentration of radiation sources in the food, water, and air.

The amount of background radiation is not the same everywhere on earth. Radiation from space is partly absorbed by the atmosphere so that less of it reaches sea level than mountain tops. People who live in the mile-high environs of the Rocky Mountain Plateau may be exposed to twice as much radiation as those who inhabit the Atlantic or Pacific coasts. There is also more intense radiation in the polar regions than at the equatorial belt because cosmic rays streaming in from space tend to be pulled toward the poles by the earth's magnetic field. The earth's magnetic field stops some of the radiation resulting in high-intensity radiation in the so-called Van Allen zone. The amount of radiation from minerals in the soil varies greatly from place to place. For example, the inhabitants of Kerala, India, are exposed to many times the average amount of environmental radiation because of the high radioactive thorium content in nearby soil.

The usual method of expressing the amount of radiation in the environment is in terms relative to the amount of x-ray exposure regardless of the kind of radiation. For this purpose the rem, *roentgen equivalent mammal* (or *man*), is used as an expression of the relative danger from different sources of radiation. For example, it is estimated that for the average person the gonads (reproductive organs) of the body are exposed to about 3 to 4 rems during 30 years. This span is taken as the time for one generation during which the human reproductive organs will undergo development plus the most active reproductive years. The National Academy of Science estimated that in the United States the average genetically significant exposure of the gonads to radiation from natural sources is 90 millirems (thousandths of a rem) per year (see Table 14.2). Of this, about 80 percent is from sources outside the body, the remainder from elements found naturally in human tissue and from inhalation of air.

Man-made radiation

Ever since the discovery of x-rays and the isolation of radioactive chemicals in the laboratory late in the last century, people have been increasingly exposed to ionizing radiation from its widespread use in medicine and industry. The nuclear age, beginning with the development of the atomic bomb in 1945, propelled the environmental burden of radiation to new heights, not only from the explosive release of radionuclides in the atmosphere which eventually invade water, soil, plants, animals, food, and the human body, but also from the greatly increased use of radioactive substances in medicine, industry, and research.

Lifetime records of exposure to man-made radiation are almost nonexistent, although the practice of keeping such records has been strongly recommended. The Federal Radiation Council estimated that of the average exposure of people in the United States to all man-made sources (67 millirems per year), 75 percent is from diagnostic x-rays. Of the remainder, about 16 percent is from radiation therapy, 6 percent from radioactive fallout, 3 percent from such miscellaneous sources as luminous watch dials, television tubes, and industrial wastes, and a small additional amount from nuclear research laboratories. Thus most of the man-made radiation—more than 90 percent—is from medical uses, and this comprises about 30 percent of all radiation to which people are ex-

TABLE 14.2   Sources of Radiation Exposure

Source	Average dose* per year, millirems	
	Whole body exposure	Genetically significant exposure
Natural radiation		
Cosmic radiation	44	
Radionuclides in the body	18	
External gamma radiation	40	
Subtotal	102	90
Man-made radiation		
Medical and dental	73	30–60
Fallout	4	
Occupational exposure	0.8	
Nuclear power (1970)	0.003	
Nuclear power (2000)	<1	
Miscellaneous	2	
Subtotal	80	
Total	182	
Radiation Protection Guide for man-made radiation (medical excluded) to the general population	170	

*Average values only. For some segments of the population, much greater exposures are experienced. (From *The Effects on Populations of Exposure to Low Levels of Ionizing Radiation*. NAS-NRC. 1972).

posed, both natural and man-made. If the higher value is taken as the exposure that might be expected in the United States, equal to 5.5 rem (5,500 millirems) for a 30-year period, the effect of man-made radiation is now nearly twice that already caused by natural, or background, radiation.

Radiation effects on the gonads   The greatest damage can occur in the gonads, where the germ cells give rise to the gametes in an ultimate process of nuclear division called *meiosis*. Genetic changes caused by radiation absorbed by the gonads can have a variety of effects, depending upon the intensity of the radiation. There may be mutations involving damage to a relatively few nucleic acid molecules in the DNA, or there may be complete sterilization of the treated organism. The damage may be so severe as to be lethal to the egg or embryo (see Figure 14.11).

Threshold   Hermann Muller and other investigators working with the effects of x-rays on fruit flies determined that the number of mutations in the progeny of irradiated flies was proportional to the amount of radiation. For example, if the amount of radiation were doubled, there would be twice as many mutations; if ten times the radiation dose were given, the number of mutations would be increased tenfold. If this relationship is plotted, as in Figure 14.12, the result is a straight line. Thus, it is said that there is a *straight line*, or *linear, relationship* between radiation dose and mutation rate. It can be assumed that as the radiation dose is reduced, the line

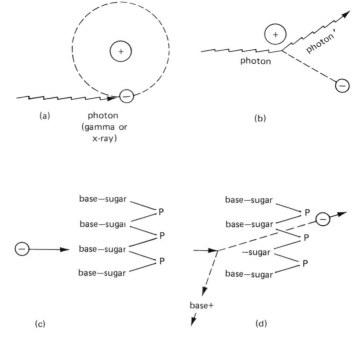

Redrawn from N. A. Frigerio, *Your Body and Radiation*, 1967 (revised), U. S. Atomic Energy Commission.

FIGURE 14.11   Radiation effect by production of secondary energetic electrons having effects similar to those of primary beta particles: (a) photon of gamma or x-ray strikes the electron of a hydrogen atom, (b) the electron is knocked out of its orbit at high speed, the energy transfer resulting in a photon of lower energy, (c) the electron strikes a portion of a molecule of DNA, and (d) disrupting it by knocking off a base. The electron departs, having lost some of its energy.

continues downward in nearly a straight line to very low radiation values. This assumption means that no matter how small the dose, even of background radiation, there will be some radiation-induced mutations.

The dose of a toxic substance below which there is no biological effect is said to be the *threshold* level. Insofar as anyone can determine with certainty, there is no exposure to radiation low enough to be without some effect (see Figure 14.14). Thus, many biologists believe that ionizing radiation has no threshold, and that the smallest amount, even that encountered in the natural background in food, water, soil, and air, does some slight damage. Any additional radiation exposure from man-made sources will add to the mutations proportionately and some of the effects on the sex cells will show up in subsequent generations.

PROBLEM AREAS

A disturbing feature of radiation safety standards is that the injurious effects of ionizing radiation have been consistently underestimated. It has been necessary to reduce the estimated safe level several times. Since most of the man-made radiation comes from the use of x-rays for exami-

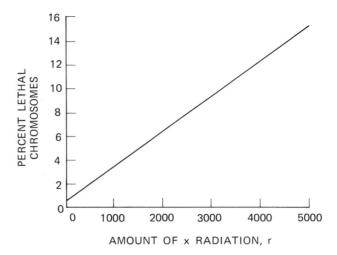

Redrawn from I. Asimov, and T. Dobzhansky, *The Genetic Effects of Radiation,* 1966, U. S. Atomic Energy Commission.

FIGURE 14.12 The number of mutations is believed to be generally proportional to the amount of radiation absorbed, shown by the straight line on the graph. The line does not come all the way down to zero mutations because of natural background radiation and mutational effects of background chemicals.

nation of teeth, bones, lungs, and other organs, it is important to know precisely the long-range genetic effects of tripling the radiation exposure of the population over that received from the natural background.

Much of the experimental work on radiation effects has been done using small animals, especially the fruit fly *Drosophila*. But in general, higher animals are many times more susceptible to genetic damage than lower animals such as insects. The mutation rates in the fruit fly *Drosophila* were well worked out during the studies by Muller and later geneticists. When radiation workers began to study the genetic effects on mice, it was found that the effect *per gene* was about 15 times that on fruit flies. Since genetic experiments cannot be humanely performed on people, it must be assumed that humans are at least as susceptible to genetic damage as mice. It is possible that the effect is more drastic. Common sense requires that we exercise extreme conservatism in estimating radiation hazards and that we avoid unnecessary exposure of humans to radiation in any form. Special precautions should be taken against radiation exposure of young people and pregnant women.

READINGS

Kinds of Radiation

CRAVEN, C. J., *Our Atomic World*. Oak Ridge, Tenn., U. S. Atomic Energy Commission, 1963 (rev. 1964).

HINES, N. O., *Atoms, Nature, and Man*. Oak Ridge, Tenn., U. S. Atomic Energy Commission, Div. of Technical Information, 1967.

KASTNER, J., *The Natural Radiation Environment*. Oak Ridge, Tenn., U. S. Atomic Energy Commission, Div. of Technical Information, 1968.

SCHUBERT, J. and LAPP, RALPH E., *Radiation: What It Is and How It Affects You*. New York, The Viking Press, 1957.

Biological Effects of Radiation

ASIMOV, I. and DOBZHANSKY, T., *The Genetic Effects of Radiation*. Oak Ridge, Tenn., U. S. Atomic Energy Commission, Div. of Technical Information, 1966.

FRIGERIO, N. A, *Your Body and Radiation*. Oak Ridge, Tenn., U. S. Atomic Energy Commission, Div. of Technical Information, 1967, pp. 1–78.

GROSCH, D. S., *Biological Effects of Radiations*. New York, Blaisdell Publishing Company, 1965.

Human Radiobiology. U. S. Atomic Energy Commission, Div. of Biology and Medicine. U. S. Government Printing Office, 1970. Reprinted from *Fundamental Nuclear Energy Research*, 1969.

KISIELESKI, W. E. and BASERGA, R. *Radioisotopes and Life Processes*. Oak Ridge, Tenn., U. S. Atomic Energy Commission, Div. of Technical Information, 1967.

Late Somatic Effects of Radiation. U. S. Atomic Energy Commission, Div. of Biology and Medicine. U. S. Government Printing Office, 1967, pp. 1–14. Reprinted from *Fundamental Nuclear Energy Research*, 1966.

Photo by Dorothea Lange, from the collection of the Library of Congress.

CHAPTER 15

POPULATION GROWTH

No one knows exactly how many people there are on earth because only a few countries have reliable census data. The best estimate is that the human population has reached the 4 billion point. The population is unevenly distributed, not only in the available surface space but also with respect to the resources in usable land, water, minerals, energy, climate, forests, and other natural flora and fauna. The growth rate has become accelerated in recent times. In a finite world that is neither expansible nor reproducible, the question becomes increasingly important, "Where are we going?"

The implosion

The early phases of the growth in human population could be referred to as an explosion, for it was characterized by a rapid bursting outward from the centers of high density. Today, because there is little room for expansion outward, the expression *population explosion* appears to be a misnomer and it would be more accurate to speak of it is an *implosion*, defined as a bursting inward, and causing compression.

The human population of the earth is estimated to be increasing at present by about 75 million people each year. This figure, taken by itself, would not seem to be alarming, for it is estimated that the earth could support at least twice the present number, if people were willing to accept certain adaptations. Even the calculation that the present rate of increase comes to at least 2 percent per year does not seem to convey any sense of urgency, although it does give us pause when we consider that this adds up to an increase in the world population every three years by a number greater than the 212 million people now living in the United States. It is only when we examine the present growth rate in relation to the human populations of the recent and distant past that we discover a trend of truly frightening proportions. In order to understand what is taking place, let us first take a look at what happens when a species of the lower organisms goes on a reproductive spree similar to that of the human population.

THE GROWTH CURVE

The population of a living colony, whether human, animal, plant, or bacteria, is dependent on the relationship between the birth rate, the death

351

Babies are being born at the rate of about 4 per second. Those who survive increase the world's population by about 75 million people per year.

rate, and the average survival time. There are times when a colony, or an individual capable of reproducing, finds itself in circumstances that are favorable for the support of an increase in numbers without a corresponding decrease by death. When we follow the growth of such a colony throughout the normal life span of the "founders" of the colony, and assume that there are no deaths during the time of our study, the population will increase at a rate determined solely by the rapidity of reproduction.

Exponential growth

Let us consider one of the simplest organisms, a yeast, which reproduces by a primitive mechanism called *budding*. In this process, a bulge appearing on one side of the yeast cell enlarges and eventually separates from the parent cell to form a new individual. If we put 10 yeast cells in a large amount of nutrient that is favorable for their growth and assume that under the favorable conditions of nutrition, temperature, etc., each yeast gives rise on the average to one more yeast cell by budding every ten minutes, we have a population that doubles in number every ten minutes. At ten-minute intervals, we could count 10 (at start), 20, 40, 80, 160, 320, and 640 yeast cells, or an increase of 64 times during the hour. This type of increase is called *geometric progression*. If we plot it, we get a growth line that, if continued, soars upward indefinitely (see Figure 15.1). This is a type of curve that is called *exponential*. An exponential increase is one in which, as the population rises, the growth rate becomes ever faster, so that the rate of increase is always accelerated.[1]

In our hypothetical yeast colony with an adequate food supply, unlimited space, insignificant accumulation of toxic waste products, and

---

[1]A curve of geometrical progression is one in which the ratio of a value to its predecessor is always the same. This can be expressed by an exponent; thus, the term *exponential curve*.

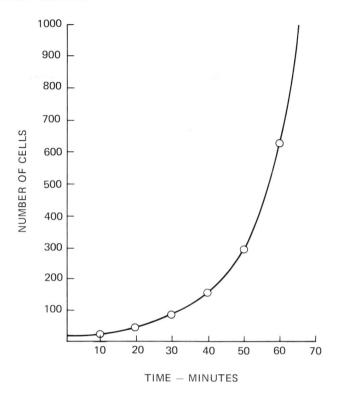

FIGURE 15.1 Population growth curve of a hypothetical colony of yeast cells, each reproducing itself every ten minutes in a microbiological "Utopia" where there is unlimited food and space.

other conditions favorable for indefinite growth, the population curve starts off innocently enough; but as time goes on, the curve bends more sharply upward, until at ten minutes after the hour, it has disappeared from the chart, careening off into space toward infinity. Thus, within a relatively short span of time, yeast would overwhelm the earth if nothing were done to stop them.

Actually, nothing of the sort ever happens. One of the things that takes place in nature was described by Raymond Pearl, an American statistician, in his epochal book *The Biology of Population Growth,* published in 1925. Pearl worked with the common laboratory fruit fly *Drosophila melanogaster:*

> My first approach to the population problem was purely mathematical. But it immediately became apparent that in its real essence the problem was a biological one.

The logistic curve

Pearl prepared a bottle containing suitable food and moisture. Then he took a few flies and enclosed them in the bottle to feed, to breed, to grow, and to die. Thus, he created a small insect "world," well equipped with everything a fly could want, but like the human world, spatially limited. In due time, the flies would produce offspring. Some would die, but others would mature and have offspring of their own. Eventually the old ones would die, but not before their descendents of several generations would have accumulated in a seething crowd of feeding, breeding, flies. Pearl counted the population of his living insect world at frequent intervals, about every second or third day. He found that his flies increased in population slowly at first and then began to increase rapidly, following a curve very similar to the first part of the exponential curve. However, the rate in population growth then slackened off and eventually reached a point at which there was no increase at all. What Pearl found for the population growth of his fruit flies was not the curve of geometric progression, though it started out that way, but an S-shaped curve sometimes called a *sigmoid* curve. Mathematicians refer to it as a *logistic curve* (see Figure 15.2). Pearl had confirmed experimentally the basic feature of the population growth curve worked out by a French mathematician, P. F. Verhulst, nearly a century earlier (1838).

LIMITING FACTORS

During the early stages of population increase when there is little or nothing in the environment to impede growth, the population increases much as the hypothetical exponential curve implies; but later, as the environment becomes less favorable from overcrowding, depletion of the food supply, accumulation of toxic waste products, and possibly the buildup of diseases and enemies, the rate of growth decreases.

Fluctuations in population density

The logistic growth curve falls short of depicting the entire picture of population growth. Studies by various investigators on a wide variety of organisms show that, in the world of nature, the population does not often become stabilized at a constant level but fluctuates between low points and high points. In some populations, these fluctuations are so extreme that near-extinction is approached before the trend is reversed to sky-rocket again to a level that cannot be sustained, whereupon the cycle is repeated. Depletion of the favorite food is sometimes responsible. Many other environmental conditions may be involved, such as the normal seasonal cycle, unseasonable weather, invasion of parasites and predators, or competition for the avilable food supply. Parasites or predators, when too successful, often build up to the point where they overwhelm the victims.

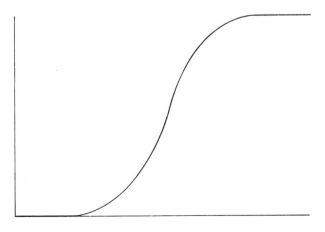

FIGURE 15.2 The logistic curve.

The maximum population growth that an organism would achieve in an unlimited environment is called its *biotic potential*. The environmental pressures that limit the organisms' inherent capacity for population growth are called collectively *environmental resistance*. The population density at which the growth curve levels off represents an equilibrium between the biotic potential and environmental resistance. The population level that the growth curve approaches as it proceeds indefinitely but never quite reaches is called the *asymptote*.

While population growth in nature does not often follow a neat logistic curve, the sigmoid feature is usually present in one or more stages of the population history of an organism. The S-curve, therefore, is useful as a generalized population growth curve because it shows graphically the opposing influences of biotic potential against environmental resistance. The biotic potential causes the population to come on strong during the early and middle phases of population growth, generally against relatively little resistance from the limiting factors that later build up. Then, as environmental resistance, including inimical biotic factors, becomes increasingly stubborn, the relative influence of biotic potential wanes and population growth can no longer make headway. The population typically reaches a point at which some limiting factor, or a combination of factors, prevents further population increase.

The upper limit of the S-curve, the asymptote, beyond which there is no further increase, has been defined by ecologists as the *carrying capacity*. When a population exceeds the carrying capacity of the environment, it can result in a sudden population decline. Thus, the environmental resistance is an important component of the concept of carrying capacity, an important consideration when trying to determine the optimum population size of an organism in a given habitat. Carrying capacity can be defined loosely as the ability of the ecosystem to support an organism or group of organisms. The carrying capacity of the earth for *Homo sapiens* has an important bearing on the future of the human population. What are the limiting factors? Some of them that affect both man and other animals are worth discussing.

Ecologists distinguish between density-independent and density-dependent limiting factors. Those that are density-independent exert an influence regardless of population density and include such factors as weather and climate and seasonal ecological cycles. Density-dependent limitations include parasites, predators, competition, nutrition, and behavioral patterns related to crowding and other effects of population density.

Crowding

Lemmings

The best-known example of density-dependent change in a population is that of the so-called Norwegian lemming[2] of Norway, Sweden, Finland, and northern Russia. Lemmings are rodents belonging to the same family as rats and mice. Five species are found in various parts of the arctic regions of the Northern Hemisphere. The Norwegian lemmings are about

---

[2]*Lemmus lemmus*.

five inches long, weigh less than four ounces, are yellowish-brown, and live on the moss and other vegetation of the rough terrain on high plateaus. They usually give birth to four or five young per litter and from one or two litters per season. But the reproductive rate sometimes builds up in successive years. At intervals ranging from 4 to 20 years there is a great surge of reproductive power. They will give birth to as many as three or four litters of eight to eleven young each. The cause of this sudden surge of reproductive power is not known, but is speculated to be either a hormone or vitamin from the food. Another theory is that over-crowding causes an increase in the activity of the adrenal glands. The result is that the population outgrows its food supply and the lemmings emigrate in vast hordes, overflowing into the lower land masses where they overwhelm every barrier in their compulsive escape. They press on into valleys, rivers, lakes, and fjords. They form a moving mass not unlike that on a freeway during the rush hour. Some of them travel west toward the Atlantic and some east to the Gulf of Bothnia. When they reach the sea, they reluctantly plunge in and swim until they drown.

Mice   Experiments in which mice were subjected to crowding in the laboratory gave results that were variable but drastic. A summary of the results of numerous experiments follows.

The effect on mice of increasing the population density was to cause an increase in the size of the adrenal cortex and collapse of the thymus gland. The mice were stunted and the reproductive functions of both sexes were suppressed. Sexual development was delayed and at higher populations the mice did not reach maturity at all. In males, the production of sperm (spermatogenesis) was delayed, and the size of the accessory sex organs declined as the population density was increased. The effect on mature females was to prolong the estrous cycle, to reduce ova production and implantation, and to increase fetus mortality. There was inadequate milk production (lactation), resulting in stunted young at the time of weaning, an effect that was seen to a lesser degree even in the next generation not subjected to crowding. Crowding of female mice before pregnancy caused irreversible disturbances in the behavior of the subsequently conceived offspring. Possibly related to this was the independent finding that increases in the corticosterone can permanently impair brain development in mice. It is theorized that the sum total effects of crowding results from a reduction in gonadotrophin secretions, which limits the population by either reducing the birthrate or increasing infant mortality, or both.

Rats   Experimental crowding had similar effects on Norway rats. Many females aborted, infant mortality was high, and mothers were deficient in carrying out their maternal functions. Among the males, a sort of hierarchy developed. Some became aggressively dominant and established a territorial dominion. Next on the status ladder was a group that could not distinguish between appropriate and inappropriate sex partners. They would attempt to mount males, females not in estrous, and juveniles. Other males became completely passive, ignoring and ignored by both males and females. Still others were hyperactive; they were both hyperactive and hypersexual, making sexual advances to the females without the customary rat courtship ritual. Some of these hyperactive rats eventually became cannibalistic. Whether the response of mice and Norway rats to crowding is an indication of what may happen or is taking place among the human population as it experiences increasing population stress is speculative.

Many animals appear to have a built-in system of population control whereby the restrictions on population density is automatic. This type of mechanism is called *homeostatic control*. In some animals density-dependent controls take the form of physiological changes that alter the growth, behavior, and reproductive patterns of the individuals. Changes in the hormone balance may result in a reduction of ovulation or resorption of the embryos.

It has long been recognized that the functioning of the endocrine systems (glands of internal secretion) is influenced by the impingement of various factors in the external environment of the organism. The effect of crowding on physiological changes is becoming of increasing importance. One theory is that there is an endocrine-behavioral feedback system that regulates the population of rodents, lagomorphs (rabbits, etc.), deer, and possibly other animals, the biological function of which is to prevent complete destruction of the food-supplying environment and consequent extinction.

Disease and Population

Disease has an immediate impact on each of the two main determinants of population growth, the birth rate and the death rate. Thus, the causative agents of disease, including parasites, microbes, and viral pathogens, are among the most important ecological factors exercising control over the population density of all organisms.

The great expansion of science and technology that began in the nineteenth century, and its further application to agriculture, made food available to more people and increased the opportunities for nutritional well-being and improved health. The mortality in western Europe and North America dropped significantly. The application of science to the understanding, treatment, and prevention of disease had even greater effects.

Secondly, one of the most important facets of medicine in improving life expectancy and therefore population growth was the development of immunization. People in early times were aware that some diseases could produce immunity to a recurrence of the same disease. There was an ancient practice in China of blowing powdered smallpox scabs into the nostrils. The slave dealers of western Asia immunized Circassian slave girls by inoculating them with a mild form of the disease.[3] The wife of the British ambassador to Constantinople, Lady Mary Montagu, brought to England the practice apparently developed by the Turks of inoculating with pus from smallpox pustules, a process known as *variolation*. This was a dangerous practice, not only because the disease might be contracted in a virulent form, but also because the donor patient might be suffering from other dangerous diseases. Immunization made an important advance when Edward Jenner demonstrated in 1796 the efficacy of inoculation with cow pox. It was developed to a science by Pasteur as early as the 1880s, but it did not have great effect on mortality until well

---

[3]Circassians were a group of tribes of the Caucasus, noted for their physical beauty.

into this century. Together with the discovery of penicillin and other "miracle" drugs, immunization and sterile techniques have dropped the death rate dramatically.

Thirdly, a growing consciousness of the importance of public hygiene stimulated the development and application of methods for maintaining the purity of water and food and the treatment of human wastes. In little more than a half century the causative agents of many of the most virulent man-killers have been brought under partial or complete control. These include cholera, malaria, yellow fever, plague, amoebic and bacillary dysentery, and tuberculosis.

A fourth product of science and technology to cause an abrupt drop in mortality was the discovery of DDT and similar potent pesticides that eradicated malaria and other vector-borne diseases from large populated areas of the world.

Epidemics   Pestilences of the past have been among the greatest disasters to affect mankind. The black plague that struck Europe in the years 1348 through 1350, which the Germans called the Great Dying, is estimated to have killed off at least one-fourth of the population of western Europe during the three years of its lethal visitation. The plague is caused by an infection with a bacillus[4] which is transmitted to man by the oriental rat flea,[5] mainly from black rats,[6] although the pathogen can be carried in other rodents, such as ground squirrels. There had been earlier epidemics of the plague. Since ancient times, the disease has swept through parts of Europe, Asia, and Africa. Thucydides wrote of the "Plague of Athens" in 430 B.C. A terrible epidemic struck Rome in 262 A.D., killing 5,000 people a day. It is believed that the Crusaders carried the disease to western Europe. From 1334 to 1351, the plague spread through Italy, France, England, Germany, Norway, Russia, India, Persia, and China.

One of the most notable episodes was the Great Plague of London in 1665, immortalized by Daniel Defoe in a fictionalized account, *A Journal of the Plague Year*. The disease struck London at least 20 times during the fifteenth century and struck Venice 23 times between 1348 and 1576.

Few disasters have caused such fear in people and induced such dissolute behavior with respect to both the living and the dead as the outbreak of plague. Boccaccio's *Decameron* is a collection of one hundred short lively tales that were supposed to have been told to a group of revel-making cavaliers and maidens who had fled the plague in the city of Florence to take refuge in a country house.

The plague mysteriously disappeared from most parts of Europe, though as recently as 1894 it appeared in Hong Kong and spread to other ports. During the next twenty years, more than 10 million people died from the plague in India alone. It has flared up sporadically in various parts of the world in modern times, but vigorous public health measures with pesticides to control the rat and the flea vector have kept the bacillus within bounds.

Influenza epidemics have repeatedly decimated the people of Europe and America as well as other parts of the world. It was recognized as a

---

[4]*Bacillus pestis.*

[5]*Xenopsylla cheopis.*

[6]*Rattus rattus.*

killer as early at 1510, and since then there have been at least 30 worldwide epidemics, called *pandemics*. One struck in 1915 and 1916 during a cold, rainy period when crops failed in many parts of Europe and the countries were engaged in the start of a world war. Fearful and weakened from malnutrition, the people suffered a frightful death rate. It returned in the form of a major epidemic in the fall of 1918. Few areas of the world escaped the scourge. In the United States more than 500,000 died of the disease and its complications, while worldwide, more than 21 million people died from its effects before the epidemic subsided. No effective prevention has yet appeared. Flu "shots" consisting of a vaccine hold out some promise of partial control, although the fact that there are several types of influenza virus complicates the problem, and their effectiveness in keeping an epidemic in check has not yet been demonstrated. A new type of flu called "Asian Influenza" appeared in 1957 and caused 86,000 deaths during three epidemics during 1957 to 1960.

For centuries, typhus, or spotted fever, took a great toll of human lives, particularly among soldiers and refugees in times of war. The disease, in former times often called "plague," took a dreadful toll of lives from the fifteenth century to modern times. The scourge struck the Serbian army during the first world war (1914–1918) with a death rate in some places of as much as 75 percent. The discovery by Charles Nicolle that the disease is conveyed by the body louse[7] came in time to prevent widespread devastation during the first world war and subsequent conflicts in the termperate zones where the scourge is most apt to strike. The timely appearance of DDT is credited with averting a typhus disaster in Italy during and following World War II.

Syphilis appeared in virulent form in Naples in 1495 and soon afterward appeared in France, carried there by the army of Charles VIII. The origin of this dangerous disease is obscure. It is thought by some people to have been brought to Europe from the New World by Columbus's sailors, but other people believe that the disease originated in the East. The first break in controlling this dreadful affliction came in 1909, after a systematic search for a cure by Paul Ehrlich, a German bacteriologist, turned up an organic arsenical chemical named salvarsan (later called arsphenamine) that would kill the spirochetes in the blood of the victims. Today a more effective drug, penicillin, is available, so there would be little excuse for not stamping out the disease were it not for ignorance and fear of reporting to physicians by those who have been victimized by sexual contact with diseased individuals.

Asiatic cholera has caused devastating epidemics in India and other parts of Asia. An epidemic broke out in London in 1854 during which John Snow, a physician, showed that victims had used water from a pump on Broad Street, whereas neighbors who worked in a brewery and drank only beer did not get the disease. He published his findings in a classic work entitled *Mode of Communication of Cholera*. It was not until 1883 that Robert Koch, a German bacteriologist, discovered the cholera germ,[8] mainly in water polluted with sewage. The disease is kept under

---

[7]*Pediculus humanus corpora*.

[8]A bacterium called *Vibrio cholerae*.

control only by strict observance of sanitation procedures and by the injection of cholera vaccine.

Yellow fever, probably originating in Africa, became through its spread by exploration and commerce one of the most dangerous maladies throughout the tropical and subtropical parts of the world. It was not brought under control until Walter Reed of the United States Army Medical Corps proved that the virus is spread by the bite of the mosquito *Aedes aegypti,* a species that adapts to domestic life and is difficult to suppress except by constant surveillance and pest control measures.

Disease and death rate

Medical advances and public health practices have had their greatest effects on child mortality. Infant mortality has an important impact on population growth because children not only add to the population but will become reproducers themselves. It is estimated that during the 1840s in western Europe only about 85 percent of the newborn survived the first year and only 65 percent lived to the age of 15. This has drastically changed. One of the best records in reducing infant mortality was made in Sweden, where there was a decline of about 90 percent in 150 years. It dropped to 21 per 1,000 live births by 1950 and further declined to about 15 per 1,000 by 1962.

There was a steep decline in mortality in many of the underdeveloped parts of the world, due largely to the control of malaria, which was the world's leading cause of sickness and death until after World War II. The death rate declined from about 28 per 1,000 during the 1940 to 1950 period to 17 per 1,000 population during 1960 to 1970. The infant mortality rate in five developing countries dropped from 165 per 1,000 live births in 1932 to 83 per 1,000 in 1968, a decline of 50 percent. In Taiwan the infant mortality dropped almost as low as that of the industrially advanced parts of the world.

The overall effect of disease on population growth is difficult to evaluate precisely. It is widely accepted that the great advance in the treatment, cure, and prevention of disease during this century is the major factor in the explosive surge in population throughout most of the world. However, we are not immune to epidemics. Moreover, devastating epidemics of the past have usually not been confined to one disease. Disruption of the economic and social life favor a variety of afflictions and related difficulties. History tells us that epidemics strike more often and are more calamitous during times of stress from other causes such as during periods of unusually severe weather or in famine and war.

War and Population

Warfare is one of man's chief preoccupations. Statistical information concerning early conflicts is not available, but we know from historical accounts that they often took a dreadful toll of life among men, women, and children. Caesar told in his own words of his campaign in the Ardennes:

> Every village and every building they saw was set on fire; all over the country the cattle were either slaughtered or driven off as booty; and the crops, a part of which had already been laid flat by the autumnal rains, were consumed by

great numbers of horses and men. It seemed certain, therefore, that even if some of the inhabitants had escaped for the moment by hiding, they must die of starvation after the retirement of the troops.[9]

According to Pliny, Julius Caesar fought 50 pitched battles and put to the sword 1,192,000 of his enemies besides the carnage of citizens in the civil wars, although Pliny avers sardonically that Caesar was notably modest about the latter aspect of his adventures.

Genghis Kahn elevated the business of annihilation to a ghastly art. During his reign beginning in 1206, and during that of his successors, the Mongols engaged in a succession of frightful atrocities against humanity. Tamerlane, one of the cruelest of the Tartars, overran Persia and is said to have decapitated 70,000 prisoners, piling their sliced-off heads into towers. Genghis sent an army of 80,000 men to put down a rebellion at Herat, and they were said to have massacred 1.6 million inhabitants, allowing only 16 to escape the horror. The Mongol leader Batu Khan replied to the Pope's entreaty for peace in the thirteenth century with the statement, "God has commanded my ancestors and myself to send my people to exterminate wicked nations." When Louis IX, who ruled France from 1226 to 1270, tried to negotiate with the Mongols, his emissaries had to travel a whole year before reaching the khan's court. En route, they found all the land under the domination of the Mongols, many cities destroyed, and "great heaps of dead men's bones."

The human proclivity for slaughter did not subside with the coming of modern civilization. A news dispatch from Riga published in the London Times of September 1, 1922, stated that according to official Russian figures, a tribunal called the Cheka executed 1,766,118 persons, a large proportion of whom were influential leaders.

The carnage of the first world war was one of the worst in history. Battlefield casualties on both sides totaled more than 37 million men including 8.4 million dead, 21 million wounded, and 7.7 million prisoners or missing. Equally disastrous were deaths by starvation of millions of civilians, including children and old people. This was partly the result of shelling and partly the result of the Allied blockade which prevented food from reaching eastern Europe and the Balkans, where millions of refugees moved helplessly from place to place looking for shelter and food in the blasted ruins.

World War II probably set an all-time record for military losses. Over 15 million men died—more than 9 million on the Allied side and about 6 million from the Axis countries. More than 50 countries participated.

The atrocity of exterminating nearly 6 million Jews by the Nazis during World War II was unsuccessful in its intended genocidal effects because of their wide geographic distribution. The North American Indians suffered disastrously at the hands of the invading Europeans. During the longest war of extermination in modern history, which lasted nearly 300 years, the Indians were reduced from a population of about one million to less than 400,000 by 1900. Possibly disease and liquor left even a smaller

---

[9]Julius Caesar, *The Conquest of Gaul*.

fragment,[10] which, however, has reexerted itself as an ethnic force so that the population may be greater now than when America was discovered.

Although the social consequences of war are far-reaching, it should not be assumed that the mortality from wars has had a significant effect on population growth. Throughout history, only a small percentage of men have served in armies. It has been estimated that the total of violent deaths, including those from murder and assault, during the period 1820–1945 was 59 million, or about 2.4 percent of the approximately 2.5 billion people alive during the period. While deaths from war, combined with civilian casualties, have often been locally devastating, the effect on global population growth thus far appears to have been minimal.

Historically, deaths of noncombatants have probably at least equaled battle deaths, due to the disruption of normal life, destruction of the food supply, and dislocations caused by ruined homes and buildings, resulting in epidemics and famines. The high population density that may be involved in future wars could be expected to intensify these effects.

Climate and Human Population

Man is an important component of the biomes of the earth. Not only have people been able to invade all the major regions and adapt to them with phenomenal success, they have also been influential in modifying large tracts of the earth's surface. The "cradle of civilization" was in the region of the Fertile Crescent, the hub of which was Babylon, and in Egypt. The Sumerians of the Tigris-Euphrates Valley and the Egyptians of the Nile Valley settled down to intensive cultivation around 4000 B.C. or earlier. A third great civilization developed in the Indus Valley by about 2500 B.C. These were ideal places for the agricultural revolution. The climate was warm and dry, the soil was rich and renewable annually by flooding, and there was a seemingly inexhaustible supply of river water for the irrigation of crops. But today, those earlier favored areas are less densely populated than many other parts of the world. We find the climates oppressive and the geography dry, barren, and dusty. The greatest growth of human population has been in parts of the temperate zone and semitropical regions of high rainfall, notably western Europe, southern Asia, and parts of the Orient.

There is no one measure of the conditions that favor human accomplishment. The vigor, productivity, and population of a family, a tribe, or a nation are determined by the input from many inherent and environmental sources: heredity, social beliefs and practices, prevalent diseases, productivity of the soil, diets, abundance of water, minerals and other natural resources, geography, and the behavior of neighbors. Among the most important are the various manifestations of climate. Temperature, humidity, rainfall, snow, ice, sunshine, and their seasonal characteristics help determine in major degree the health and vigor of the people.

Ellsworth Huntington, a Yale University geographer, made an exhaustive study of the effect of environment on human activity. He contended

---

[10]Few, if any, of the Indian tribes had any knowledge of alcoholic beverages before contact with Europeans.

that the pattern of a highly developed civilization and its culture ". . . must be due to conditions that are highly stimulating to health and activity . . . during the most important years of life." Huntington concluded that people in general worked best when the mid-day temperature is about 63° to 73°F and that they are most comfortable when at rest if the mid-day temperature is somewhat higher, about 70° to 77°F.

The internal body temperature of humans is about 98.6°F. People who wear a great deal of clothing, such as those belonging to so-called civilized groups, seem to function best at temperatures of 60° to 70°F, depending on the amount of physical activity engaged in. Efficiency drops off when the temperature goes much above 80° when the humidity is low, or above 70° during humid weather. People who habitually go nearly unclothed can function at optimum levels at higher temperatures. The dissipation of internally produced heat places an increasing burden on the heart and other organs as the temperature and humidity rise, particularly during vigorous physical effort. At such times there is always a temporary fever in some parts of the system and fatigue comes on quickly. At least in the absence of air conditioning, the customary adjustment to temperatures higher than the optimum is to "do as little as possible."

Extremely low temperatures also limit human accomplishment. Conditions may be so extreme that mere survival places first call on the resources of the body and the mind. The effect of moderately low temperatures, however, is not as great as high temperatures, due to the fact that keeping warm is comparatively easier than keeping cool, by the use of clothing, shelter, fire or other heat, and energy-rich foods. Probably most important of all is the fact that one of the easiest ways to keep warm in a cold climate is to stay active. Thus it is both physiologically and socially acceptable to be idle in the tropics and, in the words of Huntington, ". . . to at least seem to work hard in Scotland."

The excessive heat of the tropics creates problems for people from the temperate zones. Many efforts have been made to settle areas in which climate is a formidable obstacle due to its effect on industry, agriculture, and human diseases. We have seen in developing countries dramatic increases in health and longevity of individuals following the control of malaria and other diseases. It is probably too early to tell whether the surge of vigor from these successes will produce results comparable to those prevailing in more favorable climates. Past failures do not necessarily mean that such lands cannot be utilized to greater advantage. Advances in transportation, communications, disease control, agricultural methods, refrigeration, and air conditioning may open large tracts for agricultural and industrial development. If population growth or large migrations into such areas occur, it will depend more on progress in science and technology and less on individual pioneering efforts than in the great migrations of the past.

THE HUMAN GROWTH CURVE

Early human populations No information exists concerning the size of prehistoric human populations and very little is known about their distribution. People of the genus *Homo* are believed to have used fire about 400,000 years ago. A liberal assumption would be that the use of fire and related domestic activities began no earlier than 600,000 B.C. By taking that date as the practical beginning of human populations, and by making certain additional assumptions, a group working on the problem in 1962 calculated that only about 12 billion people had lived on earth by 6000 B.C., whereas the total number of people for all time, including the present number, totals 77 billion. Thus, fewer than one-sixth of the total number of people that ever lived were born before the dawn of civilization. Even if the beginning were extended another million years earlier, the number of people that lived prior to 6000 B.C. was probably only 32 billion as against a total for all time of 96 billion.

Early checks on human population

From the early Paleolithic or Old Stone Age to the beginning of the Neolithic or New Stone Age, the size of the population is still purely speculative. It has been estimated at 5 million. Much of the population existed in wandering bands. Large areas were needed during this hunting and gathering stage of man's climb up the human ladder. But the total land area of the earth is about 58 million square miles, so living space could not have been a problem, except perhaps for isolated tribes. However, numerous forces worked against the natural endowment of human productivity. The life cycle of our ancestors was on the average much shorter than it is today. The food supply was sporadic, and though they must have become inured to a "feast or famine" routine, the periodic shortage of food increased their susceptibility to parasites and diseases that ravaged the population, especially the very young before they could reach the reproductive age.

On the other hand, epidemics were kept in check by the wide dispersal of the population and the rarity of communication among the family groups. Although space favored the disposal of human waste, there was little if any awareness of sanitation. Disease-carrying vermin must certainly have swarmed around refuse heaps in large numbers. The need for mobility in harmony with the ecological cycles of the animals and plants on which human life depended made it difficult to provide protection with any permanency against the harshness of the seasons. There was little incentive to care for the wounded and sick, for whatever time and material goods were available were needed for those who could still be useful. The tough, the brave, and the productive were honored while the injured, the weak, the ill, and the aged were by reason of necessity probably either expelled or left on their own. Over the long span of prehistory, births and deaths were almost in balance, with births having a slight edge.

The primitive surge

Homo sapiens came on strong between 110,000 and 35,000 years ago (see Figure 15.3). Neanderthal men had spread throughout Europe and the Mediterranean. They were the dominant type during the last glaciation. His relatives, Solo man and Rhodesian man, were present during part of that period in Southeast Asia and Africa, respectively. The immediate ancestors of modern man, the Cro-Magnons, emerged in force beginning about 30,000 B.C. although they must have existed much earlier. By 20,000 B.C. the Cro-Magnons had produced some of mankind's greatest artistic achievements. We admire them for their engravings and carvings of bone and ivory and for their magnificently sensitive paintings and other artifacts found in the caves of southern France and northern Spain. Important ecological companions of the Cro-Magnon people were the

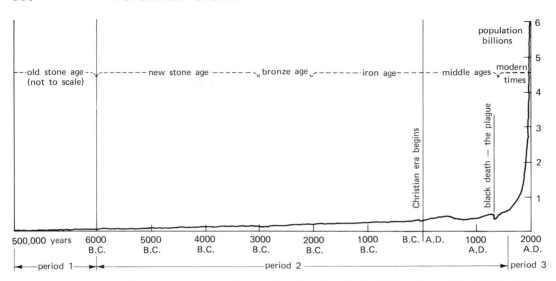

Redrawn from "How Many People Have Ever Lived On Earth?" *Population Bulletin* 18(1), 1962. Reprinted with the permission of the Population Reference Bureau.

FIGURE 15.3 Human population growth. It has taken all of man's existence on earth for his numbers to reach 3 billion. But in only 40 more years the population will grow to 6 billion unless something intervenes. If the Old Stone Age were in scale, its base line would extend 35 feet to the left.

mammoth, the bison, and the horse, on whose abundance they depended for much of their game.

Estimates of the human population at the beginning of civilization about 6000 B.C. vary from a low of 5 million to no more than 20 million. Julian Huxley thought that it did not pass 100 million until after the Old Kingdom of Egypt, which ended in 2258 B.C. There were probably 200 to 300 million people living at the beginning of the Christian era, and the population reached about 500 million during the seventeenth century (about 1650). By 1850 the population had passed the billion mark, and before 1930, it had gone over 2 billion. By then the curve was turning sharply upward.

Pearl's predictions

Raymond Pearl tried to show that the growth of the human population followed the pattern of his fruit flies. The data on world population available to him did, indeed, fit the fruit fly growth curve, but when he attempted to predict future populations by extrapolating the logistic curve, he missed the mark twice, in 1924 and again in 1936.[11] Each time he undershot the mark. Each time he assumed that growth would level off at about the point it did with his fruit flies. It did not do so and has not done so yet.

Pearl's estimate was that the "colossal total" of 2.6455 billion would be approached around the year 2100. But he modestly refrained from claiming any accuracy. He said:

It is a figure that future events and trends now wholly unpredictable may alter . . . [I] have had no inside information or revelation from Divine sources

---

[11]An extrapolation is a graphic prediction of further events by the extension of a curve that is based on previously determined information.

as to whether the upper asymptote depicted will reasonably accord with reality.

Doubling time   There emerges a picture of almost imperceptible growth in population during the first few hundred thousand years of human existence. There were probably large fluctuations caused by periods of famine, epidemics, the vagaries of weather and climate, and other ecological variations resulting in disappearance and reappearance of an adequate food supply. But throughout the span of human experience the population growth has tended to accelerate, turning ever more sharply upward, until within recent years the most convenient way to express population growth is in *doubling time*. The population doubled itself twice between 1650 and 1930. While it took nearly two centuries for the first doubling, the second took place in less than one century. From the quarter billion mark near the beginning of the Christian era, the doubling time—that is, the time for the population to reach the half billion mark—took nearly 16 centuries. It took only 200 years for the next quantum jump to the one billion mark, and only 80 years to go to the 2 billion mark. The next doubling took place in less than 50 years. Beyond that, the curve predicts that only 36 years will span the next leg of the headlong reproductive race toward "infinity" and that doubling will thereafter take place within ever-shorter time periods. (See Figures 15.4 and 15.5.)

THE SOCIOLOGICAL PROBLEM

During recorded history, at least, the population growth of humans has followed a different pattern from that of nearly all other animals. Though many species fluctuate over wide limits of population density, the long-range picture is one of relative stability except for those species that have gone into a downward spiral of extinction. Man, on the other hand, shows

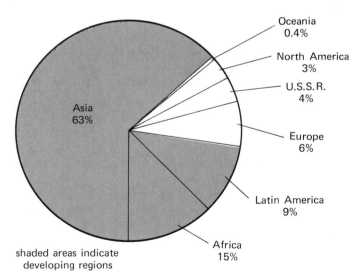

FIGURE 15.4 Distribution of world births—about 125 million per year. Based on the 1975 World Population Data Sheet of the Population Reference Bureau.

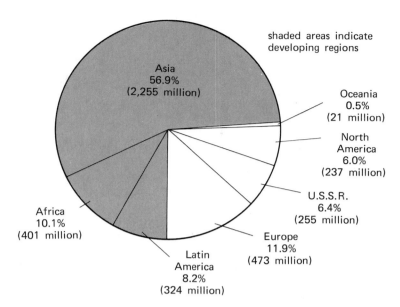

shaded areas indicate
developing regions

Asia
56.9%
(2,255 million)

Oceania
0.5%
(21 million)

North
America
6.0%
(237 million)

U.S.S.R.
6.4%
(255 million)

Africa
10.1%
(401 million)

Europe
11.9%
(473 million)

Latin
America
8.2%
(324 million)

Based on the 1975 World Population Data Sheet of the Population Reference Bureau.

FIGURE 15.5 Distribution of world population of 3,967,000,000.

a long upward trend over a period of thousands of years. It is important not only to give attention to the causes of this nearly unique characteristic of the human species but to study the probable results and ways and means to modify and, if necessary, reverse the trend.

In some primitive societies, social customs tend to restrict an increase in population beyond the limits that can be supported by the hunting range. Courting and marriage practices, strict limitations on sexual intercourse (such as by nursing mothers), infanticide, human sacrificial offerings, and head hunting often resulted in an elaborate system of taboos. Offending individuals became social outcasts and were sometimes put to death. A reversal of priorities took place when people became engaged in agriculture and urbanization. Large families were the most successful providers, large tribes were able to achieve wealth and prestige, and large cities were the most capable of defending themselves and their status against envious neighbors. A new set of social values was required.

Future prospects   Three ways by which people might bring their numbers into stable balance with the environment were theorized by scientists at the Massachusetts Institute of Technology who were working on a "Project on the Predicament of Mankind."

The first and most tragic possibility, which the Population Reference Bureau calls the "population crash curve," is that the human population would continue to double every 30 years until we populate the earth beyond its carrying capacity (see Figure 15.6). An accumulation of adverse effects would then cause a precipitous decline, along with a chaotic deterioration of nutrition, health, and economics as well as aspects of the environment associated with severely overtaxed and dissipated resources. The crash curve is typical of many animal populations and therefore may be thought of as letting nature take its course.

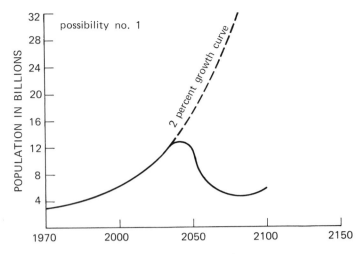

Redrawn from "Man's Population Predicament" *Population Bulletin* 27(3), April 1971. Reprinted with the permission of the Population Reference Bureau.

FIGURE 15.6 The population "crash" curve. Continued population growth at the current 2 percent rate could lead to a calamitous population crash sometime in the twenty-first century.

The second possibility, which was called the "gradual approach to zero population growth," would be characterized by a leveling off of population growth before the earth's carrying capacity is exceeded (see Figure 15.7). In this case, births would be at a replacement level that would match but not exceed deaths. The population would be in a "steady state."

The third possibility is something in between, called the "modified Irish curve" because of its similarity to the experience of Ireland following the potato famine of the 1840s, when the population declined by nearly one-fourth from death by starvation and emigration to North America (see Figure 15.8). The disaster had far-reaching effects on the economic and social pattern of Irish life. The people of Ireland reduced their depen-

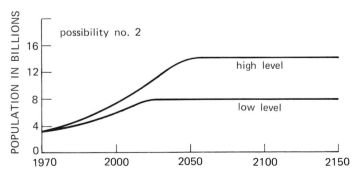

Redrawn from "Man's Population Predicament," *Population Bulletin* 27(2), April 1971. Reprinted with the permission of the Population Reference Burea.

FIGURE 15.7 Gradual approach to zero population growth. A population of 7 billion or more would present problems for agriculture and would place serious strains on resources and the environment.

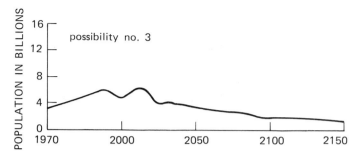

Redrawn from "Man's Population Predicament," *Population Bulletin* 27(2), April 1971. Reprinted with the permission of the Population Reference Bureau.

FIGURE 15.8 The "modified Irish" curve. Temporary declines in population could result from a number of serious disasters, followed by a determination to keep population down.

dency on a single crop by developing other types of agriculture, while they shifted their social customs to favor bachelorhood, late marriages, and fewer children. The pattern continues to be a major influence on population stability.

The first person to see that the human population was increasing exponentially, and that this would lead to disaster if not curbed, was a young English clergyman, Thomas Robert Malthus, who later turned to economics. Malthus came to the chilling conclusion that the world would outbreed its capacity to feed itself, and he published his opinion in 1798. Malthus's views on the human reproductive urge and the limitations of the world's food supply are discussed in Chapter 16.

PROBLEM AREAS

Hardly anyone who has studied population growth and decline in plants and animals can escape the conclusion that the human population problem, if left alone, would solve itself, although the solution almost certainly would not be to our liking.

Checks of one kind or another have always operated to bring a rising population to a halt. Is man exempt from this law that controls all other living things? While there is no proof that we cannot succeed where others failed, it would defy reason to think that the human population can exceed the space and other environmental resources that eventually must contain us. The question, therefore, is not if, but when? Will the human growth curve level off before it is forced into a precipitous decline by some catastrophic event? Will the population growth be held in check by conscious human effort, or will it take place by events beyond the control of man? Will it come, as Malthus predicted, by outbreeding our food supply, or by a homeostatic mechanism controlling hormonal and behavioral adaption to overcrowding, or by an orgy of mutual self-destruction after the fashion of the lemmings? The problem is that these determinations must be made years before the event takes place if disaster is to be avoided. It appears that time is short (see Figure 15.9).

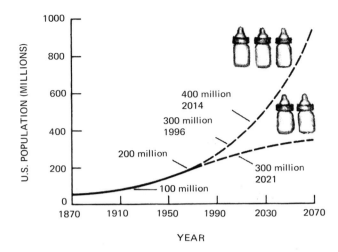

Redrawn from "Toward a U. S. Population Policy," *Population Bulletin* 27(3), June 1971. Reprinted with the permission of the Population Reference Bureau.

FIGURE 15.9 An average of three children per family would push the U. S. population up to 400 million by the year 2014; two children per family would mean 100 million less, and much greater differences in following years.

Can man beat the odds?

The exponential increase in human population has two root causes: (1) medical advances in controlling disease, and (2) an increase in the food supply derived from improvements in agricultural efficiency. But the growth in population is on a collision course with two other results of science and technology: (1) depletion of the world's resources, including those needed for food production, and (2) deterioration of the environment. The backlash from applications of science and technology intended to improve the human condition has a sting that threatens to paralyze human progress and destroy the quality of life. But we must pause to ask, "What is quality of life?" Here we are treading on dangerous ground, philosophically and sociologically. To a North American, the quality of life may mean escape to the wilderness in which to commune with nature, or perhaps to repair the psychic wounds of environmental insults from too much congestion and too many people. To others, no matter how great the need for spiritual equanimity, the quality of life may be more bluntly physical. To a Sahelian it may mean a square meal; to a resident of the Indian subcontinent it may mean a supply of pesticide to control malaria mosquitos; to a Nigerian it may mean an oil refinery to create a job and raise the standard of living above the subsistence level. Thus, the quality of life is many things to many people, and it is most productive at the outset to equate it with the most basic of human needs.

At a series of meetings held November 5–16, 1974, delegates from 430 nations gathered in Rome for the United Nations sponsored World Food Conference. The 1,250 delegates faced the grim statistic that, around the world, over 460 million people—more than double the population of the United States—were threatened with starvation and, that by year's end, some 10 million people would have died from lack of food. Most of the victims would be children under five years of age. The problems—

economic, sociological and political—are complex, and no revolutionary solutions could be expected; but the inescapable conclusion was that more would have to be done to bring the world's population and its food production more closely into balance.

While the facts of population growth are easy to communicate statistically, the numbers are meaningless unless we can translate them into terms of human values. The real-life meaning is that millions of people are receiving not daily bread but the daily ordeal of hunger. The population implosion has its meaning in the human misery of oppressive congestion, wretched poverty, starvation diets, malnourishment, and, when in children, impaired health and mental capacity of the survivors. These problems are discussed further in Chapter 16.

READINGS

BATES, M., *The Prevalence of People*. New York, Charles Scribner's Sons, 1962.

CALHOUN, J. B., "Population Density and Social Pathology," *Scientific American*, Feb. 1962. (Reprinted in *39 Steps to Biology*. San Francisco, W. H. Freeman and Company, 1968, pp. 269–276.)

CARTHY, J. D., and F. J. EBLING, *Natural History of Aggression*. New York. Academic Press, 1964.

CHRISTIAN, J. J., and D. E. DAVIS, "Endocrines, Behavior and Population," *Science* 146: 1550–1560 (1964).

COOK, R. C. ed., "How Many People Have Ever Lived On Earth?" *Population Bulletin* 18(1):1–19 (1962).

EHRLICH, P. R., *The Population Bomb*. New York, Ballantine Books, 1968.

HUNTINGTON, E., *Mainsprings of Civilization*. New York, The New American Library of World Literature, 1945.

HUXLEY, JULIAN, "World Population," *Scientific American*, March 1956. (Reprinted in *Three Essays on Population*. New York, The New American Library, 1960, pp. 62–81.)

MILES, R. E., JR., "Man's Population Predicament," *Population Bulletin* 27(2):1–40 (1971). Population Reference Bureau, Inc., Washington, D.C.

MONTAGU, A., *Man: His First Million Years*. New York, The New American Library, 1957.

MURRAY, D., "Flu—The Underestimated Enemy," *Reader's Digest,* Jan. 1965, pp. 95–98.

SIGERIST, H. E., *Civilization and Disease*. Ithica, N.Y.: Cornell Univ. Press, 1943; Chicago, The Univ. of Chicago Press, 1962.

WILLIAMS, G., *The Plague Killers*. New York, Charles Scribner's Sons, 1969.

WOYTINSKY, W. S., and E. S. WOYTINSKY, *World Population and Production*. New York, The Twentieth Century Fund, 1953.

FOOD AND FAMINE

Food is a common denominator of all life. It is foremost in the requirements of all plants and animals, and the form that it takes determines the relationship of organisms to one another. Man is the most omnivorous[1] of the higher animals. We can digest, and apparently enjoy, an immense variety of foods, including many that are so greatly modified by technology that they are almost—but not quite—synthetic.

But for many people on the earth, food is plain, monotonous, and scarce. Thousands of people are dying of hunger; millions more subsist on the ragged edge of starvation. Are they caught up in the cross current of a dwindling food supply relative to a runaway growth in global population? According to the idea first put forth by Thomas Robert Malthus, they are.

THE MALTHUSIAN THEORY

Thomas Robert Malthus, an English clergyman, was the first person to take the exponential concept of population growth seriously and relate it to the ability of people to feed themselves. In his writings of 1798 and later, Malthus contended that the human reproductive urge would outstrip the capacity of the earth to produce subsistence.

Malthus originally entered the ministry upon graduating from Cambridge University but, soon afterward, became the first professor of economics when he took a teaching post at a college established by the East India Company to train its young executives.

As an undergraduate at Cambridge University, Malthus had done a lot of thinking about the prospect of economic progress keeping ahead of population growth. Shortly after graduating, he published anonymously in 1798 his *Essay on Population,* in which he took a pessimistic view of the future of mankind. From his vantage point of prolific eighteenth-century England, Malthus contended that the struggle up the ladder of human welfare is a losing game in which every possible means of sub-

---

[1]*omnis* (L.) = all; *vorare* (L.) = to devour; hence omnivorous means eating anything; technically, an organism that feeds on both animal and plant tissue.

sistence must eventually be trampled down by the human reproductive capacity. He said:

> Population, when unchecked, increases in a geometric ratio. Subsistence increases only in an arithmetical ratio. A slight acquaintance with numbers will show the immensity of the first power in comparison of the second. . . . And the race of man cannot, by any efforts of reason, escape from it.
>
> Famine seems to be the 1st, the most dreadful resource of nature. The power of population is so superior to the power in the earth to produce subsistence for man, that premature death must in some shape or other visit the human race.

Malthus's gloomy prediction of such dire consequences from the practice of the most sacrosanct of human activities—reproduction—was highly unpopular. Malthus's theory was interpreted as an attack on the ancient admonition to "be fruitful and multiply" and instigated a storm of abuse.

Malthus also raised the ire of Karl Marx and Friedrich Engels, who took some of his comments in the *Essay on Population* as an attack on the working class. Marx contended that under socialism there would be no unemployment and population problems would disappear. During the industrial and agricultural expansionary period of the following 150 years, the Malthusian theory fell into disfavor and was not seriously revived until population pressures and the awareness of world hunger brought a renewed interest.

Let us consider further Malthus's gloomy prediction that there is no way to prevent people from reproducing exponentially and that when they do, there will eventually be no way to feed them all. His views in relation to the present state of world nutrition are best understood in historical perspective.

The hungry man   Hendrik Van Loon in *The Story of Mankind* said that history is really only the story of a hungry man in search of food. Agriculture was invented late in the evolution of man, not more than 10 to 13 thousand years ago. When people lived off the hunt, they were limited in the population that they could support. The size of their communities was small, usually a family, or at most a tribe. They were severely restricted in their cultural achievements because mobility was needed in order to survive. Their movements were determined by the availability and migrations of game and the ripening of fruit and nuts.

People did not suddenly quit hunting and gathering and settle down to farming. Agriculture was preceded by thousands of years of food gathering, shrewd observations of the most edible varieties of plants, superior locations, soils, and climate, and eventually the semi-domestication of the most favored plants and animals. At first, farming was at best a part-time occupation, supplementing the seasonal food that could be had more easily by picking, pulling, or digging wherever it could be found. Also agriculture developed independently in widely separated parts of the world. The people of the Tigris-Euphrates and Nile Valleys grew types of wheat and other cereals; farther north the people grew rye; the people of Indo-China discovered rice; and the agriculturists in the Americas cultivated maize, potatoes, squash, and several other vegetable crops.

The innovation of farming brought about profound changes in the pattern of human life. It provided a more stable source of food, thereby decreasing the need to wander in search of game and seasonal plants; it made it possible to have a greater variety of foods, improving the health of the individuals; it made it desirable to settle down in one place and establish stable communities.

The agricultural revolution made it possible for people to settle in large communities and engage in commerical specialization, trades, and professions, forming the nucleus in which civilization would germinate and flourish. The growth of industry expanded the human capacity for using the natural resources of the environment, further nourishing economic development and the growth of literature, the arts, and science. The development of science greatly expanded human control over the laws of nature, bringing about an accelerated use of the sources of energy, more highly efficient food production, and vast changes in the environment.

The Tigris-Euphrates Valley supported a flourishing population by at least 3500 B.C. The Sumerians had developed an extensive complex of irrigation and drainage canals by 2000 B.C. They or their successors ultimately failed, partly because drainage was neglected, causing an accumulation of salts in the soils of their intensively irrigated fields. Today less than one-fifth of the land in Iraq is under cultivation and the barren countryside is marked with mounds, the remnants of ancient towns. The soil is salty as a result of the deterioration of the ancient government which allowed the drainage canals to silt up. The extent of soil erosion in the upper reaches of the rivers is seen in the deposition of silt in the delta. The ancient seaport of Ur is now 150 miles from the Persian Gulf, and the annual flooding recurring for hundreds of years has buried its buildings in silt to a depth of as much as 35 feet.

The ancient Mayas are believed by some people to have failed partly because their slash-and-burn system of farming was not suited to supporting large populations and partly because the thin soils that surrounded their cities became exhausted. They looked at the sky and became superb astronomers, but perhaps they failed to study the soil with equal perception, and failed in farm management.

The fight against famine Famines have caused some of the most dramatic of the world's human migrations. The one which most profoundly affected the history of the United States was the immigration of thousands of Irish during and after the terrible Irish potato famine of 1845 to 1848. More than 100,000 Irish emigrants left their country for North America in 1847 alone. Flood, war, and pestilence during the East Pakistan tragedy of 1971 contributed to the migration of several million refugees across the border to India. Scarcity of food took a high toll.

Jonathan Swift, a seventeenth-century satirist, wrote that Irish children should be fattened, sold, and eaten to provide an income for their parents. It was not that he believed in cannibalism. Swift was born in Dublin after his father had died. Still stinging from the grinding poverty of his youth, he said that it would be better to eat children than to raise them poor, hungry, and ignorant; and that the practice would help keep down the population.[2]

---

[2]*Modest Proposal for Preventing the Children of Poor People from being a Burden to their Parents or the Country* (1729).

USDA photo.

Food in the making—picture of Minnesota farmland.

In part of Africa known as the Sahel—an immense region across the continent at the southern edge of the Sahara—lack of rain for five years, aggravated by overgrazing, left 20 million cattle dead and the people ravaged with famine. During the worst part of the drought, in 1973–1974, at least 100,000 died of starvation and millions more were struck down with disease directly or indirectly brought on by malnutrition (see also Problem Areas). In Ethiopia, where an estimated 200,000 people died of starvation in 1973, a coverup by the government helped depose emperor Haile Selassie who had ruled since 1916. Drought struck India in 1972 and again in 1974, severely affecting roughly 200 million people, a third of the country's population. Food supplies dwindled from 9 million tons to 4 million tons while the population grew by at least 25 million and food prices went up 37 percent. Fertilizer was short, and lack of power and water caused widespread industrial unemployment.

PROTEIN STARVATION

Of the basic food constituents, protein is both the most expensive and the most frequently lacking in needed amounts. Half of the children born into the less-favored segment of the population in many parts of the world die within five years after birth. It is estimated that a significant proportion of these deaths is caused by malnutrition and that the high mortality rate is the best indication of protein deficiency.

Protein deficiency in children causes a disease called *kwashiorkor* (see Figure 16.1). A more severe form of malnutrition in children is one that

Courtesy of Instituto de Nutricion de Centro America y Panama. N. S. Scrimshaw, and M. Béhar, *Science* 133:2039–2047, June 1961. Copyright 1961 by American Association for the Advancement of Science.

FIGURE 16.1 Geographical distribution of kwashiorkor, a protein deficiency disease of young children.

develops when they are deprived not only of protein but of other nutrients and adequate calories. The disease is called *marasmus;* it is often fatal. Marasmus is more common in children below one year of age, while kwashiorkor is more prevalent during the second and third years. In marasmus there is, among other signs, extreme retardation of growth and development. In kwashiorkor, there are typically severe disorders of the skin, loss of hair, edema (accumulation of fluid in the tissues), anorexia (loss of appetite), diarrhea, larger than normal stool volume indicating malabsorption, and psychic disorders involving a curious mixture of apathy and irritability. A number of sociological and environmental conditions other than protein shortage can aggravate the condition. Heavy intestinal parasitic infections, which further reduce protein absorption, and ignorance about the protein requirements for small children are often associated with protein deficiency diseases.

T.A.M. Nash, for many years director of the West African Institute for Trypanosomiasis Research, pointed out that nagana, a trypanosoma disease of livestock, is of far greater importance both economically and medically than human sleeping sickness, if the broad view is taken that a deficiency in animal protein among the masses is more crucial than a specific disease among the relatively few. Dr. Nash, who spent many years in Africa, said of the need for protein,

> Real "meat-hunger" is something which should be seen to be believed. In Tanzania, I shot a large zebra near dusk and left two or three men to guard the carcass overnight. On returning next morning with porters, I was amazed to see how little meat was left; the men had eaten and vomited, eaten and vomited, throughout the night.[3]

---

[3]T.A.M. Nash, *Africa's Bane. The Tsetse Fly* (London: Collins, 1969).

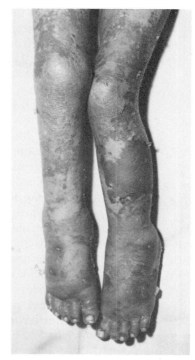

Left photo courtesy of WHO/E. Mandelmann. Right photo courtesy of Instituto de Nutricion de Centro America y Panama. N. S. Scrimshaw and M. Béhar, *Science* 133:2039-2047, June 1961. Copyright 1961 by American Association for the Advancement of Science.

Left photo: A survivor of the Sahel drought-stricken area of Africa. *Right photo*: Edema and skin lesions in a child with kwashiorkor.

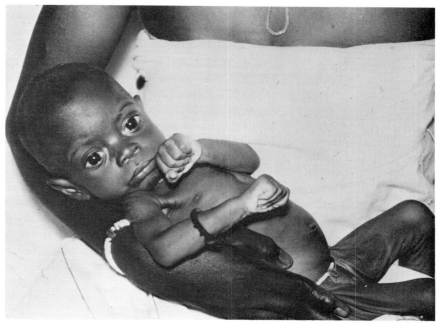

FAO photo.

Marasmus showing loss of subcutaneous tissue but bright, wide-awake eyes.

Much of the damage to children in populations suffering from protein malnutrition may occur before birth. Experiments with rats have shown that malnutrition from inadequate diets deficient in protein during the period shortly before and after birth decreases the number of brain cells. Similar effects were observed in children. Drastic changes in the biochemistry of the brain, involving the neurotransmitters (impulse transmitters) were found in the brains of rats undernourished from mid-gestation to weaning. Other experiments with rats revealed that protein-less diets during any part of the mother's pregnancy caused impaired brain development in the fetus. The damage occurred even if a full protein diet was reinstated following short periods of protein deprivation. The physiological and biochemical events in the utilization of amino acids during fetal development are evidently so tightly scheduled during the course of pregnancy that there is no chance for the fetus to catch up or repair damage, once it has occurred.

Proteins and mentality

The discovery of the effects of fetal and infant nutrition on brain development and intelligence may be one of the most important findings in environmental biology of this century. Experiments with animals give us a partial understanding of how these effects take place. In the pig, during the 13-week period between the fiftieth day before birth and the fortieth day after birth, the brain grows intensively, gaining every two weeks as much as 5 to 6 percent of its eventual mature weight, and 33 to 39 percent of its total growth during the 90-day period. Biochemical development, as in the spinal cord and other parts of the nervous system, takes place rapidly. It is not surprising, therefore, that laboratory investigators working with various animals find that the brains are smaller at maturity and that the animals mature biochemically and functionally at a slower rate when the diet is inadequate in calories and protein during a time of life when the brain is growing most rapidly.

Social customs and diet

Protein utilization is sometimes influenced by social, religious, and dietary customs. India, for example, has about the same number of cattle, including water buffaloes, in proportion to the human population as the United States. But animal protein makes up a small fraction of the Indian diet. One reason is that bullocks are needed for transportation and farm power. Famines followed the slaughter of the breeding stock during the invasions of the Mongols in the fourteenth century. The experience influenced Hindu religious philosophy, and even today in many of the Indian states, the slaughter of cattle is either prohibited by law or by social custom. Thus, India is a land of vegetarians, with animal protein derived mainly from dairy and poultry products, chiefly eggs.

What are proteins?

Proteins are gigantic molecules containing thousands of atoms of hydrogen, carbon, oxygen, and nitrogen. There is great diversity among the proteins; some contain sulfur, phosphorus, or small amounts of heavy metals. Nutritionally, for animals and humans they are the primary source of nitrogen, which averages about 16 percent of the protein content based on dry weight. Structurally, the proteins are composed of units of amino acids attached by an arrangement called peptide bonds, forming what is called a *peptide chain*.

All proteins come from the cells of animals or plants. Microorganisms such as yeast and fungi supply a fraction of human protein and caloric intake. Since neither animals nor humans can synthesize all the amino acid building blocks of proteins, some of these must be derived from plants by digesting the plant proteins into smaller units and resynthesizing them into the numerous types of proteins needed. Many special and distinct kinds of protein are needed for muscles, skin, hair, and other tissues in various organs.

Since it is possible to replace protein in the diet completely with amino acids, the so-called protein requirement is really a requirement for amino acids. There are 23 amino acids, of which 8 to 10 are called *essential* because they cannot by synthesized by the mammalian organism from substances ordinarily present in the diet, at least not at a rate sufficient to supply its physiological requirements. The number of *essential* amino acids varies with the species and even in different stages of development of the same species; eight are *essential* in nearly all animals.

Food from animals   Protein from animals is the most costly source (Figure 16.2). For example, 8 pounds of cattle feed are needed to produce 1 pound of live-weight gain as beef with efficient management. The amount of plant protein in animal feed required to produce 1 pound of animal protein is 3.9 for milk, 4.1 for eggs, 4.6 for broilers, 6.2 for turkeys, 7.1 for pork, 10.0 for beef, and 12.5 for lamb. However, vast areas of woodlands, wastelands, and rangelands in many parts of the world are suitable only for grazing, and only animals can utilize these resources. Also, meat is generally considered to be a better source of protein than plants because the amino acid content is in better balance for human nutrition.

Food from Plants   Grain has always been the basic food—the "staff of life." It directly provides about half the calories for the human population and, indirectly, much of the remaining caloric intake by conversion to meat, milk, eggs, and other animal products. About 70 percent of the world's cropland is used to grow grain, producing one-third of a ton of grain annually for each person on earth.

About one-third of all human energy comes from rice; it is estimated that one-half of the people in the world depend upon it for 60 percent of

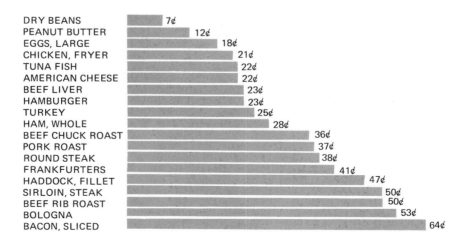

Redrawn from the Agricultural Research Service, USDA, 1973.

FIGURE 16.2 Relative cost of a day's protein, June 1973, based on one-third of the recommended dietary allowance for a 20-year-old man.

USDA photo.

Loading Texas cattle in Abilene, Kansas, in frontier days.

their nutrition. Although the cereals as a group are the most widely used food sources, their protein concentration is low: 7.5 percent in rice, 13 percent in wheat, and 9.5 percent in corn. The cereals, however, are better than potatoes, sweet potatoes, and cassava, from which tapioca is made. These are the starch foods. Both types of potatoes contain about 2 percent protein, cassava less. The 70 grams of protein needed by the

Photo by Great Plains Wheat, Inc., Garden City, Kansas. USDA photo.

Wheat is the "staff of life" for millions of people in the western world.

average person would require that he eat 8 pounds of potatoes or 25 pounds of processed tapioca daily.

The oilseeds are better than cereals as sources of protein. Some varieties of soybeans contain 40 percent or more protein, peanuts about 25 to 30 percent, and cottonseed 16 to 18 percent. Cottonseed cake (the residue from oil extraction) contains the toxin gossypol and is therefore used only for certain types of livestock feed, but efforts to develop genetically glandless (gossypol-free) cotton are underway. The quality of protein in soybeans is comparable to that from animal sources, although all of the oilseeds are slightly low in the amino acid methionine.

Food from the sea

Fish do not provide more than 2 to 3 percent of the total calorie consumption by humans, but they are an important source of animal protein, supplying nearly one-fifth of the world consumption (Figure 16.3). About three-fourths of the catch is used for human food, annually amounting to about 22 pounds per person on a global basis. This compares with 20.5 pounds of beef and 20.3 pounds of pork. Fish provide an amount of meat comparable to the production of 850 million cattle, about the total number on earth.

Fish flour is potentially one of the most abundant, cheapest, and nutritionally richest sources of animal protein. Sometimes called fish protein concentrate, it contains as much as 95 percent protein and an adequate amount of all the essential amino acids. Some people have objected to the product because it is made from whole fish and contains all the head parts, entrails, and offal. Though it is nutritious, the United States has refrained from exporting a product to needy countries that the FDA refuses to sanction for domestic consumption on aesthetic grounds. Others, however, are experimenting with concentrates made from eviscerated fish. Removal of the viscera may reduce the hazard of contaminants such as pesticides and mercury.

Although the fisheries can be an invaluable source of protein supplement, they cannot comprise more than a small fraction of the world's food supply (see Figure 16.4). The sea is not an inexhaustible source of food. Much of the ocean is a wet "desert"—a vast biological wasteland. In

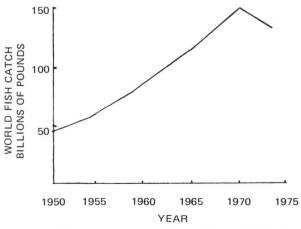

FIGURE 16.3 The world fish catch increased from 1950 until 1970, but since then, the total world catch has declined, indicating that long-range productivity of the oceans will depend on conservation and international constraints.

Left photo courtesy of Standard Oil Company of California. Right photo USDA SCS photo.

Left photo: Fish supply nearly one-fifth of the world consumption of protein. *Right photo*: Catfish are being harvested at a fish farm near Lake Charles, Louisiana.

contrast to land, where water is most often the most important limitation to production, the oceans are limited in productivity by lack of mineral nutrients.

The oceans cannot be expected to yield much more than about 100 to 150 million tons of food annually on a sustained basis. This would provide no more than 3 percent of the world population's caloric requirements. However, there are prospects of greatly increasing the present yield from the ocean. Most of the food harvested from the ocean is by refined techniques of gathering and hunting. The adoption of electronic devices and other advanced techniques has greatly increased the efficiency and the total catch.

Much so-called "trash fish" that is now ignored as human food could provide an important source of cheap protein for the human diet. Initial attempts at herding have not been successful because of the expense of catching food for the "herds." Techniques for large-scale fish farming

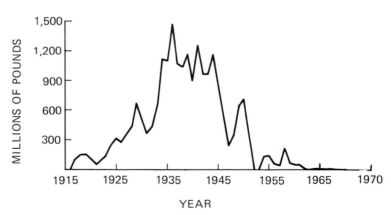

Redrawn from H. W. Frey, *California's Living Marine Resources and their Utilization,* 1971.

FIGURE 16.4 California commercial landings of Pacific sardine. It is believed that overfishing caused the virtual disappearance in commercial quantities of the Pacific sardine, but there is also evidence that there were wide fluctuations in abundance of the species over the last 1,850 years.

have yet to be developed. The major difficulty is that the ocean is three dimensional and the basic source of food in the oceans consists of microscopic organisms. The cost of harvesting it as feed for the marine "livestock" would be very great.

Paradox of plenty   The world is producing more food than ever before. Yet, more than one-half of the earth's 4 billion population subsists on an inadequate diet (see Figure 16.5). The Food and Agriculture Organization of the United Nations estimated that 2 billion people are hungry or undernourished. Malnutrition is a form of slow famine, and starvation is an ever-present reality in some of the developing parts of the world. Malnutrition claims 10,000 lives every day. Many Americans are less well-nourished than the country's livestock, most of which receives a sientifically balanced diet. Nowhere, however, is there greater contrast than between the farming industries of the highly industrialized areas and those of the developing parts of the world. The difference is a major cause of the disparity in the quantity and quality of food between the "haves" and the "have nots." In most parts of the world, men still bend their backs to the hoe for a pittance and a semistarvation diet. At most, 400 million people in the richly endowed parts of the word—little more than 10 percent of the world's population—enjoy a rich, steady, and abundant diet. They belong to what Dr. Georg Borgstrom, the eminent food scientist, calls the world's "Luxury Club." The others, at least 2 billion people, are overwhelmingly dominated by critical shortages of food and meager resources

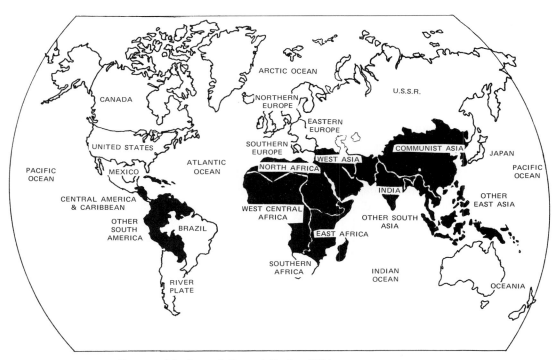

Courtesy of the Economic Research Service, USDA.

FIGURE 16.5 Diet-deficient regions of the world.

From the sixteenth-century Codex Florentino.

The Aztecs planted corn, using a digging stick, and stored the harvest for the winter.

of soil, water, and forests. Dr. Borgstrom said, "If all food in the world were equally distributed and each human received identical quantities, we would *all* be malnourished." On a global basis, the world has no surpluses, only enormous deficits.

AGRICULTURE TODAY

Food plants are cultivated on about ten percent of the earth's 33.5 billion acres of land area (excluding the Antarctic). Another 15 to 20 percent is under permanent meadows, pastures, and rangeland. The 3.5 billion acres of cropland are distributed unevenly with respect to both international geography and population densities. Nearly half of the cropland is within the borders of four countries: The United States, the Soviet Union, India, and Mainland China. With respect to acres of cropland per person, Australia, Canada, and Argentina rank first, while India and Mainland China rank low. Other factors that produce imbalances between populations and their food-producing capacity are differences in climate, soils, water supplies, crop patterns, irrigation practices, and the utilization of agricultural technology.

Agribusiness   Where agriculture has reached its highest development, man drives machines across the land into which he has fed fuels and chemicals; fertilizers are blended to meet the needs of a particular soil and crop; pesticides are pitted against enemies when they appear, and other sophisticated methods are used to supplement them; the crop is harvested, packaged, and shipped mechanically; livestock are bred by artificial insemination, kept in pens with automated feeding and fed on factory-blended feed, milked with machines, and slaughtered by production-line methods. This new face of farming is called *agribusiness*.

USDA photo.

Dependence on human labor and animal power limited food production in earlier times.

More food for more people   Production of foodstuff in the highly industrialized countries has far outdistanced that of a few years ago and is still increasing at an astonishing pace. Since 1940, cereal grain production has increased 50 percent in the United States, and farm food production is accomplished with one-third as many workers (see Figure 16.6). While this phenomenal progress was being made, the total acreage under cultivation actually decreased by about 7 percent. Agriculture is the largest source of world wealth and is the base for other segments of the economy. In most of the developing parts of the world, agriculture will have to provide the leverage for industrial growth and development. Agribusiness, if brought to its potentiality throughout the world, could profoundly change the economy of the world's population.

However, in a world devoted largely to family-size plots of land and subsistence farming, agribusiness can be socially disruptive. It is not so much the trauma of discarding ancient values in favor of greater efficiency as we have seen take place in the United States, but the fact that agribusiness can be impractical unless the financial and related resources needed to change from a subsistence economy can be sustained. We get some idea of this from the problems associated with the "green revolution."

THE GREEN REVOLUTION

The availability of plant and animal foodstuffs brought about by selective breeding are little appreciated by the average consumer. The productivity

Photo by Myrtle Maxwell.

Even until a few decades ago, animal power was needed for the best use of farm machinery.

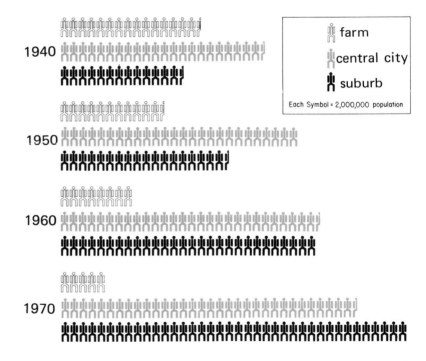

Redrawn from *Population Bulletin* 27(5), 1971. Reprinted with the permission of the Population Reference Bureau.

FIGURE 16.6 Migration from the land to the city. In thirty years suburban dwellers grew from least numerous to most numerous.

Photo courtesy of Superior Farming Company.

Mechanization and high production efficiency are a part of agribusiness.

of hybrid corn has made the growing of hybrid seed an industry in itself in the United States "farm belt." The increased yields from hybridization is a manifestation of *hybrid vigor,* well known to plant and animal breeders.

Farmers buy seeds for varieties of plants that have been selected for one or more of a large number of desirable characteristics, such as resistance to specific diseases, fast maturity, frost resistance, growth to a uniform height, uniform ripening so that the crop can be machine harvested, high protein content of cereals, gossypol-free cotton, and high vitamin A content in squash.

In livestock, we have turkeys with large breasts and small legs to give a higher proportion of white meat, cows that give milk high in butterfat, beef cattle with a high proportion of choice cuts, hogs with less fat and more lean meat, and hens that lay more than 200 eggs a year.

The development of high-yielding varieties of dwarf wheat in Mexico by the International Maize and Wheat Improvement Center in Mexico has helped to bring many parts of the world from near famine to self-sufficiency. It is estimated that 25 million acres now planted to the new varieties of grain feed 500 million people. The success of these varieties of wheat when cultivated with the use of suitable amounts and kinds of fertilizers, pesticides, irrigation, and other appropriate materials and methods has been termed the *green revolution*. The man who headed the project, Norman Borlaug, an American plant pathologist working as a staff member of the Rockefeller Foundation, was awarded the Nobel Peace Prize in 1970 for his contributions to the development and worldwide distribution of the high-yielding varieties. His educational efforts in the production and use of fertilizers and pesticides that are needed for efficient utilization of the new seeds were also recognized. A new variety of rice developed in the Philippines holds out similar promise for the rice-producing and consuming areas of the Orient.

The green revolution is more than the breeding and selection of crop plants. It is an increase in productivity resulting from the application of

USDA photos.

Left photo: A new variety of Indian millet yields more grain than older varieties—an example of plant breeding. *Right photo*: Norman E. Borlaug recording the vigor and stage of growth of wheat in his breeding plot. Borlaug was awarded the Nobel Peace Prize in 1970 for his contributions to human nutrition.

scientific methods to achieve the greatest performance from varieties of high-potential yield. These methods include fertilization, irrigation, pest control, and mechanization.

All of these create problems of their own. New varieties of plants are sometimes more susceptible to local diseases and pests than older varieties, which themselves may have become selectively adapted over a long period of years for their resistance to pestilence. Disasters could result if, given favorable circumstances, a hitherto mild disease were to sweep through large areas devoted to a single new and susceptible variety. Similar difficulties could arise if a relatively innocuous insect pest of older varieties were to suddenly find thousands of acres of new crops especially to its liking. If the compulsion to solve the immediate problem by saturating the fields with pesticides were followed, the ecological problem would be compounded with one difficulty leading to another. Thus, the age-old problem of coping with technological advance, which began with the invention of agriculture and the introduction of monoculture, becomes intensified and calls for increasingly greater attention to the ecological considerations of agricultural innovations.

FERTILIZATION

A hundred bushels of corn take from the soil 78 pounds of nitrogen, 36 pounds of phosphate (as phosphoric oxide), and 26 pounds of potassium (as potassium oxide). The stalks remove from the soil additional nutrients that are not returned when the stalks are used for fodder, equal to 52 more pounds of nitrogen, 18 more pounds of phosphate, and 94 more pounds of potash to each 100 bushels of corn. Other crops make similar demands on the plant foods in the soil. These lost nutrients must be replenished if the soil is to retain its productivity year after year. Thus the use of fertilizers not only increases yields dramatically, but the benefits are derived immediately without the wait of ten or fifteen years, which is sometimes necessary for the breeding of new varieties, the discovery of a satisfactory pest control measure, or other method for improving productivity.

The removal of plant nutrients from the soil by a growing crop is directly proportional to yield. A variety of rice, for example, that has been selected for high productivity will withdraw from the soil more plant nutrients than its low-yielding counterpart and will require correspondingly more fertilizer. Therefore, in order for developing countries to harvest the fruits of selective breeding in the forms of high-yielding varieties or strains, they must also have available other products of science and technology, such as adequate fertilizer supplies and, for some crops, an abundant supply of water. Inasmuch as the costs will thereby be increased, prudence would dictate that the crops be protected from pests and diseases, that advantage be taken of growth-regulating substances, and that maximum use be made of modern machine methods.

The plant nutrients

At least 16 elements are needed for the growth of green plants. Among these are carbon, hydrogen, and oxygen, which are abundantly available in the environment from carbon dioxide and water. The remainder are taken up from the soil. The nutrients that are needed in largest amounts are called *primary nutrients,* of which the ones that most frequently require addition to the soil are nitrogen, phosphorus, and potassium.

Nitrogen is obtained from the atmosphere and is obtainable any place in the world where the energy is available to fix it in the form of usable nitrogen compounds. The process also requires capital and technical know-how. Phosphorus and potassium, on the other hand, are found in large mineral deposits in certain parts of the world.

Other elements needed by green plants in moderate to large amounts are sulfur and calcium. Still additional elements, called *micronutrients,* are needed in very small amounts. Important micronutrients are iron, manganese, zinc, copper, molybdenum, and boron. When these are deficient or when soil conditions render them unavailable in sufficient amounts, it is necessary to add them either to the soil or directly to the plants. Various compounds of the micronutrients are applied in various ways. Others of the so-called "trace elements" that are present in green plants in various amounts are chlorine, vanadium, cobalt, iodine, fluorine, and sodium. Of these, only chlorine has been proved essential to the growth of green plants. Sodium is always present, but its role is obscure, possibly substituting for potassium, though it is indispensable for animals. Vanadium is essential for the growth of some algae but apparently has an insignificant role, if any, in the growth of higher plants. Iodine and fluorine are likewise needed by animals, but as far as we know, they are not needed by plant life.

Organic fertilizers

The use of organic fertilizer is an ancient practice. A parable of interest to the organic gardener is written in Luke 13:6–9:

> . . . let it alone this year also, till I shall dig about it, and dung it: and if it bear fruit, well: and if not, then after that thou shalt cut it down.

The first record that we have of fertilizing the soil with manure is in an Accadian tablet. Pliny attributes the discovery to an ancient king of Greece. He cites Homer, who made reference to the good king Laertes found laying dung on the land with his own hands. The ancient Romans were skilled farmers and knew how to handle and store manure so that its valuable contents would not be lost. They practiced green manuring (plowing under growing crops), crop rotation, and fertilizing with sea-weed. The North American Indians are said to have planted fish in their corn hills, but one view is that they learned it from the Europeans.

The organic cycle in nature is often taken for granted. The value of scavengers and decomposers becomes evident when we ask the question, ''How long would we be able to survive if the dead bodies of plants and animals were to accumulate without being broken down and returned to the soil?'' Such organic waste would in time cover the earth's surface to a depth of several feet. The scavengers and decomposers are essential to maintaining a balance in nature.

The most easily recognizable effect of organic matter is on the soil's physical properties. Organic matter makes heavy soils easier to work and induces a crumbly structure. Moreover, the soil crumbs are stabilized and held together so that the soil absorbs water more easily and run-off and erosion are reduced. There is better aeration in the root zone, and seed-lings emerge more easily. Another benefit from organic matter is that it moderates the tendency of the soil acidity-alkalinity relationship, called *pH,* to fluctuate when acid or alkaline materials are added. This effect is called *buffering* action. Organic matter may stabilize soil micronutrients by the formation of metal-organic complexes that prevent leaching of elements that are needed for good plant growth. Substances that are capable of forming complexes with metals are called *chelating* agents, from the Greek *chela,* meaning ''claw.'' Knowledge of chelating effects has made it possible to develop synthetic chelating agents to supply essential metals, or micronutrients, in plant nutrition.

Another way in which organic matter affects plant nutrition is through the retention and release of substances such as calcium, magnesium, and potassium ions called *cations*. Iron is sometimes present in the soil in large quantities but unavailable to the plant because of an unfavorable pH and other conditions. Reaction of soil iron with organic substances capable of forming iron-organic complexes greatly increases the availability of the element to the plant. For example, it has long been known that so-called iron humate can be used as a source of iron in solution culture.

Organic matter is a direct source of plant nutrients, especially nitrogen, phosphorus, and sulfur. Soil organic matter, sometimes collectively called humus, contains a maximum of about 5 or 6 percent nitrogen. Virtually none of it is available to plants without microbial and chemical conversion to an inorganic form. This conversion takes place slowly. Actually, there is often an initial decrease in the available nutrients to growing plants following heavy applications of organic matter due to the development of competitively high populations of microorganisms. Later, there may be a surge in plant growth derived from the nutritional value of dead microorganisms which themselves add to the organic content of the soil. Nitrogen-fixing bacteria contribute.

The liberation of nitrogen does not take place until the protein is de-

Photo courtesy of Floridin Company.

The chemical hoe has largely replaced hand hoeing, and pesticides have taken the place of manual removal of pests. Fertilizers are also applied with fast mechanical equipment.

composed and ammonia is released. This is called *ammonification* or mineralization. Ammonia that is not assimilated undergoes a second breakdown step called *nitrification,* which results in *nitrate* nitrogen, an inorganic form that is available to plants.

The phosphorus content of soil organic matter is much less than that of nitrogen, but in some soils may account for one-half or more of the phosphorus present. Inorganic phosphate is made available when the ratio of carbon to phosphorus is sufficiently small.

Sulfur is present in plant and animal residues and in microorganisms as one of the constituents of some of the amino acids in proteins. Like nitrogen and phosphorus, organic sulfur must be converted to an inorganic form (sulfate) before it can be utilized by plants.

Thus, the nutritional benefits of organic matter based on assimilation of the primary nutrients by the growing plant are ultimately identical to those derived from the addition of inorganic fertilizers. Insofar as anyone can determine, the quality or nutritional value of the food product is no different from one that has been produced from a scientifically balanced plant nutrient containing the primary essential elements in inorganic form. However, the nutrient value of organic material is usually much lower than that of commerical inorganic fertilizers; therefore, much larger quantities of organic material are needed to produce the same yield.

FOOD FOR THE FUTURE

Synthetic foods   The subsistence of man on completely synthetic food may be possible at some time in the distant future. Foods are chemicals, and the composition of some of them are well known. We already use synthetic vitamins and chemically modified products such as hydrogenated oils and fats. We add synthetic flavors, colors, and emulsifiers (digestible substances akin to fats). We can manufacture amino acids and synthesize polypeptides. Semi-synthetic foods are now widely accepted. Soybean concentrates and other plant proteins are skillfully blended with flavoring, coloring, and stabilizing agents to produce delicious food products that are reasonable facsimiles of processed ham, beef, or chicken.

Progress is being made in the artificial synthesis of amino acids. Synthetic methionine is now being used extensively in animal feed. About one-third of the nitrogen in the feed for fattening cattle can be supplied in the form of urea, a manufactured nitrogen compound. Cattle can utilize urea indirectly by the action of bacteria in the rumen.

Lower organisms

Culturing algae was at one time thought to be one of the more promising new methods for producing food. A great deal of work was done on *Chlorella,* a single-cell green alga that synthesizes carbohydrates as do the higher plants by utilizing the action of sunlight on chlorophyll. So far, the results have been disappointing because of high costs and problems of palatability, although the latter might not be significant if the product were used for livestock feed. Bacteria and yeast could also be used for a portion of human or livestock protein requirements by utilizing carbohydrate content of whey and other waste sugars for culturing the microorganisms. A food yeast called torula has achieved some acceptance.

Food preservation

In ancient times, there were two ways to preserve food: sun-drying and salting. Modern methods of preservation vastly expand the possibilities for better world distribution of food and make it technically possible collectively to deliver "fresh" food anywhere in the world with only minor losses in quality and quantity. These methods include: refrigeration; canning; freezing; vacuum drying and vacuum dehydration of frozen foods (freeze-drying); pasteurization; use of antioxidants; use of antimicrobials; scientific storage to control temperature, humidity, and atmospheric content; protective packaging with plastics and pest-proof materials; and rapid transportation by refrigerated truck, train or airplane.

Climate control

Climate control has progressed little since scientific rain-making achieved limited and questionable success. However, the possibilities have not been completely explored. The full effects of high-level explosions, the melting of glaciers and ice caps, and the diversion of waterways and ocean currents remain to be more thoroughly studied. Methods of controlling frost or its physiological effects would have a revolutionary impact on agriculture.

Maximized methods

Other technologies can be updated or used to greater effect:

- Better utilization of water resources would include reclamation of used irrigation water, desalination, use of high-salt water where possible, and more efficient irrigation practices.
- Maximum mechanization in developed countries and basic mechanization in primitive countries has yet to be achieved.
- Maximum use of the chemical means for increasing yield and quality include more efficient use of fertilizers, minerals, and pest control methods.
- Greater utilization of sewage wastes is needed.

USDA photo.

An abandoned farm in Oklahoma, showing the disastrous effects of wind erosion.

- Development and use of synthetic and natural plant hormones have undiscovered possibilities for increasing growth, controlling flowering and fruit set, and improving quality and yields.
- New methods of food production are needed, including more efficient utilization of space in farming operations.
- Leaf protein concentrate from plants such as alfalfa is an unexploited source of human food.
- Breeding and selection of new strains of plants and livestock that can thrive in the immense uncultivated regions of the tropcis are needed.

PROBLEM AREAS

The world food supply is precarious. For the first time in the history of mankind, we are faced with shortages of all four of the basic agricultural resources—land, water, fertilizers, and energy. Most of the land not now under cultivation is either marginal, unsuitable for agriculture, or useful only for grazing. Recent advances in productivity have been almost entirely from scientific and technological developments, and these may have reached a plateau beyond which further benefits become increasingly difficult and costly.

Water is in short supply in cultivable regions, and most of the water resources available for agricultural use have already been harnessed. The shortage of water, combined with the vagaries of weather in areas of marginal rainfall, may even be more critical than the availability of productive land. Fertilizers are essential if food production is to be maintained and expanded. Phosphate fertilizers are derived from a depletable mineral resource that is unevenly distributed over the earth's surface. About one-half of the fertilizer that is needed is in the form of nitrogen fertilizers, the production of which requires large amounts of energy in the form of natural gas, petroleum, or electricity (see Chapter 3). Energy is also needed at almost every step of food production from planting to processing. The increase in cost of energy has already had a devastating impact on people in food-short areas and may prove to be the most serious impediment to the success of the green revolution.

All the major crops of the world and most of the minor ones were domesticated in prehistoric times. Thus, modern agriculture, which we must use as a starting point, is basically Neolithic agriculture. The practicality of many of the potential improvements are speculative. Much research will have to be invested to find out which are true advances and which are effective on a sufficiently large scale to be economically feasible. Cultural and dietary customs remain to be overcome. Economic barriers to investment and world trade have not been solved. New sources of food that now seem promising may not materialize.

High-yielding varieties of corn, wheat, and rice are available. Whether they will perform to expectations under the wide variety of conditions of climate, moisture, soils, and pests in many cases remains to be proved. To achieve their potential productivity, high-yielding strains must have more mechanization, more processing and storage facilities, and greatly increased amounts of capital and credit.

A handicap that the developing countries have not yet overcome is that the wealthy parts of the world are drawing on the food-producing resources from areas where they are needed most. Many millions of people in the tropics restrict their cultivation of domestic food crops in order to supply the demand for export crops such as peanuts (groundnuts), cotton, coffee, tea, and cacao. These are called cash crops. They have high priority on fertilizers, irrigation, equipment, and bank credit. Not more than one-tenth of the worldwide use of fertilizer is used by the hungry part of the world, and most of it goes on cash crops.

Some of the developing countries have increased their food production. In most cases, their population has increased even faster, so that there is little or no gain. In some areas the gain in production is from marginal land, so that in terms of energy expended there is a net loss.

We have seen that drought and overgrazing caused widespread famine among the people who live along the southern edge of the Sahara desert. The area was occupied by 24 million people and the same number of animals—about a third more people and twice as many animals as the land supported forty years ago. The desert is encroaching southward at the rate of 30 miles a year, according to one estimate, and this is generally conceded to be due to the impact of increased human activity and the struggle to survive in a precarious environment. Many people are worried that the progressive collapse of the fragile steppe and savannah of central Africa is an example of what may become increasingly common in other parts of the world. Indeed, there is much to support the conclusion that the creeping corrosion of the ecosystem we see taking place in the Sahel of Africa is a continuation of the historical pattern of decline seen elsewhere from over-exploitation of agricultural resources.

It has been taken for granted for many years that people of affluence and high food-producing capacity should share the burden of food scarcity with people who are on starvation levels of subsistence. But the mouths that need feeding are increasing along with the cost in money and energy. This poses problems that are more than technical or economic. It raises the moral question of how far the affluent consumers will be willing to reduce their food consumption and at the same time bear the increased cost of food and energy in order to provide starving and semi-starving millions with more than a bare crust of bread.

Large amounts of the four basic agricultural resources are used for nonfood purposes. Lawns, parks, golf courses, cemeteries, playing fields, backyards, and landscaped highways all use land, water, fertilizers, and energy. Energy now used for many activities and consumer products is energy that could be used for producing fertilizers. Millions of people could be fed by diverting these resources to food production, but at great cost in aesthetic and other qualities of life that people value. A compromise between the desired standard of living and the obligation to feed people is being made, and may become more painfully evident as the world population increases.

We watch the population barometer and try to calculate how many people the world could feed *if* the population soared to 6, 10, or 12 billion. Yet, we cannot feed satisfactorily even half of the people on earth now. Unless increased food production is linked with population control, there are those who will ask, "What good is it to keep people alive to give birth to more people to feed?" The response to this dilemma will be the ultimate test of the ability of the human race to deal with the combined population-subsistence problem, and to answer once and for all the question, "Was Malthus right after all?" And the next question is asked, "If so, who will we keep alive and who will be allowed to die?"

READINGS

BORGSTROM, GEORG, *The Hungry Planet,* revised edition. London, Collier-Macmillan, Ltd., 1967.

BROWN, LESTER R., *In the Human Interest*. New York, W. W. Norton & Company, 1974.

EMERY, K. O., and C. O'D. ISELIN, "Human Food from Ocean and Land," *Science* 157:1279–1281 (1967).

MALTHUS, T. R., *An Essay on the Principle of Population as it Affects the Future Improvement of Society*. Printed for J. Johnson in St. Paul's Church-yard, 1798. (Reprinted in Thomas Robert Malthus, *First Essay on Population 1798*. Published for the Royal Economic Society. London, Macmillan & Co., Ltd., 1926.)

MALTHUS, T. R., *A Summary View of the Principle of Population*. 1830. (Reprinted in *Three Essays on Population*. New York, The New American Library of World Literature, 1960.)

Resources and Man. San Francisco, W. H. Freeman and Company, 1969.

SCRIMSHAW, N. S., and M. BEHAR, "Protein Malnutrition in Young Children," *Science* 133:2039–2047 (1961).

SCRIMSHAW, N. S., "Food," *Scientific American*, Sept. 1963, pp. 73–80.

WEISS, M. G., and R. M. LEVERTON, "World Sources of Protein," *Farmer's World*, the Yearbook of Agriculture, Washington, D.C., USDA, 1964, pp. 44–52.

WOODHAM-SMITH, C., *The Great Hunger*. New York, The New American Library of World Literature, 1962, p. 429.

POPULATION CONTROL

Julian Huxley believed that the two most important contributions of biology were the control of disease and the use of artificial contraceptives. But germ control and sperm control are mutually defeating. Thus far, germ control has won out (Figure 17.1).

Birth and death, the most direct determinants of population density, reach phenomenal proportions in nature. Natural mortality from enemies, lack of food, and other adverse environmental conditions take a frightful toll, often leaving a small fraction of the progeny or eggs originally produced. This great loss of life—a waste insofar as a particular species is concerned—is compensated for by a redundancy in births. Many organisms reproduce in enormous numbers in order to perpetuate the species.

Reproductive redundancy

Female Pismo clams lay about 15 million eggs a year. In less than ten miles of favorable clam-producing beach there could be more than 100 trillion eggs spawned. If they all matured into legal-size clams and were laid end to end, they would encircle the earth 300,000 times. Yet only a small fraction of one percent of the eggs ever become mature clams. An annual census showed that in some years only 33,000 clams resulted from more than 120 trillion eggs.

The male flowers of a corn plant produce more than 50 million pollen grains. Nearly all of them are lost to the wind. Plant pollen has been found as high as three miles in the air and as much as 100 miles from the point of its origin. Charles Darwin commented on the redundancy of nature and the toll of life: "... each species, even when it most abounds, is constantly suffering enormous destruction at some period of its life." He wrote of the dense clouds of fir tree pollen, "so that a few granules may be wafted by chance on the ovules."

The female of the Pacific salmon produces from 2,000 to 5,000 eggs. Even though several males may mate with the female, much of the sperm is carried away in the fast current and only a small percentage of the eggs are fertilized. Of those that are fertilized, few hatch, and even after hatching the mortality is high. It was estimated that of the young sockeye salmon that enter the sea, only 10 percent complete their cycle and return to their home river. Some die of diseases or are eaten by

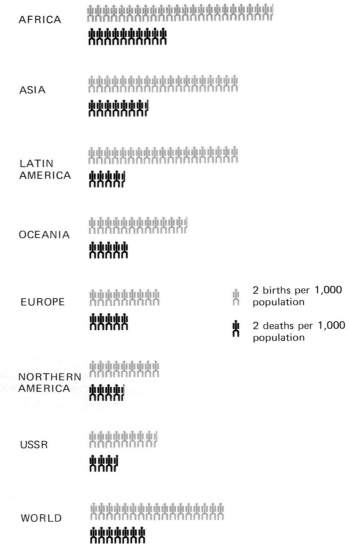

AFRICA

ASIA

LATIN AMERICA

OCEANIA

EUROPE

2 births per 1,000 population

2 deaths per 1,000 population

NORTHERN AMERICA

USSR

WORLD

Redrawn from *Population Bulletin* 27(2), April 1971. Reprinted with the permission of the Population Reference Bureau.

FIGURE 17.1 For each thousand persons, worldwide, there are 2½ more births than deaths a year. The rate is higher in Latin America, Africa, and Asia.

predators; some of the survivors fail to find their home streams; others die for lack of oxygen or from pollution. Some that enter their home streams are blocked by dams, or if there are ladders, some cannot find them. Finally, some expire of exhaustion or are too weak to spawn.

Examples of amazingly redundant fecundity are found among the insects. The queen honeybee lays 2,000 to 3,000 eggs, equal to several times her own weight, every day. A termite queen may lay as many as 3,000 or 4,000 eggs an hour for the duration of life until millions have been produced. One entomologist calculated that a pair of houseflies starting in

Flowers produce an overabundance of seeds—an example of the redundancy of nature commented upon by Charles Darwin.

April could set off a reproductive proliferation that would give rise to more than 191,000,000,000,000,000,000 offspring, enough to cover the earth 47 feet deep in one season.

In humans, only one sperm can fertilize an ovum. But an ejaculation releases an estimated average of at least 100 million sperm and in some cases as many as 500 million. Conditions in the vagina are not conducive to sperm longevity. Large numbers never get past the cervix. Millions more die in the uterus and oviduct. Others become infertile or go up the wrong oviduct. The spermatozoa probably remain fertile for no longer than 48 hours in the human female genital tract. In many other animals, however, the sperm are retained in the female in viable form for long periods, in some species for the life of the female.

Raymond Pearl studied 199 healthy married couples who did not practice contraception and found an average of 254 potentially effective copulations to each pregnancy. Thus, on an average, only 1 sperm in 25 billion or more sperm was successful in fulfilling the mission of the species. A lifetime average is probably even much less favorable to the sperm's chances, taking into account the unproductive years of life.

The balance of nature is a complex system of homeostasis whereby most species of animals, plants, and microorganisms are overproduced but kept in check by natural mortality. Each species is in balance with its environment to the extent that it is neither brought to extinction by parasites, predators, and other inimical forces in the environment nor allowed to overrun the earth to the point of calamity. Some species temporarily beat the odds with a reproductive surge almost unhampered by environmental resistance. But invariably the population growth loses momentum and declines to levels more in balance with the environment. Man has "cheated" on nature, in a manner of speaking, by technological advances in medicine, public health, and food production, so that there is no longer a balance between births and deaths. The odds are temporarily weighted in favor of births with no identifiable major obstacle to continued growth in sight, though a dark cloud of disaster looms on the horizon.

WHEN DOES LIFE BEGIN?

The ideas and definitions of when human life begins have changed from time to time, causing persistent legal problems and raging controversies over moral questions. It is beyond the scope and purpose of this book to evaluate moral problems. We will direct our attention to the biological aspects on the premise that a thorough knowledge of the facts is the best preparation for making wise decisions on human problems.

We must reject the temptation to answer the question "When does life begin?" in a legal reference, for legal rights and privileges, including the right to live or to receive recognition as a citizen, are varied and ever changing, subject to the evolving ideas, customs, ethics, and moral concepts of each society. An amusing and traditional exercise in intellectual frustration is to ask, "Which came first, the chicken or the egg?" A question that comes closer to our problem is, "When does an egg become a chicken?" Conclusions about when life begins must be based on scientific observations of life itself.

A predominant characteristic of all organisms that perpetuate themselves by true sexual reproduction—whether in primitive single-celled or more complex organisms—is a cycle that involves an alternation of at least two kinds of living entities. In all sexual reproduction a somatic (body) stage and a gamete (sex-cell) stage are required to complete one generation.

There is great variability in the reproductive methods of organisms, but even among the one-celled Protozoa, the simplest animals, examples can be found which in basic outline represent the kind of sexual reproduction that has developed in higher animals.

Single-celled sex

Small organisms belonging to the genus *Volvox* are an intermediate evolutionary step between one-cell and multiple-cell organisms. As a compromise they are usually referred to as "colonial" unicellular organisms. They are so primitive that they are regarded as animals by zoologists and plants by botanists. The latter have a point, for many of the cells contain chlorophyll, enabling them to manufacture carbohydrates with the aid of energy from sunlight as in the higher plants. *Volvox* lives in water in ball-like spherical colonies held together by a matrix of gelatinous material. Each individual in the colony is called a *zoid*. The colony forms a green ball not more than 1 millimeter ($\frac{1}{25}$ inch) in diameter containing as many as 50,000 zoids imbedded in the sticky surface of the jelly ball. Most of the zoids have whiplike structures called *flagella* (singular, *flagellum*) with which they flail the water in coordinated action, causing the ball to roll over and over in spirited gyrations that are fascinating to watch under a microscope.

Some of the zoids become specialized for nutrition and locomotion.

Other zoids have the responsibility for reproduction. While *Volvox* has the ability to reproduce either with or without sex, it is only the sexual activity of the colony that we are concerned with. The zoids destined for sexual reproduction change into sex cells called *gametes*. Some of them becomes *macrogametes* (ova) and some of them become *microgametes* (which will form sperm). The macrogametes are fewer in number and are larger, being loaded with food material for nourishment of the new colony that they will start. The microgametes divide repeatedly to form spherical bundles of tiny flagellated sperm, which leave their colony and swim about in the water seeking mature ova.

When a sperm finds and enters an egg, thereby fertilizing it, the result is a *zygote*. When conditions are favorable, sometimes long after fertilization, the fertilized ovum begins a new colony by dividing and continuing to do so repeatedly until the community of cells is mature and ready to start another colony.

The continuity of life There is no basic biological difference between the sexual reproduction in *Volvox* and that of higher animals including people, except in the details of physical accomplishment. A new generation is considered to begin upon fertilization of the ovum, which produces a zygote (fertilized egg or ovum). Are the gametes (ovum and sperm) part of the new generation or the old, or are they neither? There is no clear-cut answer to this question and perhaps it is academic anyway. When we see the futility of pointing to a particular stage of the life cycle of an organism and calling it the "beginning," the conclusion is inescapable that life as we know it is continuous.

The Japanese say that a child is one year old the day it is born. The ancient Greeks regarded the birth of a child as its first birthday, and the end of the first year after birth as the completion of its second year of life. There was some logic to this, assuming there is no particular time or stage in the development of an infant, an adult, or an aged person that marks it as a member of *Homo sapiens*. Whether we can make such distinctions biologically is controversial. It appears that we must depend on either an instinct within or a higher sense of justice to decide at what point it is justifiable to begin, to prevent, or to terminate a life.

HUMAN BIRTH CONTROL

Preventive methods that may be classified broadly as population control have been practiced for at least thousands of years. The first attempts must have been obvious, direct, and crude. Infanticide was probably an early practice. The ancient Spartans were more concerned with a form of short-term eugenics than with population control. A newborn male received a physical examination before the elders, and if it did not pass because of some hopeless deformity, it was left to die in a chasm at the foot of Mount Taÿgetus. As late as Roman times a callous view was taken of infant life. Disposition of the young by drowning or exposure was apparently commonplace and viewed casually as the prerogative of parents by writers such as Seneca, who was Nero's tutor, the historian Tacitus, and the gossipy biographer Seutonius. Infanticide has persisted through the centuries, especially during times of famine or other hardships.

Simple methods of contraception are probably equally ancient. The

Book of Genesis relates that Onan was directed by the tribal patriarch Judah to take as his wife the widow of Onan's slain brother. But Onan was resentful at having forced on him the responsibility for perpetuating the family, and in order to prevent conception practiced *coitus interruptus*. This made Jehovah so angry that he condemned Onan to death. More sophisticated but equally direct methods were developed in later times. The ancient Romans used heat in the region of the genitals. We know now that the human male gonads cannot produce viable sperm unless cooled below body temperature. Early in life they drop from the body cavity into the scrotum, which acts as an air conditioning device.

Chemical contraception was an early practice. Ancient literature contains dozens of formulas for both inducing and preventing conception. Egyptian papyri written between 1900 and 1100 B.C. give recipes such as: honey and soda ash, an unidentified substance to be mixed with mucilage, acacia tips, coloquintida,[1] dates mixed with honey, moistened fibers impregnated with crocodile dung to be placed in the opening of the uterus, and treatment of the uterus with the seed of an unidentified grain.

The Talmud contains numerous references to contraceptive methods including the use of pessaries, absorbents, and potions. Throughout history more than 100 kinds of plants are mentioned as containing substances that affect human fertility and 60 of them have been identified. They include some that are abortifacients and others that cause sterility. Some are dangerous poisons. Methods given as early as the Greek and Roman periods included potions, fumigants, vaginal suppositories, and abstinence during supposed fertile periods.

Chemical contraception was known to the American aborigenes. The Shoshone Indians of Nevada used a desert plant, Lithospermum,[2] as a contraceptive. The Paraguayan Indians of the Matto Grosso use a preparation of a native weed[3] as an oral contraceptive. The women are reported to take a daily decoction of the powdered, dried leaves and stems. In laboratory experiments it reduced fertility in adult female rats.

The popularity of various methods for limiting offspring at various times throughout early history ran the gamut from techniques for preventing entry or eliminating semen from the vagina, to physical barriers to the uterus. Methods included chemical spermacides, abortifacients, antifertility drugs, abstinence, observance of supposed sterile periods, abnormal sexual practices, castration, abortion, and infanticide.

Modern methods of birth control   Until recently, nonchemical methods of contraception dominated birth control practices in modern society. Abstinence, observance of infertile periods (rhythm method), and mechanical devices were favored. Widespread use of the condom and the pessary were made possible by Charles Goodyear's invention in 1844 of an improved method for processing rubber. These devices, however, are either too inconvenient or too costly for many people who need them. The intrauterine device (IUD), consisting of a plastic coil in one of several designs, has promise of acceptability as a

---

[1]Dried pulp of the fruit of colocynth *Citrullus colocynthis* (Cucurbitaceae), a drastic and dangerous cathartic.

[2]*Lithospermum ruderale*.

[3]*Stevia rebaudiana*.

moderately effective method, but injurious effects sometimes occur, and there are also problems that have their roots in psychological attitudes and sociocultural behavior.

Vasectomy, a surgical procedure that consists of cutting and tying off the vas deferens, thereby preventing the escape of sperm, is increasing in popularity. It involves little risk and is safer than a comparable operation on the female. Surgical abortion has had a great surge in popularity in those parts of the world that are affluent enough to afford it under suitably aseptic conditions. However, it is inconvenient and too expensive for worldwide use and, in addition, is morally objectionable to many people.

The most enthusiastically adopted of the methods that are acceptable to modern society is "the pill," which has not only a high percentage effectiveness but is easily used and peculiarly suited to our pill-oriented culture.

ORAL CONTRACEPTION

The "pill" is not a single drug. It is a euphonious expression for a variety of hormone-acting formulations that have the effect of inhibiting ovulation. The development of these substances for contraceptive use was a landmark breakthrough in biology. To understand how they work, let us review briefly the normal functioning of the human female reproductive system.

The sex hormones

Perhaps nowhere in nature can there be found more elegant manifestations of life processes than in the role of the sex hormones. The male and female gonads produce three principal sex hormones that belong to a class of chemicals containing multiring structures called *steroids* (see Figure 17.2). Each of these steroid hormones has a specific function in sex physiology.

Testosterone, produced by the testicles, is the chief male hormone. *Estradiol,* only slightly different in chemical composition and structure from testosterone, is produced by the ovaries and is called the ovarian hormone.[4] It is one of three female hormones called *estrogens* and is the most potent product secreted by the ovaries. *Progesterone,* only slightly different chemically from testosterone and estradiol, is also produced primarily by the ovaries and is especially abundant during pregnancy. Substances that have effects similar to progesterone are called *progestins,* and it is these which are of particular interest in controlling fertility. We will refer to progestins again, but first let us consider the production of the ova.

The ova (singular, ovum) are produced by the ovaries in structures called *follicles*. About every 28 days in humans an ovum matures, usually in alternate ovaries. The ovum bursts through the wall of the follicle into the body cavity and finds its way via the ovarian tubule into the uterus. There, if it is fertilized, the ovum becomes imbedded about 7 days later in the endometrium (inner lining) of the uterus where it will develop into an embryo. If this happens, profound changes take place in the secretory behavior of the ovary and in the physiology of the pregnant female.

---

[4]For the effects of estradiol on female development, see Chapter 10.

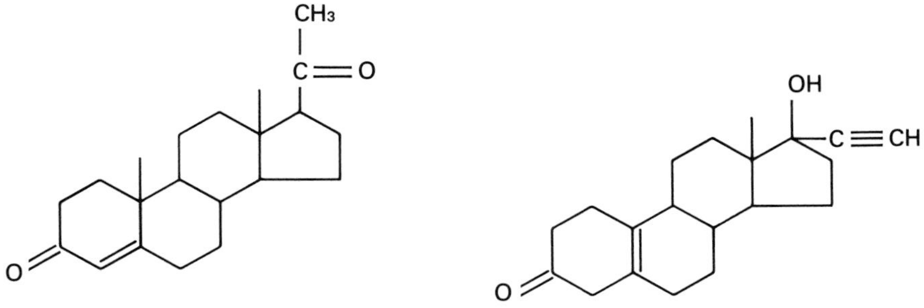

progesterone
(produced by ovaries,
especially during early pregnancy)

norethynodrel
(a synthetic progestin)

FIGURE 17.2 Minor changes in the molecules of steroid sex hormones, remarkable in their simplicity, produce dramatically different effects. Many such changes are made by nature; some can be made by man in the laboratory.

If no pregnancy occurs, the follicle, having fulfilled its function, disintegrates. But in pregnancy, the follicle, instead of disappearing, becomes swollen and reorganized into secretory tissue called the *corpus luteum*. Whereas before rupture the follicle produced large amounts of estradiol, the primary function of the corpus luteum is now to manufacture progesterone. Immediately following implantation in the uterus, the peripheral cells of the embryo start secreting hormones that stimulate the corpus luteum to sustain its activity, and it continues to do so for about 5 months into pregnancy. After that, the corpus luteum degenerates, and from then on the function of producing progesterone as well as estradiol is taken up by the placenta.

The follicle actually begins secreting progesterone just before release of the ovum, but the corpus luteum is especially active in progesterone secretion during the last half of the menstrual cycle. It is the abrupt termination in the production of progesterone by the follicle (the disintegrating corpus luteum) at the end of the cycle that is mainly responsible for initiating menstruation.

Prevention of ovulation   All present-day oral contraceptives act by preventing ovulation, that is, by preventing the rupturing of the follicle and release of the ovum. How this takes place is shown in Figure 17.3. The growth of the follicles is promoted by a hormone called *FSH* (follicle stimulating hormone). The swelling and bursting of the follicles in the release of the ova, termed *ovulation,* is stimulated by another hormone called *LH* (luteinizing hormone). Both of these hormones are released by the anterior lobe (adenohypophysis) of the *pituitary,* a small but mighty gland of internal secretion located just beneath the brain.

The anterior pituitary is stimulated to secrete FSH and LH by a hormone classed as a *neurohumor* emanating from a structure forming a part of the base of the brain called the *hypothalamus*. Thus, a chain of command is set up in which the hypothalamus issues an order for the production of follicles and eventual ovulation. But it cannot do so directly. It must go through channels. It signals, by means of its neurohumor, the gonadotropic release factor G, that the anterior pituitary should take over

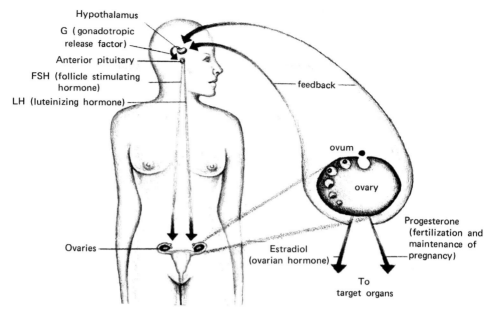

Hypothalamus

G (gonadotropic
release factor)

Anterior pituitary

FSH (follicle stimulating
hormone)

LH (luteinizing hormone)

feedback

ovum

ovary

Progesterone
(fertilization and
maintenance of
pregnancy)

Ovaries

Estradiol
(ovarian hormone)

To
target organs

Modified from G. Pincus, *Advances in Chemistry Series 45*, 1964, American Chemical Society.

FIGURE 17.3 Ovulation in the human female.

the task. Upon receiving the signal via the neurohumor, the anterior pituitary goes to work producing FSH and LH, which in turn communicate the instructions to the ovary.

To prevent the hypothalamus from getting carried away by its dominating influence on the functions of the ovary, nature has provided for built-in restraints called feedback mechanisms. The production of estradiol and progesterone by the ovary itself serves to put the brakes on the hypothalamus to prevent its ordering an excess of ovarian activity. The effect of progesterone on LH production is especially important in this feedback. The increase in progesterone production at and immediately after ovulation, through its action on the hypothalamus, inhibits the formation of LH and stops further ovulation.

The feedback action of the ovarian hormone, and especially that of progesterone in stopping ovulation, suggested that they might be used to control fertility. But there was an important obstacle. Estradiol and progesterone are not highly effective when administered orally. A major breakthrough was the discovery that effective compounds could be produced artificially by making small changes in chemical structure. These changes apparently reduced their destruction by the liver. A slight modification of the progesterone structure gave *norethynodrel,* the principal ingredient in one of the first of the commercially available oral contraceptives, Enovid.

Some of the first oral progestins to be used for inhibiting ovulation had some inherent estrogenic activity, and it was later found that preparations were contaminated with estrogen. This was a lucky accident because it was found that when estrogen was present it enhanced the effectiveness of the progestin. The most likely estrogenic contaminant, later named *mestranol,* was used in small amounts with norethynodrel.

Side effects   The control of growth, sexual maturity, and reproduction are controlled by the actions and interactions of a large number of hormones, many of them not completely understood. It is not surprising that a system so intricate and so delicately balanced can lash back in unexpected ways when one or more of the chemical messengers of the body are augmented or mimicked in a way that puts in motion a chain of events contrary to the normal physiological cycle.

The most frequent side effects from the oral administration of progestin-estrogen contraceptives have been nausea, vomiting, dizziness, headache, hypertension, and discomfort in the breasts. They are more common in the first cycle of medication than later. There were some reports of a tendency to gain weight, an increase in physical vigor, and an increase or decrease in libido. The breasts may increase in size if there is already such tendency. Premature menstrual bleeding, or so-called "break-through bleeding," has varied and seems to be related to the formulation, the dose, and the frequency of use. Prolonged use has occasionally interfered with liver function. There have been no effects on subsequent fertility except possibly for a slight increase. There is no evidence of an increased incidence of cancer either of the breast or the cervix.

The most serious complication is *thrombophlebitis,* a clotting of the blood in the presence of inflammatory changes in the walls of the veins. Because the danger of pulmonary embolism (obstruction of a lung artery) was a cause for alarm, several studies have been made of the potential danger. Both British and United States research workers found that the risk to those on the pill is about 4.4 times that of nonusers. Studies conducted under the sponsorship of the FDA indicated that the greatest risk of thromboembolism[5] was from oral contraceptives containing relatively higher doses of estrogen. On the basis of these findings, the FDA in 1970 advised physicians to prescribe oral contraceptives with the lowest estrogen content.

The so-called "morning-after pill" is essentially DES (diethylstilbestrol), a synthetic estrogen that had the unusual effect of causing vaginal cancer in the daughters of women who were administered the drug regularly when pregnant. DES is discussed more fully in Food Contaminants, Chapter 10.

In 1955, before the "pill" was known, Julian Huxley contended that if one-tenth of the money and scientific effort that was being devoted to the release of atomic energy were devoted to finding a cheap and harmless contraceptive, we would have the answer within ten years. Huxley was not far off. We have had the "pill" for several years, although its potential side effects are too great for it to be recommended without a physician's prescription.

At least 20 million women in the United States use oral contraceptives, and the number is increasing. Undesirable side effects and the legal re-

---

[5]Thromboembolism is the obstruction of a blood vessel with a clot that has broken loose from its site of formation.

quirement for a physician's prescription appear to be minor limitations. But a major obstacle to worldwide adoption of the "pill" is its cost. Control of worldwide population growth awaits a method as effective and convenient as the "pill" but cheap enough for all who want it. Though not ideal, oral contraception may point the way to a cheap, safe, and effective birth control agent of universal availability.

ABORTION

Abortion is one of the oldest, and certainly one of the most controversial, methods of birth control. Except for sexual restraint, it is also the most effective. Abortion has historically been associated with moral censure, and the public disapproval of it included rigid legal restrictions. But despite the difficulties, a flourishing trade in illegal abortions persisted, often conducted by improperly trained practitioners and under crude and unsanitary conditions, making it dangerous as well as expensive.

Beginning in the late 1950s and early 1960s, probably coinciding with a growing relaxation in the traditional inhibitions against sexual behavior, the pressures of population congestion felt in parts of western Europe and Japan led to the disappearance of much of the social objection to abortion. There followed widespread reevaluation of the disruption of personal lives and long-term social effects of having unwanted children.

In the United States, laws making it easier to obtain abortions were first passed in 1967 in California, Colorado, and North Carolina. By 1970, fifteen states had adopted similar laws. During that year, about 200,000 women obtained legal abortions in the United States. In 1971, the number had doubled to about 500,000 and by 1972 had reached 600,000. It was estimated that in 1971 between two-thirds and three-fourths of all legal abortions replaced what otherwise would have been illegal abortions. Because about two-thirds of the abortion patients were unmarried— single, divorced, separated, or widowed—there was a reversal in the upward trend of illegitimate births. Many of the women traveled outside the states of their residence to seek legal abortions in states or countries where abortion laws had been liberalized. But in January, 1973, the United States Supreme Court handed down a 7-2 ruling that, for practical purposes, removed all legal barriers to abortion and made it unnecessary for women to go outside their states to obtain legal abortions.

The Supreme Court said that states could not place restrictions on abortion during the first three months of pregnancy, called the *first trimester*. The Court noted that medical techniques have improved so much that an early abortion "is now relatively safe." During the second trimester, or between three and six months of pregnancy, when the dangers of medical complications are greater, the justices said that the states would be justified in imposing controls that are "reasonably related to the preservation and protection of maternal health," such as requiring that the abortion be performed by a physician or in a licensed hospital. Once the fetus becomes viable—capable of surviving outside the womb, usually considered to be between the twenty-fourth and twenty-eighth weeks or about six months of pregnancy—the states may *prohibit* abortion as long as they leave open the possibility of terminating the pregnancy when it is necessary to preserve the mother's life or health. Thus, a woman and her physician legally can decide whether a pregnancy should be aborted if there are no state or local laws involved and, in any case, women have an unrestricted constitutional right to an abortion during the first three months of pregnancy.

Abortion now is said to be the most common legal surgical procedure, except for tonsillectomy. It was estimated that there were at least 750,000 legal abortions in the United States in 1973 and 900,000 in 1974.

PROBLEM AREAS

The birth rate has declined sharply in recent years in the United States and in several other of the more highly developed countries. The *fertility rate*—the estimated number of children a woman will have during her lifetime—is now less than an average of 2 in the United States. One step toward the goal of a nongrowing population is a *replacement level*. This is not solely dependent on the fertility rate. If the death rate is high, the replacement level will have to be at a higher fertility rate than if the death rate is low. For example, the mortality rate in parts of West Africa during the late 1960s was such that women had to have an average of 3.5 children in order to replace the parent generation. In the United States where the death rate is lower, only 2.1 children are needed to maintain fertility at the replacement level. Since the fertility rate in the United States is under 2.1, at least temporarily, the fertility of Americans is now below the replacement level.

A fertility rate at the replacement level does not necessarily mean zero population growth, because population growth is also dependent on the age distribution. A large proportion of young people—present and future parents—can be expected to give birth to a larger number of children than a population consisting largely of old people. In the United States where fertility is below the replacement level, there are about 3 million babies born every year, so subtracting deaths, the population is growing by about one million people a year. As long as there are more babies being born than there are people dying, the population will continue to grow at a rate depending on the slowly changing age distribution, assuming no migration. Thus, the achievement of a stable population in the United States—zero population growth—will require a further reduction in fertility below the present level in order to accomplish the objective within a reasonable period of time.

On a global basis, the problem of effective birth control is formidable. The world fertility rate is almost 5 children per woman. This would have to be cut in half to bring it down to a replacement level, assuming no change in the death rate. Improvements in sanitation, medical care, and nutrition would require a further reduction in fertility to compensate for the reduction in deaths. On the other hand, there is a clear correlation between income and the practice of birth control; so the faster the economic well-being of people in poorer countries can be improved, the faster the replacement level can be reached. Still, in view of the certain lag between reducing fertility and achieving zero growth, there is practically no possibility that zero growth can be attained before at least 2 billion more people are added to the world's population, if then.

There are sociological reasons why effective birth control practices have not caught on in regions of high fertility. Personal income and economic well-being are important, but these also appear to be related to other factors such as education, the frequency and reliability of communication with sources of information, and the availability of medical and nursing services. Religious and moral convictions are influential in all strata of society, while suspicions of ethnic implications and motives have strangely emerged on both sides of the birth control argument. In poor countries where subsistence is apt to be from hand to mouth, and infant mortality is high, a large family is considered to be the only security parents will have in their old age. This is their form of Social Security. It is not easy to argue against.

One of the unforeseen problems related to birth control is the staggering increase in veneral diseases, no doubt stemming from a combination of causes, but believed to be at least partly due to the popularity of the pill and what has been called the copulation explosion. The condom is one of the most effective prophylactic agents against venereal diseases, but because of the inconvenience and a degree of unreliability, its replacement by oral contraceptives removes another of the restraints against sexual activity—fear of pregnancy. Syphilis and gonorrhea can be cured with antibiotics, though the latter disease has developed stubbornly resistant strains of the bacterial pathogen, and treatment has become increasingly more difficult. A new venereal disease has ominously appeared, caused by a virus called HSV-2 (Herpes Simplex Virus Type 2), for which no satisfactory treatment has been found.

Abortion is increasing in popularity and availability, rapidly in some countries and slowly in other parts of the world. In the United States, the Supreme Court Justices deliberated for two years on the abortion issue. While their decision answered the legal questions and solved some of the social problems, it did not alleviate all of the social tensions that surround the abortion controversy. Many people think that abortion is an act of aggression and is morally wrong. On the other hand, those who favor abortion say that the evils of having children who are unwanted and perhaps unloved should receive first consideration. The two views are not apt to be reconciled easily.

READINGS

HARDIN, G., *Birth Control*. New York, Pegasus, 1970.

HARDIN, G., ed., *Population, Evolution, and Birth Control,* 2nd ed. San Francisco, W. H. Freeman and Company, 1969.

KISTNER, R. W., *The Pill*. Pinebrook, N. J., Dell Books, 1968.

MONTAGU, A., *Life Before Birth*. New York, The New American Library, 1964.

NOONAN, J. T., JR., *Contraception*. Cambridge, Mass., Harvard Univ. Press, 1965. (Reprinted 1967, The New American Library, New York.)

SCHROGIE, J. J., *Oral Contraceptives: A Status Report*. FDA Papers, May 1970, p. 23.

SKLAR, J., and B. Berkov, "Abortion, Illegitimacy, and the American Birth Rate," *Science* 185: 909–915 (1974).

"World Population Projections: Alternate Paths to Zero Growth," *Population Bulletin* 29(5): 1–3 (1974). (A Summary of: *The Future of Population Growth: Alternate Paths to Equilibrium* by Tomas Frejka, John Wiley & Sons, 1973.

WYNNE-EDWARDS, V. C., "Population Control in Animals," *Scientific American* 211(2): 68–74 (Aug. 1964).

Upper corner and middle photos courtesy of USDA; bottom corner photo courtesy of
New York Convention and Visitors Bureau.

PART SIX

SOCIETY AND THE ENVIRONMENT

THE "DOMINION" OF MAN:
ENERGY, THE SUN, AND THE ATOM

Practically every convenience of modern life requires energy. Every year in the United States we use up 133 million tons of steel and 224 million pounds of aluminum. We dig out 913 million tons of sand and gravel for construction of roads, bridges, and buildings. We wheel down the concrete and asphalt highways in 96 million passenger cars propelled by 70 billion gallons of gasoline per year. We support 22 million trucks, which burn up 27 billion gallons of fuel and carry each year, along with the railroads, 654 million tons of cross-country freight. We eat 24 billion pounds of beef, 70 billion eggs, and 29 billion pounds of potatoes. We drink 126 million gallons of whiskey, 140 million barrels of beer, and smoke 593 billion cigarettes. We use 175 million refrigerators and run 150 million washing machines. We use up 13.7 million bales of cotton and 7.3 billion pounds of synthetic fibers in order to wear out, among other fine things, 184 million pairs of men's pants and 81 million pairs of pantyhose. These all have one thing in common—they use up energy.

Where does the energy supply come from? Petroleum and natural gas liquids supply 46 percent, natural gas 31 percent, coal 18 percent, hydroelectric power 4 percent, and nuclear fuel 1 percent. Thus, oil and natural gas furnish 77 percent of the energy, and the combined total of oil, gas, and coal—the fossil fuels—account for 95 percent of the nation's energy. What does this mean for the average American family? In a typical day, they use directly or indirectly 9 gallons of oil, 7 gallons of natural gas, 46 pounds of coal, 1 gallon of hydro-power, and ½ pint of nuclear energy. (See Figures 18.1 and 18.2). In a year's time the average American uses an amount of energy equivalent to that in 10 to 15 tons of coal.[1] (See Table 18.1.)

FOSSIL FUELS

About 95 percent of the energy used in the United States is biological in origin. Oil, coal, natural gas, oil shale, peat, and firewood are the

---

[1] Equal to 390 million Btu (British Thermal Units). A Btu is the amount of heat needed to raise the temperature of one pound of water one degree Fahrenheit.

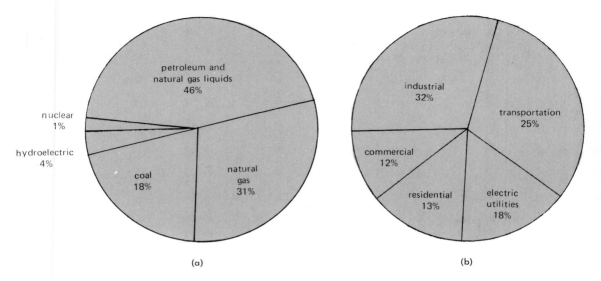

(a)   (b)

Redrawn from *Chemical and Engineering News,* November 19, 1973, pp. 14–15; *Chemical and Engineering News,* March 1974, p. 23; *Energy Index,* December 1973.

FIGURE 18.1   Energy—(a) where it comes from and (b) where it goes.

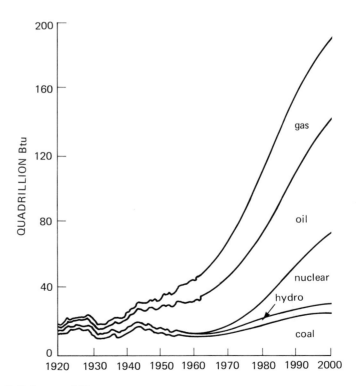

Source: U. S. Bureau of Mines.

FIGURE 18.2 The demand for total energy of all kinds in the United States will more than double in the next 20 years.

TABLE 18.1   Energy Consumption per Capita (1971 data)

Country	Kg. Per Capita (Coal Equivalent)	Energy Use as Multiple of Nigerian Use
United States	11,244	191
United Kingdom	5,507	93
West Germany	5,223	89
USSR	4,535	77
France	3,928	67
Japan	3,267	55
Italy	2,682	45
Mexico	1,270	22
China	561	10
Brazil	500	8
Philippines	298	5
India	186	3
Indonesia	123	2
Pakistan and Bangladesh	96	2
Nigeria	59	1
World average	1,927	33

From *RF Illustrated* 1(2). August 1974. Rockefeller Foundation.

remains of living material—organic substances that came from the carbon-containing compounds of living organisms, directly or indirectly the products of photosynthesis. The so-called fossil fuels—oil, natural gas, and coal—are the fossil remains of prehistoric organisms that settled to the bottom of ancient seas and became modified, probably first by bacterial action, then by pressure and temperature caused by the sediment of sand, mud, and rock deposited over millions of years.

While no one knows with certainty how the fossil fuels were formed, chemical analyses confirm the presence of substances called porphyrins, organic material that is clearly linked to the chlorophyll of plants; and recent studies also show the presence of steroid acids, more probably of animal origin. In the laboratory, bacteria can change matter of organic origin into material that resembles petroleum. It is believed that petroleum is derived primarily from plants and to a smaller extent from microscopic animal life. Some scientists believe that there is a component of nonbiological origin as well, derived from carbides deep in the earth, but this view does not have strong support. Thus, the overwhelming evidence is that fossil fuels, and therefore most of our power, derived their energy originally from the sun, and their key raw material, carbon, from the carbon dioxide of the atmosphere.

PETROLEUM

More than three-fourths of the energy consumed in the United States now comes from oil and natural gas. This may seem surprising in view of the estimate that nearly 90 percent of all fossil fuel reserves are in the form of coal. But there are several advantages to oil and gas. Liquids are easier to handle and transport than solids, especially for use in mobile equipment

Photo courtesy of New York Convention and Visitors Bureau.

Cities use large amounts of energy day and night. This is the New York midtown skyline with the United Nation's 39-story Secretariat Building.

such as cars, trucks, tractors, and locomotives. Additional conveniences of natural gas for stationary equipment and heating brought this fuel into prominence. And finally, there is a massive body of petroleum technology that has led to the use of petrochemicals in the manufacture of a vast number of industrial and consumer products, from Scotch tape, plastic teacups, and spectacle frames to wallboard, automobile bodies, and boats.

The rapidly expanding economy throughout the world demands and consumes enormous amounts of energy from petroleum. In the United States, about 75 percent of all the energy used comes from oil and natural gas. A domestic reserve ratio, at one time 2 to 1, slipped to 1 to 1 and is now apparently in an unfavorable inverse ratio. Concurrently, there has been increasing dependence on foreign crude oil, so that now 37 out of every 100 barrels of oil used in the United States are imported from foreign sources—Canada, Latin America, the Middle East, Africa, and Indonesia.

Estimates of the U.S. Geological Survey, based on a study completed in 1975, placed the nation's offshore oil resources at 10 to 49 billion barrels, and the onshore reserves at 37 to 81 billion barrels. In the same study, the Geological Survey estimated that 322 to 655 trillion cubic feet of natural gas remained to be discovered. These estimates are predictions of highly unpredictable events; some individuals and agencies have made much higher estimates—as much as 130 billion barrels of oil and 800 trillion cubic feet of gas. The Prudhoe Bay field and vicinity, on Alaska's North Slope, is the largest ever discovered in North America. Development of these sources are part of a program to become independent of foreign sources that are susceptible to volatile political conditions. A similar effort, and for similar reasons, is underway on the discoveries in the North Sea.

COAL

Coal has gradually increased its dominance as a fuel for production of electric power in the United States. In 1973, coal supplied 54 percent of the electric power market, partly the result of a decrease in the availability of natural gas and fuel oil, though petroleum continued to be the major overall source of energy.

Domestic coal reserves are enormous. The U.S. Geological Survey puts the coal reserves in the United States at 3.2 trillion tons, enough to last 5,000 years at the present rate of consumption if it could all be used. But not all of it is available. According to an estimate by George Hill of the Electric Power Research Institute, about 62 percent of it is potentially recoverable. If this were converted to oil at the rate of 2 barrels per ton, the recoverable coal in the country would be equivalent to more than 10 times the known worldwide oil reserves. Thus, coal can contribute more to the nation's future energy needs than in the past—from about 600 million tons a year used now to possibly double that rate within a few years.

However, in spite of the favorable supply of coal reserves, a combination of several problems will delay for an indefinite time the full use of coal to ease dependence on petroleum. Coal has several disadvantages: it is more difficult to handle and transport than oil, its use creates air pollution problems that are costly to prevent, and the mining of coal is apt to be so destructive to the environment that thorough studies must be made in each case before mining operations should be permitted. There are other problems as well. Mining technology will have to be improved both for economic reasons and to comply with the Coal Mine Health and Safety Act of 1969. Probably most important, if the coal is to be used efficiently, much of it will have to be converted to synthetic fuel at the mines. Such plants are as dependent on water as on coal, so in water-deficient western states, massive government programs to provide the necessary water will have to be started.

Strip mining   The effect of coal mining is often more than just subterranean removal of the coal. When deposits are near the surface, they can be mined by simply removing the overburden of earth and taking the coal out with surface equipment, a procedure widely known as *strip mining*. Strip mining for coal in the eastern United States has gouged out and disturbed the surface of 1.3 million acres, or nearly 2,000 square miles of land. Deposits of near-surface coal beds in the western United States are of greater average thickness than in the east, and the deposits are more extensive. There are 128 million acres of coal and lignite in the West, though only about 1.5 million acres, or 2,340 square miles, of this are believed to be suitable for stripping.

The huge tonnage of strippable coal poses problems that may turn out to be more costly in terms of environmental rehabilitation than originally thought. Many of the coal beds underlie arid or semiarid land, and care-

USDA-SCS photo by Wayne R. Grube.

Strip-mined area in Pennsylvania.

less stripping could severely damage supplies of native surface water and ground water that might be difficult or, in some cases, impossible to replace. Water, in fact, may turn out to be the most serious problem because the presence of moisture is critical in the establishment and growth of plants on arid land. Ponderosa pine and mixed grass prairie areas that receive more than 10 inches of rainfall a year—about 60 percent of the strippable western coal land—have a good chance of recovering if given the best possible management. But if the area receives less than 10 inches of rain a year, or if the evaporation rate is high, such as in deserts and foothill shrub areas, the chance of recovery is extremely low. Even with careful management, stripped desert lands may take centuries to recover. A panel of the National Academy of Sciences, referring to the aesthetic effects of stripping, said that strip mining of desert lands "amounts to sacrificing such values permanently for an economic reward."

The Four Corners area is an example of the magnitude of the rehabilitation problem. The Four Corners is a picturesque area near the point where the borders of Utah, Arizona, Colorado, and New Mexico come together. It covers about 11 percent of the western coal land, and the extensive coal beds there are being vigorously exploited to provide fuel for several large power plants in the area. The blue-sky country of the Four Corners has already turned brown from air pollution—the dark plumes of smoke from the power plant stacks could even be seen from space by the Apollo astronauts. But air pollution is only one of the problems. The National Academy of Sciences panel could see only two alternatives to costly and long-range rehabilitation: either "nondevelopment" of the resources or simply writing off the ravaged deserts as "National Sacrifice

Areas." For further discussion of the rehabilitation of stripped land see *Problem Areas*.

Gasification   One way to alleviate the air pollution problem of burning coal is by a process known as *coal gasification,* by which a synthetic fuel is produced comparable to natural gas. Several processes for using high-sulfur coal are under study. In one of them, the coal comes in contact with hot steam, and the carbon in the coal combines with the hydrogen in the steam to form a compound that is very similar to the hydrocarbons in natural gas. Because natural gas is in short supply, and is apt to become even scarcer, a great deal of research is being directed toward gasification technology. Underground gasification is an attractive possibility, especially in deep coal seams of the West where conventional mining is hazardous or unfeasible. The possibility of producing gas from oil shale is also being studied. but the problem is, simply, that enormous amounts of material must be removed for processing. Another scheme for improving the use of coal is by a process called *liquifaction,* which produces a product much like crude petroleum.

NATURAL GAS

The first use of a combustible gas on record was by the Chinese about 900 A.D., when they piped natural gas through bamboo tubes and used it for lighting. Gas was first produced from coal in England in 1665 and used for lighting in 1792. People soon recognized the commerical possibilities, and gas companies were organized to manufacture gas from coal and coke long before exploitation of the natural gas fields took place.

Photo courtesy of Texas Eastern Transmission Corporation.

Pipeline network for natural gas makes for ease of transport for this form of energy. Other advantages are clean burning, low pollution, and controllable heat. Disadvantages are cost and limited supply.

Natural gas has been called the Cinderella fuel—it was once dismissed as a nuisance by crude oil producers, and billions of cubic feet have been wasted by burning it off in the oil fields. But the advantages of natural gas over other forms of fuel stimulated the building of a network of high-pressure pipelines that made it possible to distribute the fuel at reasonable cost from the gas fields to major parts of the country. The enforcement of air pollution control regulations has in many areas restricted the use of high-sulfur coal and fuel oils and forced the partial substitution of natural gas in their place. Gas is now preferred by both residential and commercial users because it is convenient, clean-burning, efficient, and reasonably priced. Natural gas is favored by industries that must control heat within narrow margins because it generates heat accurately and can be "fine tuned." Gourmet chefs prefer it to other forms of fuel because it permits instant control of higher or lower heat.

The ability of natural gas to meet air pollution control standards without costly treatment or alteration, such as desulfurization, has greatly increased the interest in gas as a clean-burning fuel. The demand for natural gas, which already furnishes close to a third of the nation's energy supply, has been growing at the rate of 5.9 percent a year for the last ten years. It is predicted that the potential gas demand will outstrip the supply in the immediate future. One solution, besides expanded exploration for new sources, is to produce gas from coal. This will be a full circle return to the days when "coal gas" was manufactured, but this time it is hoped that improved technology will make it possible to produce gas of high quality (by gasification) with the same clean-burning and heating value as natural gas.

HYDROELECTRIC POWER

The cleanest energy readily available in large quantity is hydroelectric power, a form of energy that now supplies about 4 percent of the United States energy needs. While further development of hydroelectric power in the United States is limited, large resources remain to be developed in other parts of the world. A huge power project involving a complex of power plants in Quebec in the region of James Bay will generate 16 million kilowatts, the largest hydroelectric project in North America. Quebec already generates 97 percent of its electricity from water power, second only to Norway. In South America, on the Parana River, Brazil and Paraguay are underway on a project that will generate 10 million kilowatts, the largest single hydroelectric plant in the world.

The development of hydropower sometimes requires the construction of dams that can cause extensive destruction of other valuable natural resources. In such cases the impoundment of water and the inundation of natural areas is of questionable wisdom unless it is absolutely essential for flood control or conservation of critically short water supplies. One of the most controversial cases of this kind was the Glen Canyon Dam on the Colorado River near where it crosses the border between Utah and Arizona, and where its turbulent rapids have cut deep gashes in the colorful sandstone formations laid down by the ancient sea that once covered the region.

The area behind the dam includes some of the most spectacular sce-

Left photo: The Glen Canyon Dam on the Colorado River created Lake Powell, a reservoir about 186 miles long, and was designed to generate enough electricity for 1½ million people. However, many of the flooded canyons were of great scenic beauty. *Right photo:* When and if Lake Powell is full, water will flood the ravine at the base of Rainbow Bridge.

nery in the world, including the Rainbow Bridge National Monument. If the water in Lake Powell, as the body of water formed by the dam is called, reaches its maximum level, the water would back up as far as the ravine under the base of Rainbow Bridge and would come within 25 feet of its base. Whether water lapping at its base would eventually undermine the structure is debatable, but everyone agrees that the fluctuating level of the lake would leave "bathtub rings" and unsightly mudflats when the water in the lake is down. Even more important is the imminent probability that some of the most unusual, untouched landscape on the North American continent would disappear when the flooding Lake Powell backs up into the proposed Escalante Wilderness, the Dark Canyon Primitive Area, some of the rapids in Cataract Canyon, and dozens of other magnificent national treasures.

A court decision in 1973 limited the level of Lake Powell to about 100 feet below its maximum designed capacity. Since half the lake's storage capacity is in the top 100 feet, it is argued that if the court decision is made to stand, it would be disastrous to water and power users. But opponents of the dam insist that these effects would not be as serious as claimed because other means are available for correcting them. They point out, for example, that because the electric-generating capacity of the power plant is not high compared to coal-fired fossil fuel plants nearby—one of them within sight of the dam area—the electric power provided by the dam is not crucial.

This is a classic example of the trade-off that must take place whenever there is exploitation of a natural resource. At Glen Canyon Dam, the benefits in terms of electric power and water must be balanced against the possibility that there will be an irretrievable loss of a natural resource. Anyone who has ever seen this magnificent section of the Colorado River canyon before its submersion would surely and sadly agree that the electricity and water control obtained at this source is being bought at an exceedingly high price.

NUCLEAR ENERGY

The United States has slightly more than 5 percent of the world's population but consumes more than one-third of the world's electric power. Moreover, the demand for industrial and domestic power is increasing at a rate even greater than the population growth. It is estimated that the United States will need more electric power during the next twenty years than in its entire previous history and that the demand will then double in the following twenty years (see Figure 18.3). In contrast to fossil fuels, reserves of atomic fuels are comparatively large. With present technology, atomic fuel reserves are at least ten times as great as fossil fuels and would provide almost limitless energy if safe breeder reactors could be successfully developed. (A breeder reactor is one in which more fissionable material is produced than is consumed.)

Atomic, or nuclear, power is not a direct source of energy for consumer use; it is a means of producing electricity. An enormous amount of electric power can be produced by a samll amount of atomic fuel. On a weight for weight basis, uranium contains 3 million times as much poten-

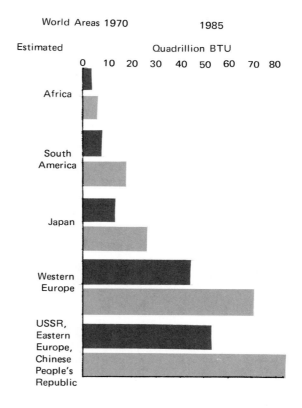

Redrawn from the USDI Conservation Yearbook Series No. 8, 1972.

FIGURE 18.3   As the United States grows, more consumers will demand more energy per person, while the rest of the world is also demanding more energy.

Photo Courtesy of the University of Chicago.

"Nuclear Energy," sculpture by Henry Moore.

tial energy as coal. One gram ($^1/_{28}$ ounce) of fissionable material releases 23,000 kilowatt hours of heat. A ton of uranium would provide as much energy as 3 million tons of coal or 12 million barrels of oil. In practice, however, only a small fraction of the potential energy in atomic fuel is utilized during one cycle in a nuclear reactor. Much of the energy is lost in the form of heat and waste products. Also, after each cycle, the fuel must be purified before it can be used further.

NUCLEAR POWER PLANTS

Nuclear power plants serve the same purpose as conventional power plants—the generation of electricity. They differ basically only in the source of energy that is used to produce heat. Conventional steam-electric power plants use a fossil fuel such as coal, oil, or natural gas. The fuel is burned in a boiler that produces steam which, in turn, drives a steam turbine generator, called a *turbogenerator*. In a nuclear power plant, the heat is produced in a device called a *nuclear reactor* instead of a boiler. (See Figure 18.4.)

The atom contains two kinds of energy. *Chemical* energy of the atom depends on the behavior of the electrons, and it is this energy that is released when a carbon-containing material such as coal is burned. *Nu-*

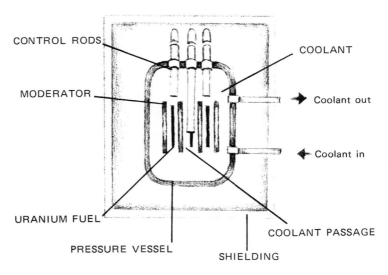

CONTROL RODS

COOLANT

Coolant out

MODERATOR

Coolant in

URANIUM FUEL

COOLANT PASSAGE

PRESSURE VESSEL

SHIELDING

Redrawn from USAEC illustration.

FIGURE 18.4 A reactor is basically an atomic furnace. This simplified diagram shows the basic components of a reactor.

clear energy is dependent on the behavior of the particles that make up the nucleus of the atom, and this energy is released when a radioactive, or fissionable, element decays. The breakdown of the nucleus of a material such as uranium-235 or plutonium-239 will, under suitable conditions, release a great deal of energy, much of it in the form of heat. Such materials are useful as fuel in a nuclear reactor.

Atomic fuel   Today's nuclear electric generating plants use pellets of uranium metal as fuel. In a typical fuel reactor, the uranium pellets are enclosed in vertical thin-walled tubes called fuel rods, constructed of a corrosion-resistant metal, in some cases stainless steel, but usually an alloy of zirconium. These metal tubes are referred to as *cladding*. The tubes are about one-half inch or less in diameter and may be ten feet or more in length. A typical reactor with a capacity of 500 MW (million watts or megawatts) will contain more than 20,000 such fuel rods. The decay of uranium in the fuel rods produces a large amount of heat as a product of nuclear fission. The heat is absorbed by water that surrounds the fuel rods and is removed by rapid circulation.

Types of reactors   There are two main types of reactors, depending on the operating temperature: the *boiling water reactor* (BWR) and the *pressurized water reactor* (PWR) (see Figure 18.5).

In the boiling water reactor the heat from the fuel rods causes the water to boil, producing steam at the top of the reactor vessel. The steam is fed directly to steam turbines which turn the electric generators.

In the pressurized water reactor, the water is under sufficiently high pressure to prevent boiling even at a temperature above the normal boiling point of water. The high-temperature water, still under pressure, leaves the reactor vessel and enters a device called a *heat exchanger,* which contains a separate, secondary water system. The lower pressure

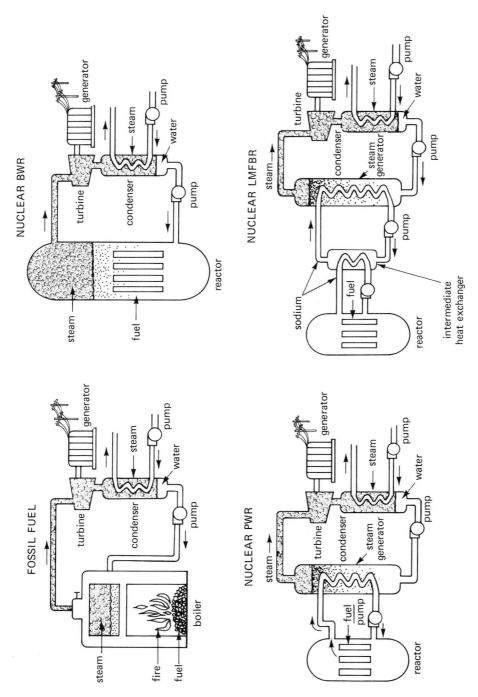

Redrawn from *Thermal Effects and U. S. Nuclear Power Stations*, 1971. U. S. Atomic Energy Commission.

FIGURE 18.5 Types of steam-electric generating plants: typical fossil fuel, nuclear BWR (boiling water reactor), nuclear PWR (pressurized water reactor), and nuclear LMFBR (liquid metal fast breeder reactor), or, simply, metal cooled breeder reactor.

in the secondary water system causes the water to boil, producing steam which is delivered to the steam turbines.[2]

FISSION

The essential ingredient of a reactor fuel is a substance that is fissionable—one that undergoes decay when struck by neutrons. The only readily fissionable material occurring naturally is one of the isotopes of uranium, U-235. This form of uranium comprises only 0.71 percent of the uranium found in nature, nearly all the rest being uranium-238. The U-238 does not undergo fission spontaneously, but it can be made into a fissionable material when bombarded with neutrons, as, for example, from U-235. Under such conditions the atoms of U-238 are changed in a two-step process of decay to plutonium-239, a man-made substance that is fissionable when bombarded by energetic neutrons called *fast neutrons* (see Figure 18.6).[3]

Runaway reactions

It is essential to have a control system to prevent the self-sustaining fission of uranium atoms from getting out of hand. This is done by using substances such as boron and cadmium, which have a high capacity for absorbing neutrons. Usually these materials are in the form of adjustable rods called *control rods,* which can be inserted into the reactor. A typical reactor has two sets of control rods, one called *regulating rods* for routine control of fission, and another called *safety rods* for the purpose of rapid shutdown in case of an emergency.

An atomic bomb in its simplest form contains two or more pieces of nearly pure fissionable material, each in itself lacking sufficient mass to be capable of sustaining a chain reaction. To explode the bomb, the pieces of fissionable material are rapidly brought together to form what is called a *critical mass*. The size of the critical mass can be reduced if the fissionable material is forced together under compression. If the critical mass is held together long enough—actually a very short time, perhaps a millionth of a second—the chain reaction accelerates to the point that a very large explosive force is generated.

The fuel in most reactors could not be made to explode, even if it were in a bomb, because it does not contain enough fissionable material in a small enough space. There is nothing in reactors to hold the fuel together. Intense heat caused by a runaway reaction would melt the fuel or cause it to come apart in some other way. When the fuel disperses, fission slows down. The design of reactors is such that upon dispersion of the fuel the reaction is automatically stopped. Most types have a built-in control such that as the temperature rises, the reaction is retarded.

The most serious hazard from a runaway reaction resulting in a "core meltdown" is the escape of radioactivity. Most nuclear power plants have a large gas-tight containment shell designed to sustain the maximum pres-

---

[2]These two systems are sometimes referred to as light-water reactors (LWR). Modifications in designs include the steam-generating heavy-water reactor (SGHWR), adopted by the British, and the high-temperature gas-cooled reactor (HTGR).

[3]Plutonium-239 is also found in uranium ore concentrate at a concentration of a few parts per trillion.

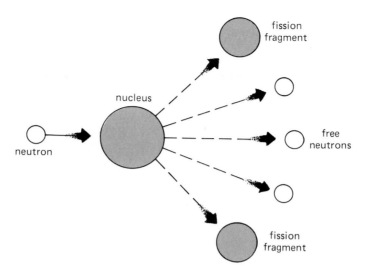

Adapted from R. L. Lyerly, and W. Mitchell III, *Nuclear Power Plants,* 1969 (revised), U. S. Atomic Energy Commission.

FIGURE 18.6   A typical fission reaction. An atom such as that of uranium-235 can undergo fission when a free neutron strikes its heavy central nucleus, splitting it into two pieces (fission fragments) that fly apart at high speed, and releasing two or three additional free neutrons. The released neutrons cause a chain reaction when they strike the nuclei of other uranium-235 atoms; and when the flying fission fragments collide with other atoms, their kinetic energy is converted to usable heat.

sure of gases that might be generated. These shells are the familiar hemispherical structures of atomic power plants. Nevertheless, accidents that might release radioactive materials to surrounding areas—remote though they may be—are of concern to many people. The danger is greatest in an experimental type of power plant called the *breeder reactor*.

In a breeder reactor, the water that would otherwise serve as a moderator by slowing down the neutrons is eliminated. Because there is little material present to impede their progress, the neutrons emitted by U-235 are faster and therefore more energetic. This, in turn, causes the production of more neutrons by the U-235, and more U-235 is converted to plutonium (Pu-239) with a net increase in the production of plutonium over the consumption of U-235. It is anticipated that the Pu-239 thus produced could be used for fuel in reactors designed to use Pu-239 instead of the present uranium-fueled reactors.

The breeder reactor must operate at a temperature that is too high for water to be used as a moderator. Instead, liquid sodium (a metal) is used in what is called a *liquid metal cooled reactor* (see Figure 18.5). Sodium is so reactive that it burns explosively in contact with air or water. Therefore, extremely strong construction and durable materials must be used to contain it. Also, the fissionable material in the breeder reactor is more concentrated and in smaller volume. The concern about breeder reactors is that an equipment failure or an accident might result in an unmoderated chain reaction, causing an extreme increase in temperature. A meltdown of fissionable material in a breeder reactor might cause it to become concentrated in a small enough volume to form a critical mass, culminating in a nuclear explosion. Such an explosion would not be on the scale of an atomic bomb, but could release large amounts of radioactive materials, and in populated areas could cause widespread disaster.

RADIOACTIVE END PRODUCTS

No nuclear power plant is contamination-proof. Leakage from pump seals and other equipment, floor drains, laboratory sinks, and laundry facilities result in liquid wastes that are radioactive. Ordinarily these are collected in liquid waste storage tanks and released to the environment if the radioactivity of the material is low. If the radioactivity is high, the material is ordinarily filtered and demineralized before being released to an outfall, usually to the ocean or a nearby stream.

Liquid effluents may contain radioactive materials in solution and as suspended undissolved substances. Both dissolved and suspended materials entering the environment can eventually be conveyed to humans by contamination of drinking and recreational water supplies, by entering the aquatic food chain, through watering of livestock, or uptake by plants from irrigation water. Dissolved radioactive gases that may be in the waste water will enter the atmosphere. Thus, radioactive wastes in the effluent from atomic power plants can eventually become widely distributed in air, water, soil, plants, animals, and humans.

Atomic power plant stack effluents contain both gases and suspended particulate matter. Particulate matter can be removed by high-efficiency filters, but a problem with the boiling water reactor plant is the lack of provision to remove or treat the high releases of gaseous waste products except by shutting down the plant. Some of the radionuclides in the stack effluent have short half-lives, most of the radioactivity disappearing within minutes or hours; others remain radioactive for years. Some, such as strontium-90, are produced in larger quantities than others. Some become distributed throughout the environment and are transmitted to humans via the food chain. When taken in by humans, certain radionuclides become concentrated in specific organs where they can be injurious to health.

Iodine-131 comprises about 5 percent of the fission products from a nuclear reactor. Radioactive iodine readily contaminates pastures and other plants, causing it to appear in milk and dairy products (see Figure 18.7). In the body, iodine becomes concentrated in the thyroid gland. Damage to the thyroid gland such as that observed to be caused by fallout from atomic testing can be serious, especially among children.

Strontium-90 finds its way into biological pathways by reason of its ready acceptance in biological systems as a substitute for calcium (see Figure 18.8). Contaminated food plants, meat, and dairy products appear to have normal nutritional properties for humans but cause radioactive strontium deposits in the bones and other tissues in place of calcium. Strontium, like radium, belongs to the alkaline earth family of elements, which also includes calcium. The physiological pathways of these nuclides are similar. About 99 percent of the long-term radioactivity from either strontium or radium taken into the human body is found in the

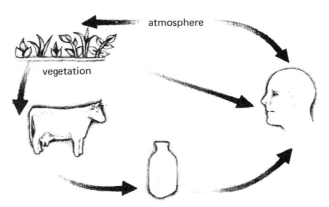

Adapted from C. L. Comar, *Fallout From Nuclear Tests*, 1967, U. S. Atomic Energy Commission.

FIGURE 18.7   Iodine-131 pathways from fallout. Iodine-131 has a half-life of only 8 days but is the radionuclide that causes the greatest radiation exposure within a short time. It is produced in large amounts by nuclear explosions and is transmitted efficiently in the food chain.

bones where damage can occur to bone cells and marrow—blood cell-producing tissue. Because of the long radioactive life of strontium-90 (its half-life is 25 years), the potential for damage is very great.

Fission products   The fission products formed by the splitting of atoms in the nuclear fuel comprise the main source of radioactive wastes, and they are intensely radioactive. Much of the fission waste material is enclosed within the "cladding" or metallic material of the fuel rod tubes which forms the first containment barrier. If the cladding holds perfectly, there is very little escape of the radioactive fission products. However, some of the tubes eventually develop fractures or pinholes even though the cladding material, usually a zirconium alloy, is very resistant to corrosion. When such leaks occur, some of the fission products can escape into the water sur-

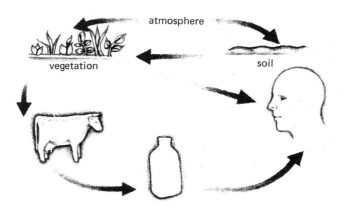

Adapted from C. L. Comar, *Fallout From Nuclear Tests*, 1967, U. S. Atomic Energy Commission.

FIGURE 18.8   Strontium-90 pathways from fallout. Because strontium-90 behaves much like calcium in the body, it reaches man primarily through foods of plant origin and dairy products. Strontium-90 has a half-life of 28 years; strontium-89, also produced by nuclear explosions, has a half-life of 51 days.

rounding the fuel rods. This comprises one of the principal hazards of environmental contamination from the operation of an atomic power plant. To guard against it, the usual practice is to shut down the plant about once a year and replace some of the fuel rods with new ones.

The gaseous wastes must be continuously removed from the primary cooling water of the reactor in order to prevent the accumulation of intolerable levels. In the pressurized water reactors, the gases are held in storage tanks for as long as 30 days or more, during which the radioactivity is reduced to low levels by natural decay of the nuclides. The gases are then released to the atmosphere via the plant stack. In the boiling water reactors, the large amount of gaseous wastes that is produced makes it difficult to hold them in storage for long periods. Only 30 minutes or so are allowed for decay before the waste gases are put through the plant stacks. Consequently a larger amount of radioactivity is released to the environment. Materials that escape include radioactive cesium, hydrogen, and krypton.

Cesium-137 belongs to the family of elements called *alkali metals,* which also includes sodium and potassium. Cesium is the member most like potassium and is physiologically analogous to it (see Figure 18.9). Most of the cesium taken into the body has a biological half-life of about 100 to 135 days, meaning that it takes that long for one-half of the cesium that is taken in to be eliminated from the body. During that time, damage to the body, especially the blood-forming tissues, can occur. Cesium 137 has an environmental half-life of 33 years.

Hydrogen-3, or tritium, sometimes called heavy-heavy hydrogen,[4] is produced in comparatively large quantities in pressurized water reactors. It is present in the reactor wastes as a component of water molecules. Because this radioactive heavy-heavy water is in mixture with the rest of the water molecules, there is no practical method by which it can be separated. Upon its release to the environment, heavy-heavy water undergoes the same environmental and biological pathways as ordinary water.

Tritium emits beta radiation, as does radioactive carbon, strontium, and other radionuclides. The beta radiation from tritium is less energetic than any of the other important reactor waste products. It is less penetrating, and theoretically, less damaging to tissue. Moreover, most of the water taken in is passed through the human system unchanged. Studies thus far show no evidence that tritium becomes concentrated in food chains. Radioactive hydrogen is important, however, because of the large amounts produced and the fact that hydrogen, both in water and as hydrogen nuclei, participates intimately in living processes, for example, photosynthesis in plants. Global contamination of the hydrosphere could have injurious effects of undetermined magnitude on the food chain. The environmental half-life of tritium is 12.3 years. It is theorized that if the tritium from atomic power plants becomes distributed in the oceans to a depth of 40 feet and in the atmosphere to an altitude of 10 kilometers (about 6 miles), the population exposure per person per year will be .002 millirem by the year 2000.

---

[4]Deuterium, or hydrogen-2, is called heavy hydrogen. When water is composed of oxygen and deuterium, it is called heavy water.

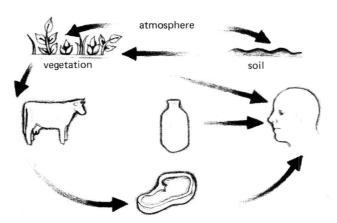

Adapted from C. L. Comar, *Fallout From Nuclear Tests*, 1967, U. S. Atomic Energy Commission.

FIGURE 18.9   Cesium-137 pathways from fallout. Cesium-137 has a half-life of 30 years and behaves in the body much as potassium does.

Another gaseous waste product, krypton-85, is a member of the family of elements called *noble gases*. They were formerly called *inert gases* because they rarely enter into chemical reactions. Although krypton-85 is highly radioactive and has a moderately long half-life (10.76 years), the absence of any known participation in biological processes causes it to be an environmental contaminant of less concern than some of the other nuclear waste products. But because krypton is not chemically reactive and does not combine with other substances, it is long-lasting in the atmosphere; and, further, because of its long half-life, it accumulates in the environment. Remember that a radioactive element damages tissue by the emission of radiations; unlike nonradioactive poisons, it does not have to be involved in a chemical reaction to be injurious. It is estimated that the annual exposure from krypton will be about .02 millirem by the year 2000. Krypton-85 and tritium are released from nuclear reactors in much smaller amounts than from nuclear fuel reprocessing plants.

TOTAL ENVIRONMENTAL POLLUTION

Fallout   Radioactive nuclides are widely distributed in nature (Figure 18.10). But their concentration in the earth's crust and in water is highly variable. Radioactivity in the environment from nuclear testing now comprises about 3 percent of the man-made radiation to which people are exposed. Two of the most dangerous materials in radioactive fallout from weapons testing are strontium-90 (see Figure 18.11) and cesium-137. Because they have long half-lives, 28 and 30 years respectively, these radioactive materials will contaminate the environment for many years. Nuclear testing also produced quantities of tritium. With a half-life of 12 years, significant amounts remain in the biosphere.

The United Nations Scientific Committee of the Effects of Atomic Radiation estimated that residents of the United States during the 1950s and early 1960s—the period of heaviest fallout—will have received a

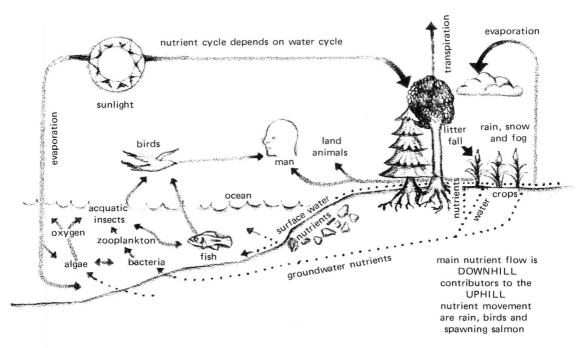

Redrawn from N. O. Hines, *Atoms, Nature and Man*, 1967, U.S. Atomic Energy Commission.

FIGURE 18.10   Movement of radioactive elements in a forest-lake ecological system.

fallout dose of 110 millirems by the year 2000. This is roughly equal to the amount they are already exposed to annually from natural background.

Atomic power pollution   If the plans of the former U.S. Atomic Energy Commission materialize, about 25 percent of the total electricity will come from atomic energy plants by 1980, and between 60 and 70 percent by the year 2000. Inasmuch as the total demand for electric power is expected to nearly triple, the amount of electricity to be derived from atomic energy will be nearly double the total amount that is presently produced by fossil fuel plants. Both long-term and short-term effects of releasing radioactive substances to the environment are considerations of major concern.

Atomic Energy Commission experience indicates that people living near a commercial nuclear reactor would be exposed to radiation in the range of 1 to 10 millirems per year. Regulations were established calling for reasonable efforts to keep radiation releases "as low as practicable." In 1971, it was proposed that maximum allowable nonoccupational exposure from nuclear sources be reduced to 1 percent of the former limit, or 5 millirems per person per year. If this is taken as the legal limit at the boundary of a nuclear power plant, the estimated dose drops off to about .005 millirems at a distance of 50 miles, and the average dose to people within that radius is estimated to be about .01 millirem per year. The average dose to all people in the population of the United States would be much smaller, probably in the order of .001 millirems per person per year.

These exposures compare with that from natural background radiation of about 126 millirems per person per year and medical-dental radiation of about 70 millirems per year. Thus, if the estimates are accurate, the

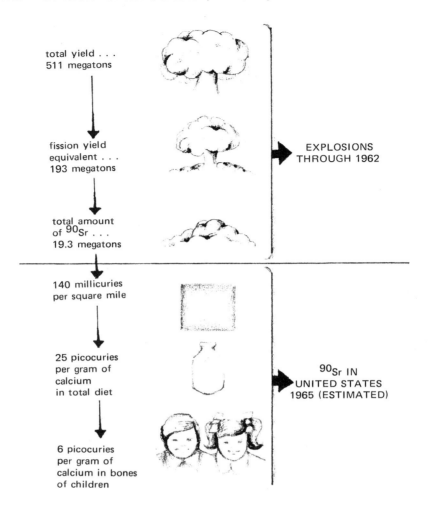

total yield . . .
511 megatons

fission yield
equivalent . . .
193 megatons

total amount
of ^{90}Sr . . .
19.3 megatons

EXPLOSIONS
THROUGH 1962

140 millicuries
per square mile

25 picocuries
per gram of
calcium
in total diet

6 picocuries
per gram of
calcium in bones
of children

^{90}Sr IN
UNITED STATES
1965 (ESTIMATED)

Adapted from C. L. Comar, *Fallout From Nuclear Tests,* 1967, U. S. Atomic Energy Commission.

FIGURE 18.11   Nuclear explosions and strontium-90 levels in the environment. These relationships may vary from time to time with conditions.

maximum radiation exposure from nuclear power sources is about 2.5 percent of that from all other sources, and the *average* exposure of the population as a whole from nuclear power sources is roughly .0005 percent of that from all other sources. On this basis, it was calculated that the additional radiation attributable to power sources will increase the spontaneous mutations and cancer rate among the general population by a fraction of a percent, taking into consideration that background radiation accounts for only a portion of the spontaneous occurrence of mutations and cancer.

The foregoing calculations have been hotly disputed by scientists who have made less favorable estimates of the radiation hazards from nuclear power production. There is great difference of opinion as to the amount of environmental contamination from nuclear power generation, the human exposure rate, and the damage that could occur by reason of mutations and cancer. These questions will probably not be resolved until additional studies have been made of the environmental impact of nuclear power production.

DISPOSAL OF RADIOACTIVE WASTES

Disposal of the huge quantities of radioactive wastes from reprocessing nuclear fuel is of great concern to all mankind. While the wastes accumulate in ever-increasing amounts, the problem remains unsolved.

Spent nuclear fuels are highly radioactive and comprise an enormously difficult disposal problem. Only a small amount of the fuel of a reactor can be used before it becomes "poisoned" by accumulated waste products. The fuel rods must then be removed and sent to a plant for repurifying. The rods are still radioactive, for much valuable fuel material remains along with highly radioactive fission products, activated reactor materials, chemicals, and corrosion products. At the reprocessing plant, the fuel is dissolved in acids and chemically reclaimed. Following purification, the fuel is used again. But huge amounts of radioactive fission products remain for disposal.

The gaseous wastes from fuel reprocessing are released to the atmosphere. The amount of radioactivity thus released to the atmosphere for disposal by dilution and dispersion is very large. A single reprocessing facility may release as much as 12 million curies of krypton-85 and 0.5 million curies of tritium per year. There are practical methods by which such gaseous wastes can be contained and they must be put into effect soon, for it is generally conceded that disposal through the stacks would be unacceptable over a long period of time.

Solid wastes are mostly of low-level radioactivity. Land burial has been used for about two-thirds of the solid wastes produced—about 1.5 million cubic feet per year since 1963—a volume that requires 20 to 25 acres of land annually. It is estimated that 3 million cubic feet of low-level solid waste will be produced annually by 1980.

The highly radioactive liquid wastes are placed in underground storage tanks for long-term decay. The heat generated by such wastes causes them to boil for as long as two years. It is estimated that about 100,000 gallons of high-level wastes were disposed of in 1970 and that the amount will increase to about 6 million gallons per year by the year 2000. About 4 billion gallons of less radioactive wastes have been disposed of during the past twenty years.

High-level liquid wastes will have to remain in storage for hundreds of years before decay will have progressed sufficiently to render them harmless. While experience indicates that some of the tanks may last for decades, there is no accurate prediction of tank service life. Therefore, it will be necessary to transfer the wastes from old tanks to new tanks periodically. Continuous inspection of the stored liquid wastes is mandatory. In view of these difficulties, it would be highly desirable to convert the liquid wastes to solids. Research is being conducted with that in mind. Salt beds are being investigated as potential underground disposal sites for the high-level solid wastes.

HEAT WASTE

THERMAL POLLUTION

Thermal pollution of water

The term *thermal pollution* means a change of temperature that adversely affects the ecosystem. The amount of heat generated in the United States from all sources is nearly 90 quadrillion BTUs (British Thermal Units)[5] per year. This enormous release of energy significantly modifies the urban climate and aggravates existing conditions. The electrical consumption of a single New York building will soon exceed 80,000 kilowatts. It is estimated that by the year 2000 the release rate in the megalopolis represented by the area extending from Boston to Washington will be nearly a third as much as the incident solar energy.

The principle of the conservation of energy, known as the first law of thermodynamics, says that "energy can neither be created nor destroyed by any means known to man." But there is also another principle—called the second law of thermodynamics—that says, "in the conversion of energy from one form to another, there is always a decrease in the amount of usable energy." Thus, the efficiency of most machines is distressingly poor, and the equipment used for generation of electric power is no exception. Heat is needed to produce steam to run turbines whether the heat is obtained from fossil fuel or nuclear fuel. After part of the energy in the steam has been converted to mechanical energy to drive the electric generators, the steam is converted back into water by passing large amounts of cooling water through the condensers. The heat that is transferred from the steam to the cooling water is waste heat, and it is the discharge of this heated cooling water that causes thermal pollution.

The most widely used method for cooling power plant condensers is to take water directly from rivers, lakes, estuaries, or the ocean. The water coming out of the condensers will be anywhere from 10° to 30°F hotter, depending on the design and the operation. Thus, the effect on the body of water to which the cooling water is returned—usually the same as that from which it was taken—can be considerable. For example, a typical nuclear reactor will discharge more than 300,000 gallons of heated water per minute.

All types of power plants release large amounts of heat to the environment. The efficiency of fossil fuel plants varies from 25 percent to as high as 40 percent in the newest coal-fired plants. The overall average is only about 33 percent. Most of the nuclear power plants in the United States—the LWR, or light water reactors—convert about 32 percent of the heat they generate to electricity, nearly all of the waste heat being lost in the cooling water. Fossil fuel plants, on the other hand, discharge part of their waste heat to the atmosphere through the boiler stacks, which accounts in part for the fact that nuclear plants discharge about 50 percent more heat to the cooling water than do the most modern fossil fuel power plants of the same electric-generating capacity. Advanced designs, when put into use, would improve the thermal efficiency of both fossil fuel and nuclear power plants.

---

[5]One BTU is the heat required to raise one pound of water 1°F.

The effluent from a nuclear power plant is about 15 to 20°F higher than the normal water temperature. This is not high enough to eliminate aquatic life but will change the amount and diversity of the fauna and flora. However, until more detailed knowledge can be gained, broad conclusions will have to be made with reservations. Some biologists theorize that thermal loading of sea water might cause an increase in plankton production, resulting in a localized improvement of the commercial and sport fisheries. But it has been pointed out by others that in the tropics, the plankton and other aquatic forms live at temperatures that are near the upper limit of their heat tolerance, indicating that the margin of safety may be uncomfortably narrow.

In contrast to tropical and arctic fish species, which have a narrow range of thermal tolerance, temperate species generally have a wide range. Tropical species live near their upper limit. But rainbow trout, a temperate species, can survive for a short period in water at temperatures ranging up to 75°F. Effluent from six to nine plutonium production reactors at the Atomic Energy Commission plant at Hanford, on the Columbia River, has caused no significant thermal fish kill of salmon. Nor were there any apparent effects on plankton. However, pending more extensive study, it is difficult to predict precisely the risks to fish in the specific ecosystem at every proposed power site.

On the coast of southern California at San Onofre, the site of a nuclear generating plant, there has been a decrease in the amount and diversity of the algae in the immediate vicinity of the outfall. However, turbidity is apparently more responsible than thermal effects. There was an increase in the number of fish species, from five to twenty-one, attributed to the presence of the cement coffer and the artificial reef formed by the outfall structure.

An example of damage to the biosphere was found in the Patuxent River estuary in Maryland at a place called Chalk Point. A coal-fired generating plant with 40 percent efficiency sits about halfway up the estuary. Oysters as far as a mile from the discharge of the plant had an abnormally green coloration due to the presence of copper in the plant's discharge water. In one episode, about 40,000 crabs were found dead near the discharge canal. The population of soft-shell clams, catfish, and hogchockers decreased, but that of striped bass increased. The overall effects appeared to result from a combination of thermal and chemical pollution.

Another example is South Biscayne Bay in Florida at a place called Turkey Point, where a fossil fuel power plant is situated. A study showed marked thermal effects near the mouth of the discharge canal and nearby areas. There was a significant shift in the types of algae present and a kill of turtle grass,[6] which is normally a dominant primary producer of considerable importance to the ecology of the bay because it serves as a habitat and as food for a wide variety of small animal life. An interesting but unexplained observation was that a ring of increased production surrounded the kill area.

During the next 30 years the total consumption and release of energy per person is expected to double, while the consumption of electrical energy alone will increase 5 times. While there is a great deal of waste

---

[6]*Thalassia testudinum.*

heat in the generation of electric power, at the present time only about 20 percent of the total heat load comes from electric generating plants. As industry sophistication increases, however, the proportion of the heat load from electric power generation and use will increase.

Atmospheric thermal pollution

Much of the waste heat is discharged to the air. An 80-megawatt electric utility plant discharges hot gases from the stack at the rate of more than 3 million cubic feet per minute (Figure 18.12). A large amount of the energy release comes from private automobiles and trucks. Even if there were complete conversion to electric-powered automobiles, the reduction in heat output would be only 50 percent.

Thus far, most studies of the energy problem have been concerned with the exhaustion of natural resources. More consideration may have to be given to the conservation of energy for another reason—heat pollution. The use of water cooling towers and heat exchangers merely transfers the heat from one segment of the environment to another without necessarily resulting in an overall improvement. Even assuming that the transfer of heat to water is the most satisfactory answer, the methods of cooling are costly in terms of money, equipment, and space. Cooling ponds require vast acreages of a diminishing supply of land. Cooling towers create secondary pollution effects by contaminating the water that is released into the waterways or by emitting a fog of chemicals in the atmosphere.

CONSERVATION OF HEAT

There is no principle of thermodynamics that says the heat from power plants must be wasted. Some of the enormous amount of heat that is released to the environment in the form of "waste" could be harnessed as useful energy. The major obstacle is cost. To bring about large increases in the efficiency of power plants, that is, the amount of power produced in relation to the heat loss extremely expensive recycling equipment is needed. As the efficiency is improved beyond present economic levels,

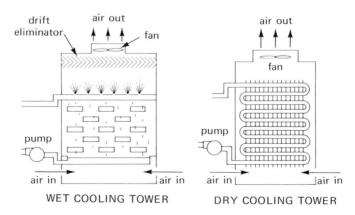

WET COOLING TOWER   DRY COOLING TOWER

Redrawn from *Thermal Effects and U. S. Nuclear Power Stations*, 1971, U. S. Atomic Energy Commission.

FIGURE 18.12 Methods of waste heat disposal to the atmosphere from steam-electric generating stations.

USDA photo.

A natural energy resource almost forgotten—elbow grease.

the cost becomes progressively more burdensome. One possibility would be to have a generating plant surrounded by a complex of hothouses for growing high-value crops such as cut flowers and subtropical ornamentals, fruits, and vegetables. Another possibility is to use the "waste" to heat warehouses for storage of commodities that require temperature regulation. Both steam and heated water could be used to supply a surrounding industrial complex. Another scheme is to use heat for large commercial farming of fish and shellfish in environmentally controlled pools. Satisfactory solutions to this problem may have to wait for the pinch of scarcer fuel and the effects of too much thermal pollution.

ALTERNATE SOURCES OF ENERGY

Other than the most important sources of energy for the present and immediate future—oil, coal, hydropower, and nuclear power—there are several promising alternate sources. They vary in the amount of potential energy they can supply and in the amount of research and engineering needed before their full use can be realized. Those under study and exploration include: oil shale, tar sands and other bituminous deposits, solar energy, geothermal energy, nuclear fusion, solid waste, wind, tidal and wave action, hydrogen fuel, thermionics and fuel cells, uranium and thorium in magmatic rocks, deuterium and uranium in sea water, and fission of boron-11. Also under study are several ways of improving the efficiency of present energy production by several methods: magnetohydrodynamics (MHD), fluidized bed combustion, potassium-steam vapor cycles, and improved gas turbine and mixed heat engines.

Fusion power   An entirely different method of producing electric power by nuclear technology is undergoing intensive research. Instead of atomic fission, consisting of the disintegration of heavy atoms, the new method would use the enormous energy released by the fusion of light elements in the formation of slightly heavier atoms. This would involve deuterium and/or tritium, isotopes of the lightest element, hydrogen. The energy of the sun is from fusion. If the kind of reaction in the sun could be controlled, energy would be available in almost unlimited amounts with the emission of very little pollution. It is referred to as *controlled nuclear fusion*. It is not yet known whether controlled nuclear fusion can become a practical source of commerical power.

Oil shale   Oil shale has been called the "ace in the hole" for the future energy needs of the United States. Exposed deposits of oil-bearing rock ranging up to 2,000 feet thick extend through northwest Colorado into Utah and Wyoming (see Figure 18.13). They hold an estimated 1.8 trillion barrels of oil, six times the oil reserves in the Middle East. The potential value of oil from these shales has been known since the pioneers noticed American Indians using the "burning rock" to warm themselves. But how to get it out economically has puzzled oil producers and scientists. Oil shale does not contain oil, but an organic substance called *kerogen,* which can be changed into an oil resembling crude petroleum when heated to around 900°F. The technology is known, but there remain at least three serious problems: cost, water, and the huge tonnage of waste. The kerogen in shale is only about 20 percent of the bulk. The remaining 80 percent, called processed shale, must be dumped or returned to the land, creating environmental problems at least as great as those associated with strip mining. Under study is the possibility of obtaining shale oil *in situ*—in place—by fracturing underground shale formations with dynamite, heating them, and pumping the released oil to the surface.

Geothermal energy   Geothermal energy is the utilization of the heat of the earth's interior. There are many places where the heat is vented to the surface in the form of hot water, steam, or geysers—a mixture of steam and boiling water (see Figure 18.14 and Table 18.2). Hot springs have been used for bathing ever since the ancient Romans built their luxurious thermal baths at Vicus Calidus (now Vichy) in what is now France, and Aquae Solis (now Bath) in Britain. The Maoris made practical use of the geysers in New Zealand for cooking their food. The first geothermal well was obtained in Hungary in 1867 in an attempt to bring mineral water to the surface for drinking. In the 1930s, natural hot water was piped into a section of Budapest to heat houses, and in Reykjavik, Iceland, wells have been drilled more than a mile deep to reach hot water, which is then used to circulate through the walls of concrete houses and for heating greenhouses. Similar use is made of geothermal waters in other parts of the world. But heating is only the simplest application. The most important future for geothermal energy is to convert it into electric power.

The first geothermal electricity was generated in 1904 at Lardarello, Italy. It was here that the sulfurous fumaroles and hissing geysers were supposed to have inspired Dante's *Inferno*. The Italians first tried lighting

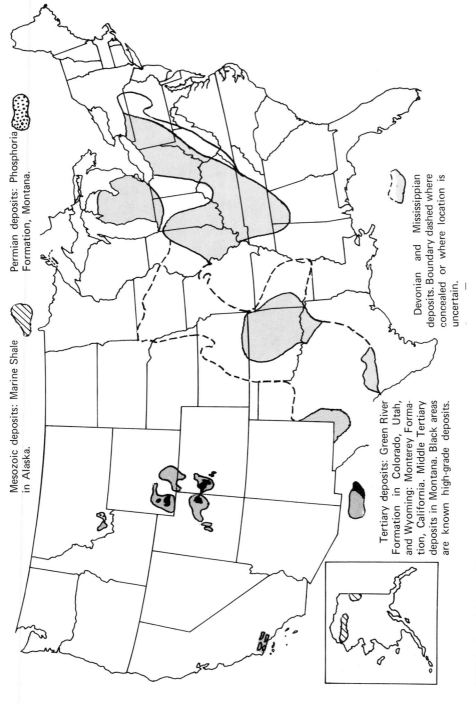

Permian deposits: Phosphoria Formation, Montana.

Mesozoic deposits: Marine Shale in Alaska.

Devonian and Mississippian deposits. Boundary dashed where concealed or where location is uncertain.

Tertiary deposits: Green River Formation in Colorado, Utah, and Wyoming; Monterey Formation, California. Middle Tertiary deposits in Montana. Black areas are known high-grade deposits.

USDI Geological Survey, 1974.

FIGURE 18.13   Principal oil shale deposits of the United States. The most extensive high-grade oil shale deposits are in the Rocky Mountain region, named the Green River, Formation. Oil shales of this region were formed from the sedimentary materials that accumulated in the bottoms of two vast shallow lakes about 50 million years ago.

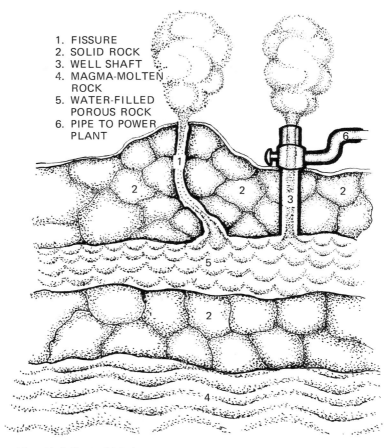

1. FISSURE
2. SOLID ROCK
3. WELL SHAFT
4. MAGMA-MOLTEN ROCK
5. WATER-FILLED POROUS ROCK
6. PIPE TO POWER PLANT

Adapted from U. S. Dept. of Interior.

FIGURE 18.14   Geothermal power is a promising source of supplementary energy. Heat for geothermal steam comes from molten rock, called magma, deep in the earth. Since magma is under great pressure, it is extremely hot (1,380°F to 2,000°F).

electric bulbs by a small generator driven by a steam turbine. The blades were turned by natural steam piped from a fissure in the ground. The next year they built the world's first geothermal power plant, which still furnishes electricity to help propel Italy's electric trains.

Geothermal electric power is still considered a potential resource that is largely untapped. Geothermal power plants are in operation in Italy, Iceland, Mexico, New Zealand, Japan, Russia, and the United States, but they produce only an extremely small fraction of the world's electricity. The production of pollution-free energy from natural thermal sources is not always as simple as it might seem. The hot water and steam are in some locations contaminated with mud, corroding salts, and poisonous gases. Hydrogen sulfide, a highly toxic gas, and radon, a natural radioactive material that can cause lung cancer when inhaled, have been found in underground steam sources used for power plants. Drilling wells that draw off large amounts of steam under pressure might cause underground

TABLE 18.2   Known Geothermal
Resource Areas

State	Acres
Alaska	88,160
California	1,051,533
Idaho	21,844
Montana	12,763
Nevada	344,027
New Mexico	152,863
Oregon	84,279
Utah	13,521
Washington	17,622
Total	1,786,612

From U.S. Department of Interior Conservation Yearbook Series No. 8. 1972.

disturbances and, as in the area of the Salton Sea, might cause the intrusion of salt water into the subterranean basin. The great geysers of the Yellowstone cannot, by law, be exploited for commercial purposes, and the environmental effect of exploitation of similar natural resources must be considered in other areas as well.

Wind power   Wind power was once relied upon heavily in rural areas, especially for pumping water. But with increased availability of fossil fuels, windmills in

Left photo: Geothermal energy is an undeveloped resource in many areas. *Right photo*: Wind power, once used widely around the world, was nearly abandoned when fossil fuels became plentiful. There is now renewed interest in this abundant natural resource.

many areas were replaced with pumps powered by electric motors. The picturesque windmills of Holland have all but disappeared, and elsewhere this practical device is hard to find. As the cost of energy from fossil fuels increases, there is renewed interest in ways of harnessing the enormous latent energy in the winds.

Solar energy   It was said earlier that the energy in fossil fuels came from living organisms. Because green plants are the direct users of solar energy, all organisms in the food chain derive their energy basically from sunlight, and this is the source of food-energy for all consumers including man. Windpower comes from the sun, for the sun's heat creates winds that sweep around the globe. The sun's heat also evaporates water that condenses and falls as rain and snow, making it possible to generate hydroelectric power. Even the tides are caused in part by the gravitational force of the sun. Thus the sun is the overwhelming original source of all usable energy.

The sun is a huge hydrogen-fusion reactor containing 99.85 percent of all the material in the solar system and occupying nearly 1⅓ million times as much space as the earth. It radiates out into space more than 90 billion calories of energy every second. But the earth is a small cosmic speck; less than two-billionths of the sun's radiation ever strikes the earth, and more than one-third of that is reradiated back into space. And of the sunlight that falls on the earth, only a small fraction of one percent is used by green plants.[7] Still, the amount of solar radiation that reaches the surface of the earth is prodigious. If we knew how to use the sun's energy efficiently, each square yard on the earth's surface could capture one-half horsepower of energy.

Solar energy has been used for many years on a small scale for heating houses in back country where fuel supplies are a problem. There has been more interest in its use in undeveloped regions than in the industrialized countries where fossil fuel supplies have been plentiful. A major disadvantage is that, for most applications, the cost of converting solar energy to a usable form is higher than conventional energy sources because the initial investment in heat-collecting and heat-distributing equipment is high. If industry and government start moving soon on solar heating and cooling of buildings, it has been estimated that by 1985 the cost of hardware will come to $1.3 billion annually.

A great deal of research will be required to bring solar energy to the point of contributing a major amount of the future heat and power requirements. But even with present technology, much greater use could be made of solar heating to supplement other sources. It has been estimated that in the continental United States, there is an average 24-hour solar energy supply of 410 thermal watt-hours per square foot, about twice the amount of energy needed to heat and cool an average house.

The Solar Energy Panel of the National Science Foundation and National Aeronautics and Space Administration guessed that by the year 2020, solar energy could provide as much as 35 percent of the total heating

---

[7]Only 0.023 percent of the solar energy received at the earth's surface is used in photosynthesis, but this amounts to 5.5×10^{17} kilocalories per year (1½ billion billion calories per day).

and cooling of buildings, could replace 30 percent of the U.S. gaseous fuel, 10 percent of the liquid fuel, and 20 percent of the electrical energy requirements.

Direct conversion of solar energy to electricity has been demonstrated, but practically all solar conversion applications have been for spacecraft power systems where high cost could be justified. If low-cost systems could be devised, they would provide an abundance of energy for billions of years without the environmental problems of fossil fuels and nuclear power. Thus, solar energy is not only one of the cleanest "fuels," but its potential availability is nearly unlimited, and its development is one of the great technical and social challenges for the immediate future.

PROBLEM AREAS

Fossil fuel resources are running out faster than new discoveries can replace them. When heavy demands for fuel began to be felt during the early years of industrial expansion, many areas quickly cut down their forests and ran out of wood. There was a switch from wood to coal, then from coal to petroleum, from petroleum to natural gas, and now there is an effort to switch from fossil fuels to nuclear energy. Each step except the first was thought to bring cleaner fuel, and the last step—to nuclear fuel—promised to greatly extend the period of time over which the human race could continue to enjoy a high level of energy consumption. But nuclear energy is more costly and is developing more slowly than its proponents had hoped for, while petroleum resources are running out faster than new discoveries can replace them. Clearly, something must be done to bring supply and demand more in balance.

Coal

Coal is by far the most abundant, readily available, energy resource in the United States. The United States has more coal than any other country in the world, about 3.2 trillion tons, enough to last at least the next 300 years if suitable energy policies are followed. This is more than three-fourths of the known worldwide fossil fuel reserves. Unfortunately, the burning of coal is a serious air pollution problem, and coal mining is another environmental problem that is just as hard to cope with.

Stripping the good earth

Coal mining is not the only operation that disturbs the surface of the land, but it is the most widely destructive. Wherever the coal lays near the surface, man and his machines of steel have flayed the skin off the naked earth. Almost half (46%) of the coal produced in the United States is strip mined—nearly 300 million tons of coal obtained by ripping off the overburden of soil with giant machines. It is more economical to get the coal from near the surface by strip mining, with equipment that can be run by one or a few workers, than by extracting it out of deep shafts by crews of miners working in perpetual danger. In the past, most of the strip-mined tonnage came from the abused hills of eastern Kentucky, Ohio, Pennsylvania, West Virginia, and other states of Appalachia. But coal is also strip mined in many other parts of the country and with the shortage of energy and the greater demand for coal, strip mining is on the increase throughout the country. The facts may prove that strip mining only seems to be cheaper. The cost of renovating the devastated land, replanting the bare soil, and reclaiming polluted streams will either have to be paid

Photo courtesy of Occidental Petroleum Corporation.

High-pressure water jets break up phosphate ore used in the manufacture of fertilizers and many other products. Similar, but destructive, procedures were used in the early days of hydraulic gold mining operations on riverways.

for in the price of coal or the environmental and social costs will be imposed on future generations.

Nuclear problems unsolved

Serious questions have been raised about nuclear power that relate only indirectly to the immediate air and water pollution problems. The disposal of nuclear waste is a nagging problem that so far has defied a well defined solution. The Nuclear Regulatory Commission[8] plans to concentrate the wastes in solidified form and put them in large concrete containers that will be buried in guarded locations. But even these will have to be watched over for thousands of years. A British report in 1972 said, "We are committing future generations to a problem which we do not know how to handle." Many people have grave doubts that the amount of energy that can be obtained from nuclear energy can even come close to justifying the costs in time, money, manpower, and risks to the environment both now and in the future.

---

[8]The Nuclear Regulatory Commission (NRC) performs some of the functions that were formerly under the jurisdiction of the Atomic Energy Commission, an agency that was dissolved and replaced, January 1975, by the U.S. Energy Research and Development Administration (ERDA).

The development of nuclear energy engendered bitterly divisive arguments over unresolved problems, including those involving costs, long-range availability of nuclear fuel, plant sites, thermal pollution of bodies of water, environmental pollution with radioactive emissions, and even the necessity of nuclear power after all. By the end of 1974, there were 53 nuclear power reactors licensed to operate and 2 others authorized in the United States, having a total generating capacity of about 3 percent of the electricity used (see Figure 18.15). A great debate continues among citizens and scientists alike over the reliability of safety devices and the danger of a nuclear explosion. We are accustomed to planning ahead a year at a time, occasionally for five years, and rarely for ten years. To plan ahead for the thousands of years needed for decay of some of the radioactive wastes seems to have been beyond the grasp of those who were eager to proceed in the belief that the answers would somehow be found when needed. But the final answers are not yet in. Major proposals include postponement of further construction of nuclear plants until the problems can be resolved, and accelerated research on development of fusion power, a form of energy that would be practically nonpolluting.

One of the more vexing problems is the possibility that nuclear materials will be stolen by hostile governments or terrorist groups for the manufacture of nuclear weapons. Nuclear technology has become so widely disseminated that some nuclear experts believe almost any well-financed organization could put together a crude bomb starting with enriched reactor fuels. Considerable quantities of missing nuclear material are still unaccounted for—enough to make several atomic bombs. The possibility of nuclear theft will be enhanced if high-temperature gas-cooled reactors come into widespread use because these require a more highly enriched form of uranium fuel, suitable for use in weapons. The danger will also be increased when plutonium is introduced instead of uranium as a nuclear fuel. Plutonium is separated as a by-product of processing uranium for the most commonly used fuel, and in the past has been used for weapons manufacture or stored for future use.

Will threats by terrorist groups to use nuclear weapons be used to replace kidnapping and other violent activities? According to a study by the Atomic Energy Commission itself in 1974, such a thing is possible and the precautionary measures were still "entirely inadequate."

Energy conservation Many people think that the immediate solution to the energy problem lies elsewhere than finding new reserves of fossil fuels or the development of nuclear energy. Cutting down on the use of energy and the consumption of material goods and discarding the economic philosophy of *planned obsolescence* is advocated. The "average" American uses 5 times as much grain as the Indian, 100 times as much steel as the Indonesian, and 191 times as much energy as the Nigerian. Much of it is excessive, profligate, and wasteful. During the fuel crisis of 1974, many power companies found that consumption of electricity fell by as much as 15 percent by voluntary action of users. A consciously planned campaign to conserve energy and manufactured materials would buy time to develop relatively nonpolluting alternate sources such as solar energy, geothermal power, nuclear fusion, utilization of heat waste, recycling solid waste as fuel, and more efficient use of present fuels.

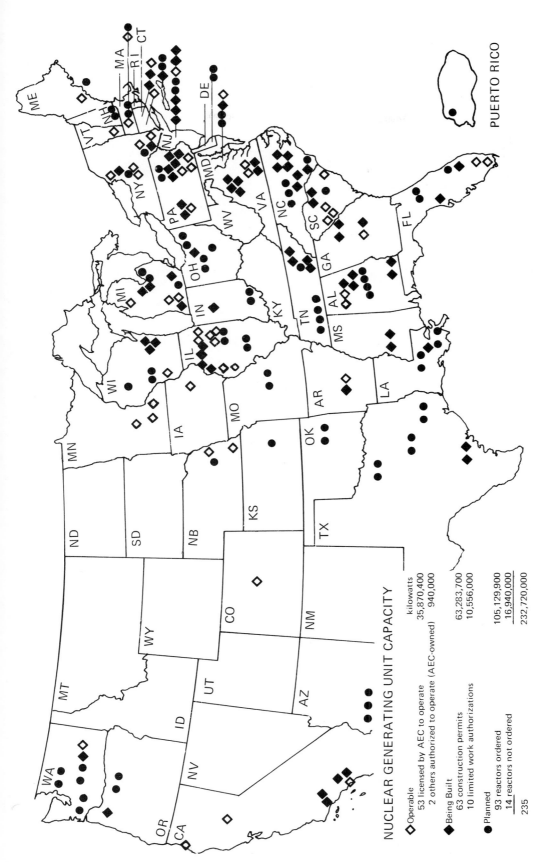

NUCLEAR GENERATING UNIT CAPACITY

	kilowatts
◇ Operable	
53 licensed by AEC to operate	35,870,400
2 others authorized to operate (AEC-owned)	940,000
◆ Being Built	
63 construction permits	63,283,700
10 limited work authorizations	10,556,000
● Planned	
93 reactors ordered	105,129,900
14 reactors not ordered	16,940,000
235	232,720,000

FIGURE 18.15   Nuclear power reactors in the United States.

Photo courtesy of Standard Oil Company of California.

"Fast lanes" for exclusive use of buses and car pools are one way to save energy—and nerves. This is a view of the toll plaza of the San Francisco–Oakland Bay Bridge.

READINGS

ASIMOV, I., *Electricity and Man*. Oak Ridge, Tennessee, U.S. Atomic Energy Commission, 1972.

BROWN, T. L., *Energy and the Environment*. Columbus, Ohio, Charles E. Merrill Publishing Company, 1971.

The Economy, Energy and the Environment. Joint Economic Committee, Congress of the U.S., U.S. Government Printing Office, Washington, D.C., 1970.

Energy and the Environment. Electric Power. Pp. i–vi, 1–58. Council on Environmental Quality, U.S. Government Printing Office, 1973.

Energy and Power. A Scientific American Book. San Francisco, W. H. Freeman and Company, 1971. Also in *Science* 184 (4134):245–389 (April 19, 1974).

The Energy Outlook for the 1980's. Joint Economic Committee, Congress of the U.S., U.S. Government Printing Office, Washington, D.C., 1973.

FENNER, D. and J. KLARMANN, "Power from the Earth," *Environment* 13(10):19–34 (1971).

FOREMAN, H., ed., *Nuclear Power and the Public*. Minneapolis, University of Minnesota Press, 1970.

FOX, C. H., *Radioactive Wastes*. Oak Ridge, Tenn., U.S. Atomic Energy Commission, Div. of Technical Information, 1969 (rev.).

GOFMAN, J. W., and A. R. TAMPLIN, *Poisoned Power*. Emmaus, Pa., Rodale Press, 1971.

HAMMOND, A. L., "Dry Geothermal Wells: Promising Experimental Results," *Science* 182: 43–44 (1973).

HAMMOND, A. L., "Geothermal Energy: An Emerging Major Resource," *Science* 177:978–980 (1972).

HAMMOND, A. L., W. D. METZ, and T. H. MAUGH II, *Energy and the Future*. American Association for the Advancement of Science, Washington, D.C., 1973.

HOGERTON, J. F., *Nuclear Reactors*. Oak Ridge, Tenn., U.S. Atomic Energy Commission, Div. of Technical Information, 1967.

NELKIN, D., *Nuclear Power and its Critics*. Ithaca, N.Y., Cornell University Press, 1971.

United States Energy. A Summary Review. U.S. Dept. of Interior, 1972.

MAN AND NATURE: CONSERVATION OF LAND AND WILDLIFE

We are by nature and tradition creatures of the earth, the rain, and the sky. Neither skyscraper canyons of cities nor the high-rise caves of apartment existence are a congenial environment for the human need to be in harmony with the things that are natural, wild, and free. A mere century ago, most Americans lived on farms, and most of those who did not were close enough to the wide open spaces to come in direct contact with that part of nature that was still beyond the touch of man. Today, only 4 percent of the people remain in rural areas. The vast majority live in the "concrete jungle" or its suburbs.

But the rush to cities, in itself, is not always the most important influence on the quality of life. The way we plan and manage the resources of the landscape determines the degree to which they can be preserved for permanent use. The state of Alaska has a human population of only 0.6 per square mile, and its resources and natural beauty are close at hand for most residents. By contrast, the Netherlands has the highest population density on earth, more than 1,000 per square mile, yet the citizens of that small country manage their resources better than most people in less populated parts of the world. It is not always the number of people but, instead, their disrespect for the environment and their careless and callous use of machines that have the greatest destructive effect on natural resources.

LAND USE

All around us the landscape changes with startling suddenness as each new construction project is unleashed on the available open space—more streets, more buildings, sprawling suburbs, airports, power plants, transmission lines, and factories. Trees and shrubbery disappear and pastures fade away in the murky environs of industrial encroachment as the demands increase for more land for more people. Land development in the United States swallows up the open landscape at the rate of one million acres every year (see Figure 19.1). The effects are more than local. For example, Los Angeles has grown into a megalopolis with an

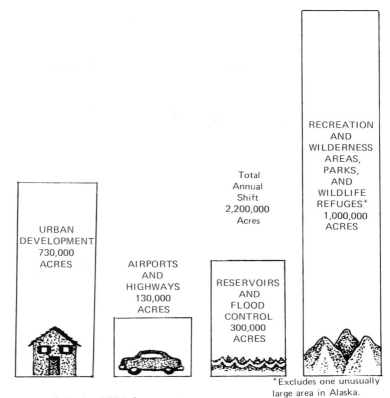

Economic Research Service, USDA data.

FIGURE 19.1   Rural land in the United States lost annually to other uses, acres, 1959–69.

insatiable appetite for resources, space, and energy. The need for more water for its thirsty people and industries has made it necessary to dig canals to the far reaches of the state and to build dams for temporary impoundment of the water. The need for electricity, combined with anti-pollution standards, has forced the construction of new power plants across state lines a thousand miles away in New Mexico and Arizona. These, in turn, generate problems in connection with strip mining for needed fuel, transmission lines, and air pollution, thus spreading the dark mantle of metropolitan engulfment over the open spaces of other states.

In Florida, two miles of concrete runway were built for a planned 2 billion dollar jetport sprawling over a 34-square-mile site in the lush swamp of the Everglades before construction was halted at the insistence of conservationists. The runway came into use as a training facility, but the major impact, that of support facilities employing 10,000 or more people, was avoided, at least temporarily.

As population grows, the congestion in recreational areas becomes critical. Beaches and related recreational facilities have been the first to feel the brunt of the encroachment of more people than they can accommodate. Many state beaches and national parks require reservations several months in advance for a place to camp, and some of the mountain areas are just as congested. In short, there are too few readily ac-

Photo courtesy of the City of Huntington Beach and The Huntington Beach Company.

Oil-drilling derricks following an oil strike displaced a community of beach-front homes on the coast of Southern California. This picture of the conflict between energy and recreational needs includes vacationers, many of whom arrived in automobiles powered by gasoline derived from crude oil obtained from wells similar to these.

cessible outdoor recreational areas for too many people, many of whom are entrapped for most of the year in the pressure-cooker of urban life. People have an increasing need to get out of city apartments, condominiums, and subdivisions. Rapid expansion in the number and size of outdoor recreational areas is one of the urgent needs brought about by the combination of population growth and urban concentration. But, too often, recreational and commercial developments are in conflict with good conservation practices. This means, in effect, that desirable open spaces must be selected and preserved before their full usefulness can be nullified by the destruction caused by planless commercial enterprises, timbercutting, highways, and the blight that inevitably follows the irretrievable effects on the landscape of narrowly purposeful exploitation of natural resources.

Most of the land area of the United States, about 57 percent, is used to produce crops and livestock.[1] Much of it is grazing land, some of it forested. The portion cultivated for crops is only about 21 percent of the United States land area, and not all of that is in crop production each year.[2] In any given year, part of it is used for crops, part is used only for

---

[1]The land area of the United States is about 2,264 million acres, or about 3.5 million square miles.

[2]The nation's total cropland totals 475 million acres.

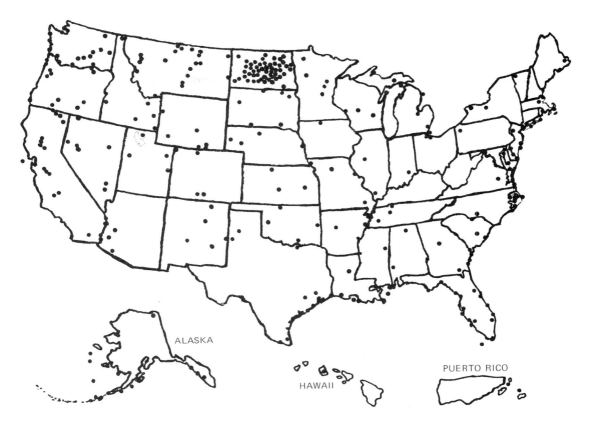

FIGURE 19.2   The world's largest network of managed wildlife areas is in the United States. Pelican Island, off the coast of Florida, was established as the first National Wildlife Refuge in 1903, and since then more than 30 million acres have been set aside, now consisting of 356 refuges.

pasture, and part of it is idle. Harvested cropland averages roughly about 300 million acres, or less than one-seventh of the nation's land area. "Special use" areas (178 million acres) take up 8 percent of the land area of the country; cities and land occupied by transportation facilities account for one-third of that, or 61 million acres, while about 28 million acres have been set aside for defense and atomic energy purposes. National and state parks, and related recreational areas, occupy 49 million acres, about 2 percent of the nation's land area, while an additional 32 million acres are reserved for wildlife propagation and protection (see Figure 19.2).

Public domain

Three-fifths of the land area of the United States is privately owned and two-fifths is owned by federal, state, and local governments. The *public domain* consists of otherwise unappropriated lands that have been ceded by the states to the federal government to be held in trust for the people of the United States. Well before the end of the last century, the United States government had become the proprietor of more than 1.8 billion acres of land, "to be disposed of for the common benefit." But many people in and out of the government viewed the public domain as a gigantic piece of cake on the public-handout menu. By the time what historian Vernon L. Parkington called "The Great Barbecue" was finished, millions of acres had been gobbled up by the railroads, corporate farmers and ranchers, miners, lumber companies, and speculators. Vast areas were given to the railroads and by 1900 the railroads alone owned more land in the continental United States than any other organization other than the government itself. The public domain has now dwindled to one-fifth the nation's land area.[3] Today's 460 million-acre public domain is administered by the Bureau of Land Management, whose management has at times been severely criticized by conservationists.

Forests   One-third of the United States is forested (Figure 19.3). Two-thirds of this, about 500 million acres, is commercial forest land available for growing continuing crops of industrial timber products. Probably more will be classified as commercial after the interior of Alaska has been completely surveyed.

There are 154 national forests in the United States consisting of 187 million acres (about 290,000 square miles). The national forests are used for both recreation and commercial purposes including water resources, timber, and livestock grazing. Thirty percent of the nation's softwood sawtimber comes from national forest lands and they sustain more than 7 million head of livestock. Sales of timber to the lumber and pulp industries, plus fees for the other uses, bring in a revenue of about 300 million dollars a year, of which one-fourth is returned to the states for use in the counties having national forest land.

National parks   The idea for national parks originated a little over 100 years ago. Yellowstone was set aside as the first national park in 1872, most of it in the northwest corner of Wyoming, and it is still the largest of the national parks. But the story of Yellowstone goes back further than that—to the days of the earliest explorers. Early in the 1800s some strange and decidedly tall tales came out of the region of the Yellowstone. A man named John Colter, who had been with the Lewis and Clark expedition, visited the area in 1807 and came out with incredible stories of boiling mud, belching caverns, steaming craters, spouting springs, multicolored waterfalls, exquisitely beautiful lakes with splashing trout, an abundance of wildlife, and mountain scenery whose grandeur was beyond description.

Artful lying was a common and sometimes admired avocation in the Old West and Colter's stories were far too fanciful for anyone to take very seriously. John Colter became accustomed to winks and knowing smiles—the Marco Polo of his day. But the rumors kept filtering out. Jim Bridger, another explorer who visited the area in 1830, and other adventurous wanderers and trappers who penetrated the region, told stories that made a few people suspect there were some strange things going on that needed investigating. So in 1870 a party of nineteen skeptical people—nine civilians, six soldiers, two packers, and two cooks—officially called the Washburn-Doane-Langford expedition, set out for the high country where things were happening beyond belief to get the straight story.

"Never was a party more completely surprised," said a report, despite their almost certain expectations of the spectacular. What the Washburn

---

[3]All federal agencies, including both civil and defense, are estimated to own and administer 765 million acres of land of which more than one-fourth is in Alaska. (Watkins, T. H. The Public Domain. Sierra Club Bulletin. 1973. p. 9–13, 41.)

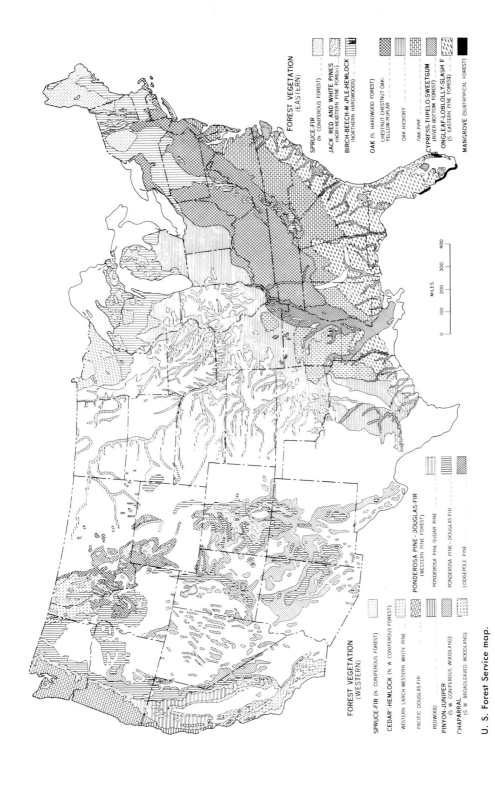

FOREST VEGETATION
(EASTERN)

SPRUCE-FIR ----------------------
(N CONIFEROUS FOREST)

JACK RED AND WHITE PINES ------
(NORTHEASTERN PINE FORESTS)

BIRCH-BEECH-MAPLE-HEMLOCK -----
(NORTHERN HARDWOODS)

OAK (S HARDWOOD FOREST) --------

CHESTNUT CHESTNUT OAK- ---------
YELLOW POPLAR

OAK HICKORY --------------------

OAK-PINE -----------------------

CYPRESS-TUPELO-SWEETGUM --------
(RIVER BOTTOM FOREST)

ONGLEAF-LOBLOLLY-SLASH F -------
(S EASTERN PINE FOREST)

MANGROVE (SUBTROPICAL FOREST) --

FOREST VEGETATION
(WESTERN)

SPRUCE-FIR (N CONIFEROUS FOREST) --

CEDAR-HEMLOCK (N W CONIFEROUS FOREST) --

WESTERN LARCH WESTERN WHITE PINE ---

PACIFIC DOUGLAS-FIR ----------------

REDWOOD ---------------------------

PINYON-JUNIPER ---------------------
(S W CONIFEROUS WOODLAND)

CHAPARRAL --------------------------
(S W BROADLEAVED WOODLAND)

PONDEROSA PINE - DOUGLAS-FIR --------
(WESTERN PINE FOREST)

PONDEROSA PINE SUGAR PINE ----------

PONDEROSA PINE - DOUGLAS-FIR --------

LODGEPOLE PINE ---------------------

MILES

0   100   200   300   400

U. S. Forest Service map.

FIGURE 19.3   Forest vegetation of the United States.

party found seemed more like a storybook picture of natural wonders. Talk around the campfire turned to speculation of what would happen to the region. Certainly it would be a profitable venture for a syndicate to acquire large tracts of land in the area—an easy proposition—and build a road or railway into the area to attract the droves of tourists who would visit the region when its true wonders became public knowledge. But Cornelius Hedges, a member of the expedition, said no, they should never let any part of it get into the hands of private owners but, instead, all of it should be set aside as a great national park. It was a startling idea, for nothing like it had ever been done before, but the other members of the party enthusiastically agreed. Thus a new idea was born, for the law that set aside Yellowstone National Park, signed by President Grant after two years of hard work with Congress, created not only the first national park in the United States, but the first in the world. The act provided for the "preservation, from injury or spoilation, of all timber, mineral deposits, natural curiosities or wonders, within the park, and their retention in their natural condition," and fish and game were to be protected from "wanton destruction."

Setting aside a natural area as a national park does not, in itself, assure the best use of an area. Campgrounds in Yellowstone National Park, one of the most spectacular natural wonders of the world, turn away hundreds of campers during the summer season, as do many of the other more popular tourist attractions. Yosemite National Park, a magnificent but comparatively narrow ice-carved pocket in the Sierra, became so congested that it was called the "garbage can of the country's park system." More than 2½ million people visited the park in 1975 and the numbers have been increasing steadily. Tents were pegged on the next tent's pegs and "people were sleeping next to everybody." Campfire smoke covered the valley like a fog. In desperation, a program was started that took people out of their cars and put them on bicycles and butane-powered trams, or simply on their feet, and the road system was altered to create a free-flow pattern, much of it closed completely to private cars. Campsites have been improved and campers are urged to use portable self-contained stoves. The master plan calls for the eventual elimination of all private cars from Yosemite Valley. There are similar plans for other parks, such as Grand Canyon, Point Reyes, and Everglades National Parks.

Wilderness   Many people have a yearning for the peace and solitude that can be found only in wilderness areas far from parks and other formalized recreational areas. Theodore Roosevelt expressed this feeling about Africa:

> . . . there are no words that can tell of the hidden spirit of the wilderness, that can reveal its mystery, its melancholy, and its charm. There is delight in the hardy life of the open. . . . Apart from this, yet mingled with it, is the strong attraction of the silent places . . . where the wanderer sees the awful glory of the sunrise and sunset in the wide waste spaces of the earth, unworn of man, and changed only by the slow change of the ages through time everlasting.[4]

Because wilderness has vanished from much of America, that which remains, about 2 percent of the land of the nation, is a priceless treasure. The idea of protecting wilderness areas originated in 1924 when the

---

[4]Theodore Roosevelt, *African Game Trails* (New York: Charles Scribner's Sons, 1909).

U.S. Forest Service photo by Clint Davis.

Elbow Lake in the Bridger Wilderness Area.

Forest Service set aside the Gila Wilderness, a special area of the Gila National Forest of southwestern New Mexico. It now consists of 427,817 acres, about 670 square miles of canyons and mountains at the headwaters of the Gila River. But for many years there were no laws to protect the country's wilderness areas. Finally, in 1964, Congress took action with passage of the Wilderness Act, which established the National Wilderness Preservation System. The System originally consisted of 54 wilderness units, plus 34 units previously set aside as primitive areas by administrative action to be temporarily managed for wilderness purposes. These 88 units in 14 states consist of a total of 14.3 million acres, or about 22,000 square miles, an area nearly the size of West Virginia. By January, 1973, 23 of the primitive areas had been added or recommended for addition to the System. The remaining primitive areas, as well as other national forest lands, remained under study for wilderness classification.

Even the designation of a natural scenic resource as a wilderness area is no guarantee that it will escape damage. The 10.7-mile-long trail to the summit of Mt. Whitney, the highest point in the United States outside Alaska, goes through the John Muir Wilderness. In 1973, 15,700 people used the trail, up from 12,800 in 1971. Camping areas have been denuded of vegetation, firewood is nonexistent, green trees have been chopped down, and human waste is left in the open. Such intensive use was outside the objectives of the Wilderness Act, which was for the preservation of natural resources. Two critical problems—sanitation and the gradual loss of natural wilderness quality—forced the Forest Service to place a daily limit on overnight campers based on a carrying capacity of 150 to 175 and to place a daily limit of 50 on those continuing on the trail into Sequoia National Park. It is now necessary to make a reservation with the Forest Service in advance to camp overnight in a wilderness area that has turned out to be too popular.

Meanwhile, there is danger that untouched wilderness will be taken over for commercial development. In addition to mining and timber, two of the most serious impediments to setting aside wilderness areas, there is the ever-increasing pressure for new vacation resorts. A long-standing controversy developed over the proposal to build a major summer and winter resort in a remote section of the High Sierra in California called Mineral King Valley. The valley is in one of the most spectacular settings on the North American continent. As originally planned, there would be a highway into the valley in place of the present road. Later, when opposition developed, the plan was changed to build a cog-railway, with parking outside the valley. The Forest Service issued bids for the resort and awarded the bid to Walt Disney Productions in 1965. If everything had gone as planned, there would now be thousands of skiers there every winter. But in 1969, before the project could get under way, the Sierra Club, a national conservation organization, strongly opposed it and filed a lawsuit to block the project. The suit went all the way to the U.S. Supreme Court, which sent it back to the federal district court in San Francisco. The litigation, which threatened to tie the project up in the courts for years, combined with the new need for environmental impact studies and the approval of various government agencies, appears to have brought the project to a dead halt, leaving Mineral King Valley still relatively undeveloped. The Sierra Club wants Mineral King Valley to beome sequestered as part of Sequoia National Park, but most likely, Mineral King will remain as part of the larger Sequoia National Forest, already a national wildlife refuge, though parts of it are used heavily by campers and hikers.

WILDLIFE

"Wildlife," in its broadest sense, includes all living things except man and domesticated plants and animals. In this sense, wildlife includes organisms such as pathogenic bacteria, parasites, pests, and a few predators that are inimical to human welfare and, therefore, are not wanted. Considerable human effort is directed toward stamping them out. But in the more common use of the term, wildlife refers to those organisms— totaling nearly 2 million species—that are beneficial and desirable for various reasons or, at most, only insignificantly harmful.

A widely recognized principle of biology is the concept of the balance of nature—the checks and balances of the ecosystem that maintain a dynamic equilibrium in the populations of plants and animals in their natural environment. Throughout the history of life, various natural conditions have upset the balance many times, at least partially being responsible for the extinction of millions, perhaps billions, of species during geological time. Climatic changes, disappearance of the food supply, increase in effective predation, epidemics, or the inability to compete, reproduce, or adapt to new conditions brought about the natural death of entire populations and their replacement by species that had more suitable qualities for adaptation. But of all the forces that have upset the balance in the recent past, man has been by far the most disruptive element. Human manipulation of the environment and mindless aggression against once-abundant species of wildlife has had effects that have been appropriately called catastrophic.

During the last 200 years, about 600 species of animals have declined nearly to the point of extinction; and in the last 2,000 years, 110 species of mammals have disappeared from the earth—never to be seen alive again. An exhibit at the Smithsonian Institution in Washington, D.C., described 225 species of fish and wildlife that have become extinct in the past 400 years. The commencement of modern extinction is taken to be the year 1600 by the International Union for Conservation of Nature and Natural Resources. Since then, 75 percent of all species extinctions have been those in which human beings played a major role.

WILDLIFE VICTIMS

The great auk   The human species, by nature or cultural adaptation, is a destructively aggressive predator, a quality that has been especially evident since the nineteenth century. This period in human history is noted for its great destruction of birds and mammals at the hand of man. The great auk was one of the first birds to become extinct, gradually dying out from the effects of savage human predation. The *great auk* was a powerful diver and a swift swimming bird. It had to be, since it could not fly. It had been known in Europe for centuries before it was discovered in North America in 1497 or 1498 by adventurous French fishermen who had begun to fish on the banks of Newfoundland. On land, the bird was a "sitting duck," or *Pingouin,* as the French called it. The bird could be taken easily in such numbers that the fishermen did not have to provision their vessels; they found all the fresh meat and eggs they needed by visiting the bird islands. One report said that a crew filled two boats in less than half an hour. The bird was destroyed first by the demand for eggs and meat and secondly by the use of the bird's feathers for eiderdown beds and coverlets. The bird had probably disappeared from North America by 1840, before its extinction in Europe where the last specimen was taken off Iceland in 1844.

Buffalo   The best known case in North America of the wholesale slaughter of wildlife is that of the American bison, known in the lingo of the plains as the buffalo. The first European to see one of these huge animals is said to have been Hernando Cortez, who visited a zoo in Mexico City at the time of his conquest of the Aztec empire. Many years later, the Englishman Sir Samuel Argoll, subsequently a deputy governor of Virginia, reported seeing one of the animals about where Washington, D.C., is now located. Few people realize that the buffalo, generally thought of as a plains inhabitant, also inhabited the eastern woodland areas in numbers estimated as high as 5 million.

It is hard for anyone now to comprehend the numbers of buffalo in America when the Europeans first invaded the plains. Alvarado, Coronado's captain, could only compare them to the fish in the sea. He and his company were the first white men to hold a buffalo hunt, and for the purpose, they learned how to use horses—animals that indirectly proved to be the downfall of the buffalo. Just before the Indians acquired the horse, the naturalist E. T. Seton estimated that 75 million bison ranged from northern central Canada to northern New Mexico. General Sheridan, in 1860, figured 100 million bison roamed the southern plains alone. One traveler on a journey in 1834 related, ''The whole plain, as far as the eye could discern, was covered by one enormous mass of buffalo. . . .''

People had plenty of reasons for shooting the buffalo. The plains Indians depended on the animals for meat, hides, and bones, used in dozens of ways for food, tepees, weapons, and household items. The military wanted the buffalo out of the way because the animals supported the Indians that the soldiers were trying to round up and confine to reservations. The ranchers wanted the range for their cattle, and the farmers wanted to be rid of the bison because they trampled their fields and crops. But in the end, it was none of these that wiped out the buffalo—it was the hide hunters. When it was discovered in 1870 that buffalo hides were commercially valuable, especially for making industrial drive belts, hide hunters swarmed over the plains by the thousands and attacked the animals with indiscriminate ferocity. The terrible destruction that followed is almost beyond belief. A good hunter could kill as many as his skinners could handle, as many as 150 to 200 animals a day. They worked fast, using horses to tear off the hides. The carcasses rotted where they lay and

Photo by L. A. Huffman.

About 20 million buffalo roamed the plains and forests in 1850; but by 1889 there were only 551 left. There is now a protected stable population of about 10,000. In this 1882 photograph titled ''After the Chase,'' the severed head between the two downed bulls was taken from a medium-sized bull, and served as a skinning wedge to be slipped under a half-skinned carcass with the help of a horse and lariat so it would be easier to get the heavy hide from underneath.

the stench of putrifying flesh over great parts of the plains was overpowering, attracting buzzards, coyotes and other animals that gorged on the carrion. Later came the "bone pickers," mostly settlers, who gathered the skeletons in wagons and shipped them to fertilizer companies in the East. Often the bones were the farmers' first cash crop. Finally, much of the buffalo's range was plowed up, destroying forever the greater part of their natural habitat.

By 1889, Dr. William T. Hornaday made a survey of the continent and could find only 1,091 buffalo left. When an effort was finally made to keep the American bison from becoming extinct, only 551 of the big animals could be found alive. By 1895 the American Bison had been virtually exterminated as a wild animal and today it exists in semidomestication only in highly protected areas.

Sea otter The sea otter was also fortunate in being resurrected from practical extinction.[5] This seagoing relative of the weasel was once so numerous that in 1811 Russian fur traders collected 1,200 skins from San Francisco Bay. At one point, a sea otter pelt brought as much as $1,700. But the supply ran out and for many years the sea otter was thought to be extinct. It has been protected since 1913 and is making a comeback, but at the expense of the abalone, one of its favorite foods. For this reason, commercial abalone fishermen and some scuba divers are unhappy with complete protection of the sea otter.

Carolina parakeet Other species of wildlife were not as fortunate. The great auk had been wiped out a half century earlier, never to reappear. The Labrador duck, victimized by feather hunters, had gone under before anyone realized that its existence was threatened. The last one ever seen alive was killed by a man on Long Island in 1875. The Carolina parakeet, a beautiful orange, yellow, and green bird, was once abundant over a wide range in the southern United States. It had two strikes against it: milliners liked the feathers for trimming hats, and farmers disliked the birds because of their destructive feeding on fruit orchards. When shot at, the scattered survivors would quickly return to their fallen companions again and again so that often an entire flock would be wiped out. The birds gradually disappeared; the last flock ever seen was in the Florida Everglades in 1904.

Passenger pigeon The most unrestrained slaughter of birdlife on record was the wanton destruction of the passenger pigeon, once the most abundant bird in North America. During their migrations, flying very close together, the pigeons passed over in great sky-darkening flocks. Alexander Wilson, the father of American ornithology, described one flock that was 240 miles long and at least a mile wide. He estimated that the flock contained about 2¼ billion birds. The birds at times were atrocious pests, breaking branches of trees by their prodigious numbers and wiping out crops. The Indians used both the meat from adults and the fat from squabs as food, but it was not until the European settlers discovered a market for the birds that the passenger pigeon became threatened. Pigeon hunters used every weapon they had—guns, clubs, rocks, sticks, bait, and nets. Squabs were also collected by the ton and sent to markets in New York and elsewhere. Despite the relentless campaign against the pigeons, it took nearly 200 years to drive them to extinction. After 1887, the comparatively few remaining flocks declined quickly. A flock was seen in 1895; by 1900 only

---

[5]*Enhydra lutris* (Family Mustelidae: weasels, skunks, badgers, minks, and otters).

Left photo by Bill Beebe. Right photo by Fred Sibley, courtesy of U.S. Fish and Wildlife Service.

Left photo: Sea otters, once nearly extinct, are making a comeback, but abalone fishermen don't like them. Sea otters use paws like hands to crack shellfish, using rocks they bring up from the bottom as hammers. *Right photo:* The California condor—can it win its struggle for existence?

50 birds were reported. The last surviving passenger pigeon died in the Cincinnati Zoological Park in 1914.

Condor   An example of a species of wildlife that has slowly declined in numbers to the point of near-extinction is the California condor,[6] North America's largest soaring land bird. This bird is a huge dark brown vulture with a prominent white patch under each wing, a bare orange head, and a wing span of over nine feet. It is a relic of the Ice Age; prehistoric remains have been found over much of western North America as far east as Texas. In the days of early explorers and settlers, the condor was known from the Columbia River region of Oregon, south to northern Baja California, in Mexico, and ranging east to southwest Utah and Arizona. Today it is found only in a small area of rugged mountain ranges in south-central California, although there are unconfirmed reports of a few of the birds still surviving in the remote mountains of northern Baja California, Mexico. The total population is estimated at no more than 50 to 60 birds, but the number may be less.

The California condor has always attracted attention because it is a spectacular bird, but it was probably never the most abundant vulture during historical times, and it may already have been in a state of delicate balance with its environment. William L. Finley, writing in 1917, said that he searched for the birds for years without success. But by then their population had greatly declined, due at least partly to the intrusion of man into their range. When cattle-raising was taken up in California by the early Spanish settlers, cattle were raised for their hides, and most of the carcasses were left for the buzzards and other wild animals. The abundant supply of ready-made food may have helped the condor build up to unusual numbers. Later, following the gold rush when livestock-raising be-

---

[6]*Gymnogyps californianus.*

came more extensive, the ranchers commonly drove their livestock into the remote mountain areas where the pasture was better during the dry summer months. But there they encountered coyotes, mountain lions, and even grizzly bears, all more or less effective predators that the ranchers thought were their mortal enemies. The easiest way to get rid of these large predators was to bait carcasses with a poison such as strychnine. Since the condors normally feed on carrion, large numbers of them were probably poisoned to death. The great birds were also shot for their feathers; the quills were used by early miners as containers for their gold dust; the eggs and feathers were prized by collectors; and many of the condors were killed out of curiosity by people who simply wanted to examine the large birds. The condor's slow rate of reproduction—no more than one young every other year—was not enough to enable it to hold its own. By the turn of the century only a few remained in the ring of rugged mountains at the southern end of California's San Joaquin Valley.

As early as 1901, students of wildlife saw that the condor was on its way out unless something was done to reverse the trend, and laws were passed to protect it. But it may have been too late. The National Audubon Society and the University of California made a survey in the late 1940s and estimated that there were only 60 condors left. By 1964, their numbers had dwindled to about 40. It is too early to tell whether the condor can make a comeback. Two sanctuaries, the largest one covering 53,000 acres, have been set aside to protect the nesting, roosting, and resting areas of the condors. The sanctuaries are closed to public use, firearms and drilling rights are prohibited, and even air traffic is restricted above the condor sanctuaries. Experiments are also being conducted with the closely related South American condor to determine its potential for breeding in captivity, a possibility that now looks good.

Brown pelican The brown pelican[7] is another spectacular bird that is highly susceptible to adversities imposed by human intrusion. This was recognized early in the century when a wildlife refuge was established in 1903 on Pelican Island, off the east coast of Florida. The population of brown pelicans in south Florida now appears to be thriving, but the population in other parts of the country are not doing as well. Drastic declines in recent years throughout most of its range have caused wildlife experts to be concerned about its survival. Due to declines in South Carolina, Louisiana (the "pelican state"), Texas, and especially in California, the bird has been placed on the endangered species list, even though it is still present in fairly large numbers. A survey along the Pacific Coast in 1972 revealed a total population of the California subspecies of about 100,000 birds. Their main nesting area is on the remote islands off the coast of Mexico, in the Gulf of California and as far south as Las Tres Islas Marias, a string of three islands about 200 miles beyond the tip of Baja California. But the pelicans also nest as far north as Anacapa Island, a few miles off the coast of southern California, and it is the recent failure to reproduce effectively on this rookery that has stimulated an investigation into the causes.

Ornithologists have learned that the pelican suffers grievously from disturbances caused by visitors to their nesting and roosting habitats.

---

[7]*Pelicanus occidentalis*.

Photo by Daniel W. Anderson, U.S. Fish and Wildlife Service.

California brown pelicans *(Pelecanus occidentalis californicus)* at Isla Mejia, Gulf of California.

Eggs are destroyed and the big birds are frightened, causing them to behave in ways that are not conducive to reproducing. Another problem is that pelicans get hooked accidentally by fishermen. One ornithologist, Ralph Schreiber of Seabird Research, Inc., said that 80 to 85 percent of the pelicans he handled had a hook or line attached. However, the most obvious cause of the decline in reproduction on Anacapa Island is the failure of the young to hatch due to collapse of the eggs because of thin eggshells. Only one young pelican was produced in 552 nesting attempts in 1970 and only seven young from 600 nesting attempts in 1971, but reproduction has improved since corrective measures were taken. Public access to West Anacapa Island is prohibited during the nesting period, and the use of DDT was banned throughout the state for garden and household use in 1971. It was also discovered that large quantities of DDT were being inadvertently deposited in the waters off the southern California coast by the only manufacturer of DDT remaining in the United States. Since DDT is believed to be the main cause of thin eggshells in pelicans as well as some of the other predatory birds, correction of this condition is expected to improve the pelican's chances, although another widely distributed class of pollutants called PCBs (polychlorinated biphenyls), synthetic chemicals used in the plastics industry, are known to have similar effects and remain largely an unknown factor with respect to their long-term effects in the environment.

EXTINCTION AND SURVIVAL

The "mystery" of extinction

The disappearance of the passenger pigeon and other extinct birds was often referred to as a "mystery." People could not understand how anyone could kill every last one of a species of animal. Surely there would be some individuals who could escape into the wilds and never be found. But it is not necessary to kill every specimen to bring about the extinction of a species. The passenger pigeon was a prolific breeder; more than 100 nests were observed in one tree; one nesting area covered 100,000 acres; and the birds nested over a wide geographic range. But many nests, eggs, and young dropped to the ground where they were easy pickings for scavengers and small predators; and winds, weather, and other abiotic factors probably took a large toll. We have already seen that biotic potential and environmental resistance operate to hold the population density in a state of relative stability. But this does not hold below a certain critical level.

A species is in jeopardy when the population is drastically reduced to the point that reproduction can no longer offset the limitations imposed by environmental resistance. The critical level in the population for each species, below which it can probably no longer survive the onslaught of adverse weather, predators, and disease, is comparable in some ways to the critical mass of an atomic explosion. If the amount of reactive material is above the critical level, an explosion takes place; below that, nothing happens. Reproduction is often impaired when a population is under exceptional stress; physiology and behavior are changed in ways that may adversely affect mating and care of the young; and the population may become so sparse that few mates find each other. The occasional natural catastrophe that sometimes wipes out local populations would normally leave surviving pockets to reproduce and perpetuate the species, but at critically low population densities the species dies out and becomes extinct.

In the final analysis, the quickest way to wipe out a population of plants or animals is to destroy their habitat. You have started a species on its way to oblivion when you remove its source of food, destroy its breeding place, or alter the habitat so much that it cannot thrive. This may have been an important factor in some of the great extinctions and near-extinctions of the past, such as those of the passenger pigeon, the Carolina parakeet and the buffalo. (Where would the buffalo roam on today's intensively farmed prairie?) The hazards to species survival are becoming more widely recognized, and increasing attention is being given to several hundred kinds of wildlife throughout the world that are under exceptional stress from one or another of man's activities around the globe.

ENDANGERED SPECIES

Within the space of a few years we have seen a distressing decline in the prevalence of some species of wildlife in the United States. The brown pelican has disappeared from Louisiana, the "pelican state," and hatching of young birds has declined at its principal rookery on Anacapa Island off the coast of California, probably due to contamination with DDT and the plasticizer PCB (polychlorinated biphenyl) which cause thin eggshells and poor reproduction. The Guadalupe fur seal was twice thought to be extinct, the first time after heavy exploitation for its fur during the last century when several scientific expeditions found and collected what they thought were the last remaining specimens; and, again, when a disgruntled fisherman and zoo collector in a frenzy swore to kill the whole herd

Photo by Carolyn Austin.

This sea lion was washed ashore on San Martin Island after being shot in the head with a high-powered rifle. Wanton killing such as this nearly drove a remnant of the Guadalupe fur seal to extinction.

and reportedly did so. The grizzly bear—the official symbol of California—can no longer be found in California except on the state flag. And the bald eagle—the official symbol of the United States—is in danger of extinction in many parts of the country, though still abundant in Alaska. Plants are no exceptions. The weather-carved bristlecone pines cling tenaciously to their last stronghold in the White Mountains where some of them have held out for 5,000 years, but now need protection.

The United States is not the only part of the world where wildlife is suffering from human encroachment. Many of the species of big game in Africa are now being protected in game preserves. The Indian Tiger has been hunted so aggressively that its numbers have dwindled from 40,000 at the turn of the century to fewer than 2,000 which now need protection by the Indian government. Even monkeys, used extensively in medical research, are in critically short supply. Nearly 60,000 monkeys are used each year in the United States alone to keep medical research going.

The great mammals of the sea, the whales, also face extinction. Eight kinds of whales are in danger, and the United States government has denied U.S. whalers licenses to hunt them. Those in most danger of extinction by whaling are the blue whale and the humpback, but the finback, sei, and sperm whales are also threatened. The blue whale is the largest animal that ever existed. Modern methods of hunting whales with the aid of spotter helicopters and explosive harpoons has so depleted their numbers that of all the nations only Russia and Japan continue the slaughter, even though many of the products derived from whales can easily be obtained from other sources. Another sea mammal, the porpoise, suffers a terrible toll by being inadvertently caught in tuna nets. About 200,000 porpoises were destroyed every year during 1969–72, but improved techniques could greatly reduce the loss.

At least 400 species of animals the world over are facing the threat of extinction. These are officially classified as *endangered species,* those

U. S. Forest Service photo.

Bristlecone pines at 11,000-foot elevation in the Inyo National Forest, California. Some of them, still alive, have lived for 5,000 years.

whose prospects for reproduction and survival are in immediate jeopardy. They include nearly 300 species of foreign fish and wildlife and about 100 species of fishes, reptiles, amphibians, birds, and mammals native to the United States. The endangered species and subspecies in the United States include 31 fishes, 8 reptiles and amphibians, 53 birds and 17 mammals. This listing does not include endangered mollusks, crustaceans, or other invertebrates.

There is no exact criterion for deciding that a species is "threatened with extinction." The Endangered Species Preservation Act of 1966 required the Secretary of the Interior to judge what was endangered and to publish lists regularly of such animals. The sequel of this law, the Endangered Species Conservation Act of 1969, directs the Secretary of the Interior to seek the counsel of specialists and agencies with expertise on the subject and to rely on their combined judgments. The actual number of animals is only one criterion. In some cases a declining population is all that is needed to indicate that a species is in danger. But some species that still exist in large numbers may be in peril because they are under some form of stress such as a polluted environment, disease, predation, excessive hunting or fishing, over-exploitation for commercial purposes, or destruction of their habitats by human intrusion. The law authorizes the federal government to conduct research and to take measures to preserve and bring about the recovery of endangered wildlife.

Photo by Camp Wells Russell.

The *big trees*, or *Sierra redwoods (Sequoiadendron giganteum)*, were once distributed over most of the northern hemisphere, but now grow in fewer than 50 groves on the western slopes of the Sierra Nevada mountains. Some of the big trees are several thousand years old.

CONSERVATION AWARENESS

Concern for the loss of natural resources is not new, nor exclusively the invention of environmentalists. Plato (*d*. 347 B.C.) called Attica a skeleton of the heavily forested mountains, cultivated trees, and pastures making up the original landscape. He spoke of the visible traces of earlier forests and of buildings still standing with roofs hewn from the timber, and he pointed to the shrines on the sites of extinct water supplies as evidence that abundant rainfall had been received by the country instead of being lost in his time by flowing over the "denuded surface to the sea." This sense of the fragility of nature's endowment to man came to be a moral obligation in the Greek tradition. There was an oath taken by the young people of Athens, ". . . I will leave to my fatherland not less, but more and better than was left to me . . ."

Some people are apt to think because so much has to be done in America to defend and protect our natural resources that very little has ever been done. Actually, giant steps were taken at the opening of this century. The most distinguished and influential conservationist of our time was Theodore Roosevelt. The term "conservation" in reference to natural resources came into use in the American vocabulary during his administration (1901–1908), but the word has since then been abused by some people and avoided by others because of the resulting bad connotations. Too often, economic gain was the overriding thought, with the effects on the ecosystem receiving secondary, if any, consideration. But despite such problems, there were accomplishments of great importance. Theodore Roosevelt was a great outdoorsman and big-game hunter, and a competent naturalist. He intended to become a biologist, but while a student at Harvard, he became disenchanted with the emphasis on laboratory work and turned to the study of law. But he never lost his interest in nature. As an avid conservationist, Roosevelt promoted steps to preserve our natural resources "for the good of the greatest number of people over the longest possible period of time." Gifford Pinchot became the Chief Forester of the United States and, under the team of Roosevelt and Pinchot, more than 200 million acres of land were withdrawn from the public domain for conservation purposes. The first federal wildlife refuge was established at Pelican Island, Florida, in 1903, the Forest Reserves became the U.S. National Forests, and new national parks were established. More than 125 million acres were added to the national forests and 25 irrigation and water reclamation projects were started. Other important steps were taken that set the stage for a series of subsequent resource management measures of which the following, in chronological sequence, are some of the more important ones:

- The Byrne Law of 1910 in New York State prohibited the sale of wild game and the possession or use of wild bird plumage.
- The International Seal Treaty of 1911, protecting the Alaska fur seal, was signed by the United States, the United Kingdom, Russia, and Japan.
- The Weeks-McClean Act of 1913 authorized the Bureau of Biological Survey to make regulations concerning migratory birds.
- A Federal Tariff Act in 1913 stopped the importation of wild bird plumes into the United States.
- The United States Park Service was established in 1916.
- The Federal Migratory Bird Treaty Act in 1916 terminated the spring shooting of migratory game birds.
- The Alaska Game Law was passed in 1925.
- The International Whaling Convention of 1937 prohibited the taking of gray whales and right whales and established whaling grounds and seasons for commercial whaling on the high seas.
- The Fish and Wildlife Coordination Act of 1958 required federal agencies such as the Corps of Engineers to consult the U.S. Fish and Wildlife Service on water resources projects.
- The Wilderness Act of 1964 established the National Wilderness Preservation System. It limited the developments and uses of resources within the system and provided for additions of wilderness areas by act of Congress.

- The Endangered Species Protection Act of 1966 required the Secretary of the Interior to judge what species of native wildlife were "threatened with extinction," to publish a list of endangered species, and to conduct research on such animals.

- The Endangered Species Conservation Act of 1969 broadened the scope of the 1966 act. Among other things, it included all vertebrates, mollusks, and crustaceans.

- The National Environmental Policy Act of 1969 required environmental impact reports, including effects on wildlife, for all federal agencies proposing projects or legislation. The statement must include an evaluation of the environmental impact, any unavoidable adverse effects, short-term and long-term effects, alternatives, and any irreversible effects or irretrievable commitments of resources.

- The Coastal Zone Management Act of 1972 has been called the "second generation of environmental concern by Congress." It required federal agencies to build into their decision-making process consideration of the environmental aspects of proposed federal actions.

- The Coastal Zone Conservation Act (California) of 1972 was one of the more comprehensive and far-reaching of numerous state laws. It established one statewide and six regional commissions to control construction projects within 1,000 yards of the coast, pending formulation of an overall plan extending farther inland.

- The Land Use Policy and Planning Assistance Act of 1973 encourages the states to regulate land use for areas where local decisions would have more than a local impact.

PROBLEM AREAS

A trend is emerging to make ecological surveys before decisions on land use are made. Current planning in some of the more farsighted communities is not for growth *per se* but for *environmental quality*. Because it is sometimes costly in terms of time, effort, and money, thereby withholding these resources from other useful purposes, decisions are hard to make and are often controversial. Above all, it requires a perspective of how to take advantage of the intrinsic values in the natural environment.

Recreation and wilderness

The Mineral King controversy raised sociological issues that go beyond the mere necessity for preserving open spaces for recreational use. Conservationists say that if wilderness areas are not set aside and protected from easy access, they will inevitably be overrun with swarms of vehicles packed with pleasure-seekers, trampled down by hordes of visitors, and strewn with garbage by transient picknickers, few of whom are apt to have any real love for the wilderness but will visit such areas in uninhibited droves for parties and sports that are equally available elsewhere. Others say that there is already plenty of wilderness area but that much of it goes practically unused while the more highly publicized places are overused and abused.

Still other people question whether remote public lands in spectacular settings should be for the exclusive benefit of the relatively few people who are willing and physically able to get there. They argue that there is no valid reason to exclude the many people whose trip would be easy and comfortable, who would have resort facilities available, and who, otherwise, would probably have no access at all in winter. Most likely this question will arise time and again as population pressure creates more demand for outdoor recreation and the philosophy of "the wisest use for the good of the greatest number of people over the longest possible period of time" is put to the test. At present, the wisest course seems to be to preserve exceptional wilderness areas from the incursions of too many people, while at the same time permitting carefully planned development of other remote areas for the enjoyment of the larger number of people who must have convenient access in order to benefit from it.

The conflict between recreation and other uses of natural resources is not confined to wilderness areas. We see it in almost every aspect of daily life where cities have engulfed rural areas, covering meadows with concrete, filling ponds and streams, removing trees and other natural vegetation, and changing the landscape so drastically that people seeking relaxation are crowded out. Mountains, forests, deserts, and beaches all share similar problems.

Wildlife management

Aldo Leopold, a professional forester, founded the profession of wildlife management and wrote extensively in the field of wildlife conservation. He told how he once went hunting as a boy and emptied his rifle into a mother wolf.

> We reached the old wolf in time to watch a fierce green fire dying in her eyes. I realized then, and have known ever since, that there was something new to me in those eyes—something known only to her and to the mountain. I was young then, and full of trigger-itch: I thought that because fewer wolves meant more deer, that no wolves would mean hunters' paradise. But after seeing the green fire die, I sensed that neither the wolf nor the mountain agreed. . . .[8]

Many people who live close to nature come to a similar understanding about plant and animal inhabitants of the earth. Edmund C. Jaeger, a noted naturalist, said it this way:

> It is always a wrongheaded attitude to assume that only man has rights; the wild animals have some, too.[9]

There is no such thing as being a little bit extinct. The finality of the end of a species is complete. Once a species has dwindled past the point of no return, the genetic resources in its chromosomes that took millions of years to develop in that particular form will have been lost forever. We can expect this to happen more often in the future, and there will be doubts about how much effort is justified to preserve the hundreds of species of wildlife that appear to have lost their place in a world dominated by man. Take, for example, the gorilla, a remote relative of man. He lives in the rain forests of central Africa where he leads a peaceful life. Though subject to diseases and parasites, he has no real enemies except humans, who hunt him for game, cut down his forests, and capture him for display in zoos. But the gorilla seems to be playing a losing game in the contest for survival. His comparatively low reproductive vitality suggests

[8] Aldo Leopold, *A Sand County Almanac,* Copyright © 1949, 1953, 1966 by Oxford University Press, Inc.

[9] Edmund C. Jaeger, *Our Desert Neighbors* (Palo Alto, Calif.: Stanford University Press, 1950).

that his string is playing out and that he cannot much longer hold his own in a life that is no longer adaptable to the conditions of the present rain forests. The species may soon join its ancestors in the now-extinct primate lineage. But it would be a mistake to underestimate the value of knowing everything there is to know about this great anthropoid ape. Some people believe that man himself has gone a long way down the road to oblivion. It may be of vital importance to understand the ramifications of the process of extinction in order to make sensible plans for our own survival.

READINGS

At the Crossroads. A report on California's endangered and rare fish and wildlife. Sacramento, Calif., California Dept. of Fish and Game, 1972, pp. 1–99.

BOUILLENE, R., "Man, the Destroying Biotype," *Science* 135: 706–712 (1962).

FISHER, JAMES, et al., *Wildlife in Danger*. New York, Viking Press, 1969.

FORBUSH, E. H., "Great Auk," in *Birds of America*. New York, The University Society, 1917, pp. 29–30.

FRIGGENS, P, "Last Chance for Yellowstone?" *Reader's Digest,* March 1971, pp. 190–195.

GODDARD, M. K., "Healing Strip Mining Scars," *Outdoors USA*. Yearbook of Agriculture, 1967, pp. 259–262.

HAMMAN, R. E., "Legislation and Regulation of Insecticide's Impact Upon Economic Entomology Industrial Viewpoint," *Bul Entomol. Soc. America*.

Major Uses of Land in the United States Summary for 1969. Agricultural Economic Report 247. Pp. 1–42. U.S. Dept. of Agriculture, Economic Research Service, Washington, D.C., 1973.

National Forest Vacations. Forest Service PA 1037. U.S. Dept. of Agriculture, Washington, D.C., 1973.

Plato: Collected Works, Oxford Text, Vol. IV: Critias, 111 A.D. In A. J. Toynbee, *Greek Historical Thought*. New York, New American Library of World Literature, 1952.

"The Plight of the Pelicans," *Sea Secrets* 18(1): 8–9 (1974).

The Third Wave . . . America's New Conservation. Conservation Yearbook No. 3. U.S. Dept. Interior, Washington, D.C., 1967.

Threatened Wildlife of the United States. Resource Publ. 114, March 1973 (Revised Resource Publ. 34). Bureau of Sport Fisheries and Wildlife, Fish and Wildlife Service, U.S. Dept. of Interior, Washington, D.C., pp. 1–289.

WATKINS, T. H., *The Public Domain*. Sierra Club Bulletin, 1973, pp. 9–13, 41.

WELLMAN, PAUL I., *Glory, God and Gold*. New York: Doubleday & Co., 1954.

GLOSSARY/INDEX

This index is intended to serve also as a glossary. **Bold-face numbers** indicate a definition or description. *Italics* indicate a figure or table.